DNA Science

DNA LEARNING CENTER

Mark V. Bloom, Scientific Editor
Susan Zehl Lauter, Art Director
Carrie Abel, Art and Production Assistant
Kelly Flynn, Editorial and Photo Research
Henrietta Dold, Editorial Assistant
J. Jiji Miranda, Laboratory Photographer

COLD SPRING HARBOR LABORATORY PRESS

Nancy Ford, Managing Director
Annette Kirk, Assistant Managing Director
Dorothy Brown, Technical Editor

DNA Science
A First Course in Recombinant DNA Technology

David A. Micklos
DNA LEARNING CENTER
COLD SPRING HARBOR LABORATORY

Greg A. Freyer
COLLEGE OF PHYSICIANS & SURGEONS
COLUMBIA UNIVERSITY

ILLUSTRATED BY SUSAN ZEHL LAUTER

CAROLINA BIOLOGICAL SUPPLY COMPANY

COLD SPRING HARBOR LABORATORY PRESS

DNA Science
A First Course in Recombinant DNA Technology

All rights reserved

Text © 1990 Cold Spring Harbor Laboratory Press and Carolina Biological Supply Company
Art © 1990 Carolina Biological Supply Company

Printed in the United States of America

9 8 7 6 5 4 3 2 1

Cover design by Leon Bolognese

Cover (clockwise from top left):

Supercoiled plasmid DNA (magnification, 20,500×) coated with the protein UvsX to make it thicker for visualization. Each bead-like unit contains exactly four turns of the DNA molecule. (Courtesy of J. Griffith, University of North Carolina, Chapel Hill.)

James Watson (*left*) and Francis Crick with an early demonstration model of DNA, Cambridge University, 1953. (From J. Watson. 1968. *The double helix,* p. 125. Atheneum, New York. Courtesy of Cold Spring Harbor Laboratory Archives.)

Scanning tunneling micrograph of single-stranded DNA (magnification, 25,000,000×). Shown are two strands of deoxyadenine, each with four nucleotides. (Courtesy of D.D. Dunlap and C. Bustamante, University of New Mexico, Albuquerque.)

Scanning electron micrograph of human chromosomes (magnification, 9,500×). (Reprinted, with permission, from L. Nilsson. 1985. *The body victorious,* p.44. Delacorte Press, New York.)

Library of Congress Cataloging-in-Publication Data

Micklos, David A.
    DNA science : a first course in recombinant DNA technology / David
  A. Micklos, Greg A. Freyer ; illustrated by Susan Zehl Lauter.
      p   cm.
    Includes bibliographical references and index.
    ISBN 0-89278-411-3
    1. Genetic engineering—Laboratory manuals. 2. Recombinant DNA—
Laboratory manuals. I. Freyer, Greg A. II. Title.
QH442.M54 1990
575.1'0724—dc20   90-42300
                                                                                                                                                   CIP

WARNING

Individuals should use the laboratory sequence of this course only in accordance with prudent laboratory safety precautions and under the supervision of a person familiar with such precautions. Use of the laboratory sequence by unsupervised or improperly supervised individuals could result in serious injury.

Additional copies of this book can be purchased under catalog number 21-2211 from

Carolina Biological Supply Company
2700 York Road
Burlington, North Carolina 27215

Call Toll Free 800-334-5551. In North Carolina call 800-632-1231.

DEDICATION

For my parents
Charles and Charlotte
and
Gideon Lyman and ancestors all
whose genes I am.
D.M.

For my parents
Harry and Annelee.
G.F.

CONTENTS

Dedication, v
Acknowledgements, viii
Foreword: Exposing Novice Biology Students to DNA Science, xi

CHAPTER 1	**Themes in the Development of DNA Science, 1**	
CHAPTER 2	**Questions in the Development of DNA Science, 9**	
CHAPTER 3	**Basic Tools and Techniques of DNA Science, 39**	
CHAPTER 4	**Advanced Techniques of DNA Science, 61**	
CHAPTER 5	**Gene Regulation in Development, 87**	
CHAPTER 6	**Molecular Biology of Cancer, 111**	
CHAPTER 7	**Applying DNA Science to Human Genetics, 139**	
CHAPTER 8	**Applying DNA Science in Agriculture, Medicine, and Industry, 169**	

LABORATORIES

Lab Safety and Adherence to National Institutes of Health (NIH) Guidelines, 197

LABORATORY 1	**Measurements, Micropipetting, and Sterile Techniques, 199**
LABORATORY 2	**Bacterial Culture Techniques, 215** A. Isolation of Individual Colonies, 217 B. Overnight Suspension Culture, 229 C. Mid-log Suspension Culture, 237
LABORATORY 3	**DNA Restriction Analysis, 247**
LABORATORY 4	**Effects of DNA Methylation on Restriction, 277**
LABORATORY 5	**Rapid Colony Transformation of *E. coli* with Plasmid DNA, 293**

LABORATORY 6	**Purification and Identification of Plasmid DNA, 311**	

A. Plasmid Minipreparation of pAMP, 313
B. Restriction Analysis of Purified pAMP, 325

LABORATORY 7	**Recombination of Antibiotic Resistance Genes, 341**	

A. Restriction Digest of Plasmids pAMP and pKAN, 343
B. Ligation of pAMP and pKAN Restriction Fragments, 353

LABORATORY 8	**Transformation of *E. coli* with Recombinant DNA, 361**	

A. Classic Procedure for Preparing Competent Cells, 363
B. Transformation of *E. coli* with Recombinant DNA, 371

LABORATORY 9	**Replica Plating to Identify Mixed *E. coli* Populations, 383**	
LABORATORY 10	**Purification and Identification of Recombinant DNA, 393**	

A. Plasmid Minipreparation of pAMP/KAN Recombinants, 395
B. Restriction Analysis of Purified Recombinant DNA, 403

APPENDICES

APPENDIX 1	**Equipment, Supplies, and Reagents, 421**
APPENDIX 2	**Recipes for Media, Reagents, and Stock Solutions, 429**
APPENDIX 3	**Restriction Map Data for pAMP, pKAN, and Bacteriophage λ, 449**

I. Restriction Maps of pAMP and pKAN, 450
II. Restriction Enzymes That Cut Once in pAMP or pKAN, 451
III. Construction of pAMP, 452
IV. Construction of pKAN, 453

Bibliography, 455
Name Index, 461
Glossary/Index, 463

ACKNOWLEDGMENTS

The list of people to whom we are indebted has grown over the five years this book has been in the making:

To Jenny Flanagan, Ellen Skaggs, Courtney Armstrong, and Anne Zollo who ably assisted a boss with too much to do;

To Christine Bartels and Laurel Vanderkleed, early interns in our program;

To Henri Dold, Nancy Baldwin, John LeGuyader, and Kelly Flynn, who did much of the laboratory teaching;

To interns Jeff Mondschein, Steve Malloy, Sol Chen, Jeff Diamond, Tom Hyland, Ken Montaigne, and Chris Inzarillo, our summer teaching assistants;

To Mark Bloom and John Kruper, whose expertise was invaluable to both laboratories and text;

To Susan Zehl Lauter, our artist and creative advisor;

To Dave Smith and Lane Conn, who used our experiments with large numbers of teachers in North Carolina and California;

To Mary Jean and Henry Harris, who made possible a quiet sabbatical on Golden Pond;

To Neil Patterson, who introduced us to the world of publishing;

To Ed Powell, who put the resources and reputation of Carolina Biological Supply Company solidly behind this project;

To Dan James, Ray Flagg, Ray Gladden, Dave Dupont, Lonna Hall, and others at Carolina Biological Supply Company, who provided the mechanism to support instructors using this course;

To Rich Roberts, who supported the initial development of the lab course and gave us lab space in which to tinker;

To Bob Pollack, Rich Roberts, Nancy Wexler, Jim O'Brien, and Conrad Gillam, who took time from their lab work to review the text;

To Jim Watson, who provided us a stimulating environment in which to work and who turned his head when we were absent from our real jobs to chase this dream;

To the foundations, corporations, and institutions (listed on facing page), which gave us just enough funding to keep us honest and creative;

To the school systems on Long Island (listed on facing page), which have led the way to DNA literacy;

And to the many biology teachers we have worked with who are busy reshaping biology education in this country.

David Micklos
Greg Freyer

New York, Spring 1990

Development and testing of the laboratory sequence was made possible through the support of:

Citicorp/Citibank
Brinkmann Instuments
Josiah Macy, Jr. Foundation
National Science Foundation
J.M. Foundation
Richard Lounsbery Foundation
The Banbury Fund
Esther A. and Joseph Klingenstein Fund

Amersham Corporation
Argonne National Laboratory
Bethany College
Biology Teachers' Organization, Winnipeg, Manitoba
Center for Biotechnology, State University of New York at Stony Brook
Cleveland Clinic Foundation
Cooperating School Districts of St. Louis Suburban Area, Inc.
Dorcas Cummings Memorial Fund of the Long Island Biological Association
Samuel Freeman Charitable Trust
GIBCO/BRL Research Products, a division of Life Technologies, Inc.
Fotodyne Incorporated
Harris Trust
Eli Lilly and Company
New England Biolabs Foundation
North Carolina Biotechnology Center
Pioneer Hi-Bred International, Inc.
San Francisco State University
University of California at Davis

and THE COLD SPRING HARBOR CURRICULUM STUDY:

Cold Spring Harbor Central School District
Commack Union Free School District
East Williston Union Free School District
Great Neck Public Schools
Harborfields Central School District
Half Hollow Hills Central School District
Herricks Union Free School District
Huntington Union Free School District
Island Trees Union Free School District
Irvington Union Free School District
Jericho Union Free School District
Lawrence Public Schools
Lindenhurst Public Schools
Locust Valley Central School District
Manhasset Public Schools
Northport–East Northport Union Free School District
North Shore Central School District
Oyster Bay–East Norwich Central School District
Plainedge Public Schools
Plainview–Old Bethpage Central School District
Portledge School
Port Washington Union Free School District
Sachem Central School District at Holbrook
South Huntington Union Free School District
Syosset Central School District

FOREWORD
Exposing Novice Biology Students to DNA Science

Powerful methods to isolate, amplify, manipulate, and analyze DNA sequences have, in effect, created a new science that has transformed biological research. The inauguration of the Human Genome Project in 1988 marked a new era in the national commitment to apply this DNA technology toward understanding human health and development.

During the last 15 years, biologists have gained the extraordinary ability to dissect essentially any of the 100,000 human genes that compose the human chromosomes. The dissection of the molecular pathway through which hereditary information flows among DNA, RNA, and protein molecule has added rich detail to our understanding of how human life develops and changes—from fertilized egg to adulthood. It has also enabled scientists to map the chromosomal positions of genes for hundreds of genetic illnesses and even to isolate the actual DNA sequences responsible for several diseases, including muscular dystrophy and cystic fibrosis.

The excitement of the Human Genome Project offers an important opportunity to substantially reorganize biology education to include new emphasis on molecular and human genetics. It is clear that the science of DNA will increasingly generate important public policy issues, including the proper uses of and access to human sequence data. In the coming decades, when a battery of DNA diagnoses are a likely part of a routine visit to the family doctor and DNA fingerprints are the definitive form of personal identity, a basic understanding of human genetics must be considered as important as a basic understanding of hygiene and nutrition. If we indeed believe in the Madisonian concept of an informed citizenry that participates in public decision making, then DNA literacy cannot be considered an esoteric pursuit. Only through widespread education can we ensure a society that shepherds the benevolent use of genetic technology for the good of all its citizens.

Periodic studies conducted since 1959 show that scientific illiteracy is a persistent problem in the United States. According to major National Science Foundation (NSF)-sponsored studies conducted over the last several years, only about 1 in 20 U.S. citizens can be considered functionally literate in scientific matters. On topics relating to molecular genetics, a 1987 Office of Technology Assessment survey found that 64% of Americans claim to have heard little or nothing about genetic engineering. The NSF-sponsored studies show that only 10% of the U.S.

population can explain what a molecule is, and only 22% can correctly define DNA. However, 57% of the population are able to understand a simple problem dealing with the inheritance of a genetic illness, suggesting that they find this topic relevant to their personal lives.

As applications of molecular genetics leave the laboratory, trained personnel from nearly every segment of society must interface with this new technology. Young people entering the medical, agricultural, manufacturing, and even legal professions will be expected to have a basic command of DNA science. At the same time, there is growing concern within the scientific community that a predicted shortage of U.S.-born biologists during the next several decades could diminish our leadership in health-related research and development. Survey data show that U.S. students become increasingly disinterested in science during the middle and high school years and that our high school students finish last in biology achievement among their peers in major industrial countries. The "pipeline" of students destined to become science professionals constricts to but a trickle by college; in 1989, only 3.4% of college freshman intended to pursue biological science majors. Introductory college biology courses, which are taken by a cross section of U.S. youth, occupy a critical position in the biology education pipeline. It is our last chance to inculcate basic tenets of scientific literacy that are essential for college graduates as they assume their roles as opinion leaders in government, industry, education, medicine, and law. It is also our last chance to lure bright minds to the challenge of careers in science. Beginning biology students need to be stimulated by the most interesting topics in contemporary biology, rather than numbed by ponderous reviews of the phyla.

Molecular genetics is typically construed as the culminating experience of the biology major's academic career. To date, the complexity of text and laboratory materials has confined hands-on instruction in molecular biology to upper-level courses, which are typically taught by research biologists with direct experience in recombinant DNA techniques. We believe that laboratories in DNA manipulation should be an initiating experience—introduced in general survey courses and comprehensively instituted in sophomore-level electives. In this way, the molecular genetic perspective that has served so well to integrate biological phenomenon for researchers can play a similar role for students.

In our view, existing college laboratory manuals in this field are aimed more at upper-level undergraduates and presuppose that the instructor is skilled in molecular biological techniques. This is in line with the trend toward authoritative "supertexts," which are excellent references for upper-level undergraduate and graduate students, but are inappropriate for novice students of biology. Supertexts contain more than enough material for several semesters' study and require instructors to select a minority of chapters for inclusion in their curriculum.

Pedagogically, we have taken a very different approach—that less is more. Since student interest and understanding are our primary goals, no attempt has been made at an exhaustive treatment of molecular biological topics and techniques. Rather, we have structured a learning experience that integrates the theory, practice, and applications. Combining the unique viewpoints of a science educator (D.M.) and a practicing molecular biologist (G.F.), *DNA Science* integrates up-to-the-minute examples drawn directly from the research literature. Unlike supertexts that can only be sampled by even the most ambitious instructor or student, this text is designed to be read from cover to cover.

The eight text chapters are written in a semijournalistic style and adopt a historical perspective to explain where DNA science has come from and where it is going. The first chapter alerts students to themes and research traditions that contribute to the molecular perspective. Classical genetics and DNA structure are reviewed in a chapter that traces the development of molecular biology. An initial techniques chapter explains the theory behind methods used in the laboratory sequence; a second techniques chapter discusses advanced methods for analyzing complex genomes. The final four chapters discuss how the molecular approach is applied in basic and applied research in the fields of development, cancer, human genetics, medicine, and agriculture. Designed to follow from basic principles of biology, the text chapters are suitable for introducing recombinant DNA in science and society courses.

Truly a first course in recombinant DNA technology, the laboratory sequence is a cumulative experience that presupposes no prior experience on the part of instructor or student. Because every major technique is repeated at least twice, students build mastery by evaluating and correcting mistakes throughout the course. The entire lab set can form the basis of a first course in a molecular biology sequence, or sets of appropriate labs can be integrated as a genetics/DNA structure component of a general biology course or used as a unit within a microbiology or genetics course. DNA restriction analysis and rapid colony transformation can function well as stand-alone laboratories in general biology or science and society courses.

Currently used by thousands of Advanced Placement/college teachers and students in 30 states and Canada, the 10 laboratory experiments cover the basic techniques of gene isolation and analysis. We have endeavored to make this manual user friendly, while avoiding a perfunctory cookbook style. Extensive prelab notes explain how to schedule and prepare, while flow charts and icons make the protocols easy to follow. A discussion section reviews each experiment in a rhetorical style, analyzing controls and showing both ideal and "less-than-ideal" results. A final section suggests simple research projects that extend the techniques learned in each laboratory and require few, if any, additional reagents.

Our goal has been to modify current research protocols to minimize expense while maximizing safety and reproducibility in the teaching laboratory. However, we strive to maintain the integrity of research methods so that novices need not relearn techniques as they progress to advanced lab work or to a research setting. Our extensive contacts within the biological research community give us an ideal window on emerging techniques. For example, personal communications with Doug Hanahan in 1987 alerted us to a relatively obscure method he had recently published for rapidly introducing purified plasmid into *Escherichia coli.*

Dr. Hanahan had developed "colony transformation" as a means to circumvent the need to maintain large stocks of frozen competent cells, but we saw it as an ideal way to introduce transformation to novices. Within months, we began testing an adapted protocol with teachers at professional meetings in New York and other states. The protocol quickly became popular among lab educators, and by 1989, the colony method was recommended by the Educational Testing Service as a transformation method of choice in the Advanced Placement laboratory curriculum.

To make the *DNA Science* course as foolproof as possible, one of us (G.F.) developed the teaching plasmids pAMP and pKAN. Restriction digests of these plasmids yield fragments of markedly different sizes, making gel interpretation straightforward. Derivatives of the pUC19 expression vector, these plasmids transform well, are highly amplifed in *E. coli,* and give consistent yields in plasmid preparations. They are, to our knowledge, the only DNA molecules specifically engineered for educational purposes.

Many instructors have indicated that time for setup and preparation is the most serious constraint to teaching DNA laboratories. Their need for "one-step" shopping and quality-assured reagents led us to collaborate with Carolina Biological Supply Company, which distributes all reagents and equipment necessary to perform the experiments in the DNA Science. Appendix 1 lists the range of product options offered—from bulk reagents, to multi-use reagent systems, to throw-away kits—that make it easy to set up a lab-teaching program from scratch.

The first working title of this course was Recombinant DNA for Beginners. True to this ideal, we hope this text will allow significant numbers of students to participate directly in the biology of the future.

Dave Micklos
Greg Freyer

New York, Spring 1990

DNA Science

CHAPTER 1
Themes in the Development of DNA Science

Gene cloning, genetic engineering, gene splicing, and recombinant DNA are some of the buzz words of perhaps the most exciting era in human and scientific history—the biotechnology era. Literally translated, biotechnology means "life technology"—applying knowledge about living things for the practical use of humankind. The age-old uses of yeasts in making bread and alcoholic beverages and of bacteria in making cheese are, in the broadest sense, biotechnology. However, the modern biotechnology revolution is based on a deep understanding of the technology of life, the mechanics of living machines.

The technical aspects of life involve the complex chemical interactions that take place among the several thousand different kinds of molecules found within any cell. Of these, DNA (deoxyribonucleic acid) is the master molecule in whose structure is encoded all the information needed to create and direct the chemical machinery of life. Analysis of the flow and regulation of this genetic information among DNA, RNA (ribonucleic acid), and protein is the subject of molecular genetics. However, the techniques of molecular genetics are now being applied to nearly every major field of biology, from neurophysiology to botany and from immunology to paleobiology. In a broad sense, the terms molecular genetics and molecular biology have become nearly synonymous.

Egyptian Hieroglyph of Wine Production
The fermenting of grapes to make wine is one of the earliest examples of biotechnology. (Courtesy of the Metropolitan Museum of Art.)

This change in our understanding of life has launched a dramatic and seemingly sudden biological revolution. However, it is prudent to remember that embedded in the word *revolution* are both *revolve* and *evolution*. These remind us of the long-term, cyclical, and historical undercurrents of revolution—however sharp its break from the past appears. As we trace the development of the concepts that led to this biological revolution, it will serve us well to keep in mind some themes and trends that can help organize our understanding of molecular biology.

Molecular Biology Is a Hybrid Discipline

Too often, we are led to think that the three sciences are quite separate from one another. Physics is physics, chemistry is chemistry, biology is biology—and never the three shall meet. Unfortunately, this artificial division still persists even in the minds of some scientists.

Molecular biology is the antithesis of this notion of separateness. It arose from a confluence of disciplines from both the physical and natural sciences—notably, genetics, physical chemistry, X-ray crystallography, biochemistry, microbiology, bacteriology, and virology. In the beginning of the 20th century, physics and chemistry were united by quantum theory, which explained the fine structure of matter. Beginning in the

Max Delbrück (Top Right) with Students, 1949
(Courtesy of California Institute of Technology.)

third and fourth decades of the 20th century, biology, in turn, began to benefit from an influx of ideas from physical chemistry.

Two quantum physicists were especially influential in breaking down the thought barriers between the sciences: Max Delbrück and Erwin Schrödinger. Delbrück, who is rightly called the intellectual father of molecular biology, was trained under Niels Bohr, the great atomic physicist who deduced that electrons occupy discrete energy states (orbitals) surrounding the atomic nucleus. Schrödinger's wave equation defined the movement of electrons within the orbitals. Although Schrödinger never made the switch to biology, his brief book *What is Life? The Physical Aspects of the Living Cell* (1944) influenced a generation of physical scientists to take a closer look at biological systems. Both men thought that the biological mystery of self-replication could be explained in quantum-mechanical terms. They were tantalized by the prospect that there might be some new physical principles to be discovered by studying living things.

Although no new universal laws were uncovered, Delbrück and other newcomers from the physical sciences showed that the axioms and methods of the physical sciences apply equally well to biology. Molecular biology could not become a rigorous discipline until this notion was firmly established. It would have been impossible to study how molecules interact in even the simplest of organisms without a basic understanding of how they react on the chalkboard or in the test tube.

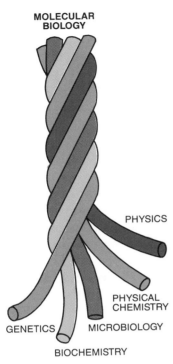

Molecular Biology Is a Synthesis of Several Disciplines
(Art concept developed by Lisa Shoemaker.)

Molecular Biology Becomes a Rigorous, Experimental Science through Application of Physical and Chemical Principles and Use of Abstract Model Systems

Physical and chemical principles underlie all biological phenomena; living things abide by all the laws of physics and chemistry. The physical and chemical behavior of elementary particles ultimately defines the parameters of behavior of any complex biological system—a bacterium, a plant, a frog, or a human being. This explanation of "life" is a direct result of biology's cross-fertilization by chemistry and physics.

Historically, biology was based on direct observation of complex natural phenomena in the real world. Molecular biology borrowed from the physical sciences the rigorous use of model systems—simplified abstractions of reality in which variables are limited and experimental situations can be controlled. The development of molecular biology was in large part driven by the quest to find increasingly purer and more powerful abstractions of essential biological processes. Experimental systems progressed from complex, multicellular organisms (such as pea plants and fruit flies used in the early 1900s) to simple, one-cell organisms (bacteria and viruses, beginning in the 1940s) to purified cellular components (in vitro systems, beginning in the 1960s).

Molecular Biology Arises from the Structure-Function Tradition

Natural scientists have always tried to find relationships between structure and function in living things. Molecular biology is the culmination of this tradition, which has led biologists to peel back successive layers of organization to ultimately reveal the molecular interactions that take place within the living cell. This pursuit began with the examination of obvious physical attributes.

Physicians from the time of the earliest civilizations tried to relate their knowledge of the human body to the treatment of illness. Thus, anatomy and physiology became the classic expression of structure-functionalism in the natural sciences. The 17th century anatomist William Harvey, for example, showed that a number of physical structures—organs including heart, lungs, veins, arteries, and valves—work together as a system to circulate blood throughout the body. The heart functions as a pump, and the blood vessels function as pipes.

Cell theory, advanced by Matthias Schleiden and Theodor Schwann in the late 1830s, was an important milestone; it moved structure-functionalism beyond systems directly observable with the naked eye. Schleiden and Schwann proposed that microscopic cells, defined essentially by the presence of individual nuclei, are the basic units of structure and function in both plants and animals. Organs were then seen to be composed of various tissues—groupings of cells with similar structures that perform a specific function. For example, cilia on epithelial cells that

William Harvey, about 1640
(Courtesy of the Granger Collection.)

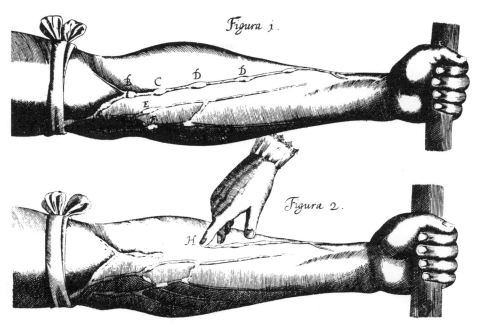

William Harvey's Demonstration of One-way Valves in Veins
(From *De Motu Cordis*, 1628.)

line the respiratory passages oscillate to help eject foreign particles such as smoke and dust. Cells, in turn, were seen to possess organelles with their own specific functions: mitochondria for energy production, vacuoles for storage, chloroplasts for photosynthesis, and ribosomes for protein synthesis.

The stage was set for structure-functionalism to move to the level of biologically important molecules during the 1930s, when physical chemist Linus Pauling codified the physical laws that govern the arrangement of atoms within molecules. During this same period, J. Desmond Bernal showed that the structures of giant molecules, such as proteins, can be studied using X-ray crystallography.

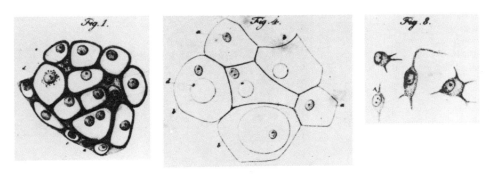

Theodor Schwann's Drawings of Vertebrate Cells Showing Nuclei
(From *Microscopical Researches into the Accordance in the Structure and Growth of Animals and Plants*, 1839. Sydenham Society, London.)

Molecular Biology Arises from the Quest to Define the Nature of Heredity

Reproduction, or autonomous replication, is perhaps the most distinctive attribute of life. To explain replication of cells and inheritance of traits over successive generations is, in large measure, to define life. During the development of molecular biology, scientists sought an increasingly

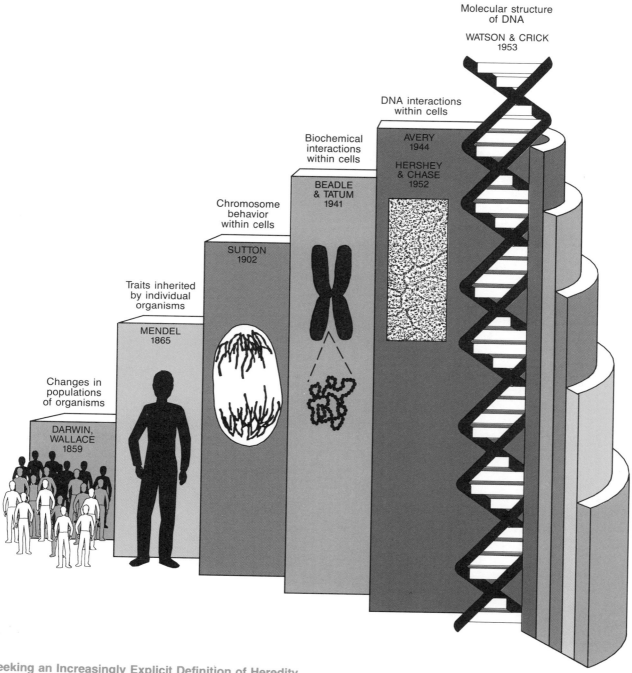

Seeking an Increasingly Explicit Definition of Heredity
(Art concept developed by Lisa Shoemaker.)

explicit explanation of the nature of heredity. The hereditary process was successively explained in terms of:

- Changes in *populations* of organisms (Darwin, Wallace).
- Traits inherited by *individual* organisms (Mendel).
- *Chromosome* behavior within *cells* (Sutton, Morgan, McClintock).
- *Biochemical* interactions within *cells* (Beadle and Tatum).
- DNA *interactions* within *cells* (Avery, Hershey).
- *Molecular structure* of DNA (Watson and Crick).

For Further Reading

Cairns, John, Gunther S. Stent, and James D. Watson, ed. 1966. *Phage and the origins of molecular biology.* Cold Spring Harbor Laboratory, Cold Spring Harbor, New York.

Fischer, Ernst Peter and Carol Lipson. 1988. *Thinking about science: Max Delbrück and the origins of molecular biology.* W.W. Norton and Co., New York.

Moore, Walter. 1990. *Schrödinger: Life and thought.* Cambridge University Press, New York.

Schrödinger, Erwin. 1944. *What is life? The physical aspect of the living cell & mind and matter.* Cambridge University Press, New York.

James Watson (Left) and Francis Crick with DNA Model, 1953
(From J. Watson. 1968. *The double helix*, p. 125. Atheneum, New York. Courtesy of Cold Spring Harbor Laboratory Archives.)

CHAPTER 2
Questions in the Development of DNA Science

The biotechnology era formally began on April 25, 1953, when James Watson and Francis Crick announced in the British journal *Nature* that they had determined the three-dimensional structure of the DNA molecule. Their gently curved structure, the double helix, was the key to unlocking the technology of life.

Beyond the double helix lay the question that would fuel the biotechnology revolution: Can DNA, and therefore life, be manipulated? Before addressing this question—which is the focus of this book—we must understand the questions that came before the double helix. The answers to these "pre-DNA" questions provide a review of basic biology and a perspective on the DNA revolution.

Hopefully, many questions will come to mind as you read through this book, and, surprisingly, they may be the same sort of questions that led to important developments in molecular biology. Bear in mind that as little as 80 years ago, there was no explanation as to why some siblings have brown eyes and some have blue eyes. As little as 55 years ago, the physical structures of simple organic molecules were not known. As little as 45 years ago, it was not firmly established that DNA is the genetic material. We still do not know enough about the molecular basis of cancer to offer systematic, rational treatments. There are lots of good questions left to ask.

How Do We Account for the Diversity (and Similarity) of Species?

The Swedish biologist Carolus Linnaeus set up a systematic hierarchy for classifying all living things based on the unit *species*. The cataloging of a multitude of organisms by Linnaeus and his disciples during the 17th century pointed up the incredible diversity of life forms. At the same time, his taxonomy emphasized the strong to relatively subtle similarities that exist among members of the same *genus, family, order, class, phylum,* and *kingdom*.

Even so, the question of how to account for diversity and similarity among living things was not one worth asking in the 18th and 19th centuries. It had already been answered, by inference, through interpretation of the biblical story of creation, which stated that each of the diverse life forms was present in perfect form at the Creation. Nowhere

Carolus Linnaeus' Drawing of *Dryas octapetala*
(From Carolus Linnaeus. 1735. *Systema naturae*. B. De Graff, The Netherlands. Courtesy of the Granger Collection.)

in the Bible is the age of the earth, or human beings, explicitly given. However, a 17th century Irish clergyman, Archbishop James Usher, concluded through study of biblical genealogies that the earth was created in 4004 BC. That left little time for any major alterations to living organisms, and similarities *between* species could be explained by minor changes that had occurred within fixed forms.

So it was presumptuous, if not downright blasphemous, to ask questions about the variety of life without mentioning the Creation in the same breath. However, questions did stir in the 19th century minds of English naturalists Alfred Wallace and Charles Darwin. Independently, they arrived at the same impudent conclusion: All living things had evolved from preexisting forms through a process of incremental change over millennia.

The publication in 1859 of Charles Darwin's epic book *On the Origin of Species* marked the first step in the biological revolution that culminated nearly a century later in the Watson-Crick structure of DNA. The theory of evolution described how heredity operates in large populations of living things:

Charles Darwin, about 1859
(Courtesy of the Museum of Natural History.)

1. There is a natural selection, over great periods of evolutionary time, for the "fittest" forms of life.
2. Natural selection arises from competition for limited food and other resources among members of the same species. Only the fittest members of a population survive to reproduce.
3. On rare occasions, a random physical change increases an individual's ability to adapt to environmental conditions and/or exploit new food resources. This "adaptive" change increases the individual's chances to survive and to reproduce.
4. Adaptive changes are passed on to offspring as part of their hereditary endowment. These individuals are, in turn, fitter than their peers and survive to pass on their physical characteristics to succeeding generations.
5. Through the process of adaptive radiation, populations of organisms evolve to exploit specialized food resources, thus limiting competition and increasing chances for survival.

Darwin was profoundly influenced by the British geologist Sir Charles Lyell and his doctrine of uniformitarianism, which stated that the earth's physical features can be accounted for by ordinary, observable climatic and geological processes—such as sedimentation, coral deposition, and erosion—operating over very long periods of time. This was in direct opposition to the biblical interpretation that all geological phenomena

Flightless Cormorant

Archaeopteryx

Darwin observed several mainland species, such as the flightless cormorant, that had adapted in specialized ways to life on the Galapagos Islands. Fossils, such as the birdlike reptile *Archaeopteryx*, provided evidence of the gradual evolution of life forms over long periods of geological time. (Courtesy of Taurus Photos; the American Museum of Natural History.)

are the products of relatively brief, catastrophic events, such as the Flood. Lyell argued that sedimentary rock layers (strata) are a geological time line extending back millions of years into the earth's history. He demonstrated that one could date the emergence and extinction of species by observing their fossilized remains in rock strata.

Darwin avidly studied Lyell's *Principles of Geology* on his world-circling voyage aboard the *Beagle*, during which he made many of the observations on which he based his theory of evolution. By tracing the distinctive anatomical changes in related fossil forms, Darwin showed the operation of evolution over long periods of prehistory. His studies of the comparative anatomy of finches that colonized the volcanic Galápagos islands, for example, vividly illustrated adaptive radiation during relatively recent history.

How Are Traits Passed from One Generation to the Next?

Evolution did not explain how adaptive traits are passed from generation to generation. It was the Austrian monk Gregor Mendel who brought the hereditary process down to the individual organism and provided a mechanism to drive evolution. Mendel's paper "Experiments in Plant-Hybridization," published in 1865, provided a basis for the mathematical analysis of inheritance. From the results of controlled crosses of garden peas, he showed that traits are inherited in a predictable manner as discrete bits of genetic information, or "factors." (The American biologist Walter Sutton later substituted the word "gene" for factor.)

Mendel was the first person to relate the outward appearance of an organism (phenotype) to its inner genetic constitution (genotype). He showed that the expression of contrasting physical characteristics—for example, red or white flower color in pea plants—is controlled by a pair of distinct genes. In this case, the dominant gene (C) specifies red flowers, whereas the recessive gene (c) specifies white flowers. Three genotypic combinations are thus possible: CC, Cc, and cc. The dominant phenotype (red flowers) is expressed when two copies of the gene are present, CC. Red flowers also result from the mixed genotype Cc, where the effect of the C gene dominates the recessive c gene. The recessive phenotype (white flowers) is expressed only by the genotype cc, where two copies of the recessive gene are present.

Now imagine simultaneously following a second trait, flower position, where axial flowers are dominant (P) and terminal flowers are recessive (p). Given two parents with genotype CcPp, Mendel hypothesized that during gamete formation (meiosis):

Gregor Mendel, about 1860 (Courtesy of the Austrian Press and Information Service.)

1. Parental genes for each trait *segregate* so that each sex cell contains only one sort. Thus, each contrasting member of a gene pair is equally likely to occur in gametes: C or c and P or p.
2. Genes for different traits *assort* into gametes *independent* of one another. Thus, each combination of genes is equally likely to occur: CP, Cp, cP, and cp.

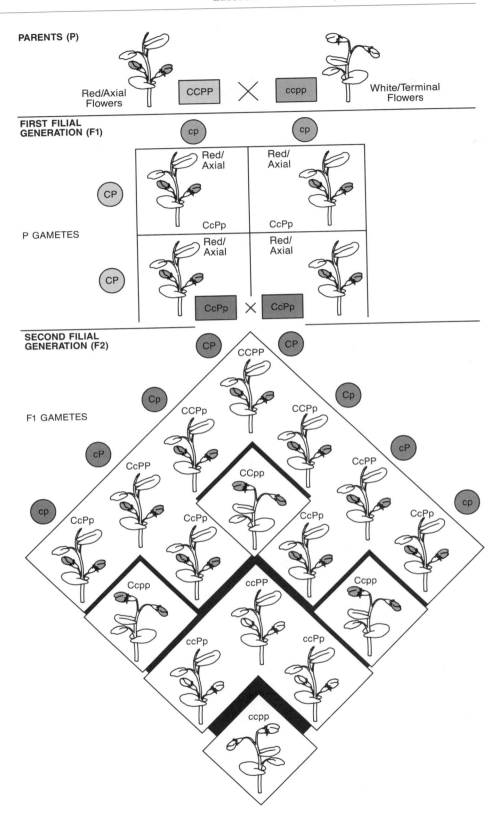

Dihybrid Cross
A dihybrid cross of garden peas showing the inheritance of two traits: flower color (C) and flower position (P).

Where Are Genes Located?

Hugo de Vries, about 1920 (Courtesy of Cold Spring Harbor Laboratory Archives.)

Edmund Wilson, about 1925 (Courtesy of American Society of Zoologists.)

Although Mendel was a contemporary of Darwin, his work lay fallow, unrecognized until the beginning of the 20th century. In 1900, the Dutch scientist Hugo de Vries and the German scientist Carl Correns rediscovered Mendel's paper and published research data that confirmed his earlier work.

In 1902, Walter Sutton, a student at Columbia University, was the first to relate genetics to the study of chromosome behavior (cytology). His analysis of chromosome movements during meiosis, in the grasshopper *Brachystola*, formed the basis of the chromosomal theory of heredity. Sutton showed that the grasshopper genetic material consists of 11 *pairs* of homologous chromosomes and that gametes (sex cells) formed during meiosis receive only one chromosome from each homologous pair. This behavior paralleled exactly the segregation of Mendel's hereditary factors and suggested that genes are physically located on the chromosomes.

The behavior of sex chromosomes, described independently in 1905 by Nettie Stevens and Edmund Wilson (Sutton's mentor at Columbia), provided the first direct evidence to support the chromosomal theory of heredity. They showed that sex is determined by separate X and Y chromosomes. Femaleness is characterized by two copies of the X chromosome (XX), and maleness is determined by a single copy of each type of chromosome (XY). The movements of X and Y chromosomes during formation of sperm and egg cells are exactly as predicted by Mendelian genetics: each egg receives a single copy of an X chromosome; each sperm receives either an X or a Y chromosome.

THE FLY ROOM

Evidence that traits other than sex are located on the chromosomes came during the second decade of the 20th century. During this period, Thomas Hunt Morgan and his astoundingly bright cadre of students—Alfred Sturtevant, Calvin Bridges, and Hermann Muller—established the

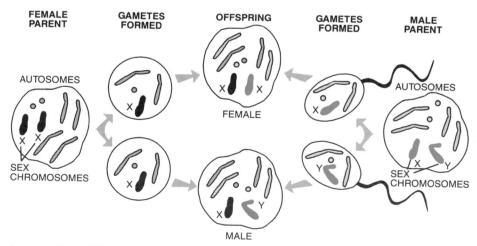

Segregation of X and Y Chromosomes in *Drosophila*

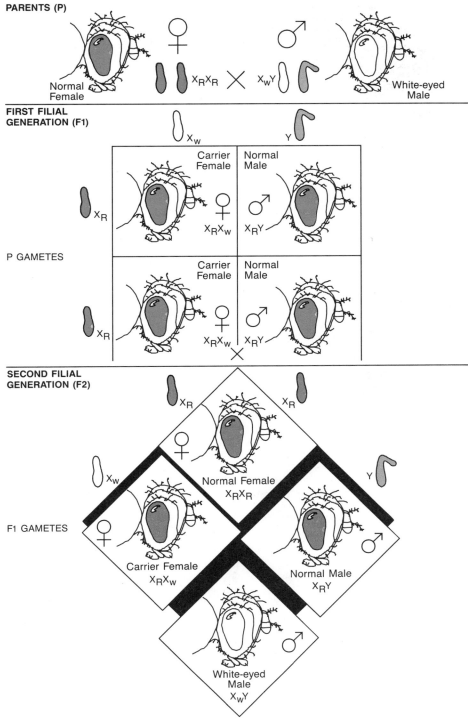

Sex-linked Inheritance of White Eyes in *Drosophila*

Thomas Hunt Morgan, about 1917, and Calvin Bridges, about 1926
(Courtesy of American Society of Zoologists.)

intellectual basis of modern genetics. They also established the humble fruit fly, *Drosophila*, as the organism of choice for genetic research. Morgan's modestly sized laboratory in the zoology department at Columbia University became known simply as the Fly Room.

Like many scientists of that era, Morgan was interested in reproducing evolutionary processes in the laboratory. The intellectual shift that Morgan and other "experimental evolutionists" underwent shows the direct historical linkage between Darwinian evolution and Mendelian genetics. It also illustrates the focusing of hereditary questions from the level of populations to the level of genes acting within individuals.

Morgan wanted to test whether the variations that result in new species happen gradually or occur in abrupt fits and starts (continuous vs. discontinuous variation). He chose *Drosophila* as his experimental model because it had a short generation time, produced numerous offspring, and was easy to culture. In addition, he lacked sufficient funds and facilities to maintain higher organisms, such as mice.

The appearance of a single white-eyed fly in Morgan's laboratory in the spring of 1910 was clearly a fluke—wild fruit flies have red eyes. Luckily, Bridges, then an undergraduate student, recognized its potential importance and saved that unique male fly. The white-eyed variant might have been viewed only as a tiny, nonadaptive evolutionary step. Instead, it became the cornerstone upon which Mendelian, chromosomal, and sexual inheritance were built into a cohesive whole. The white-eyed male established forever the importance of an observable mutation, or variation from the norm, as the starting point for genetic analysis.

The white-eyed male fly was mated with its red-eyed sisters, and the appearance of the trait was followed through several successive generations. Mendelian analysis revealed that, in most types of crosses, white eyes appeared in males only. Morgan's group showed that white eyes (like baldness and hemophilia in humans) is a sex-linked recessive trait. In so doing, the gene for eye color was localized to the X chromosome.

During the following decade, Morgan and his disciples expanded the cytogenetic analysis of heredity and provided other evidence for the chromosomal locations of genes. For example, they demonstrated linked genes—those located in such close proximity on a chromosome that they are almost always inherited as if they were a single genetic unit. In his doctoral thesis, Sturtevant used linkage data to construct the first gene map of chromosomes. He provided evidence for Morgan's hypothesis that crossing-over, the exchange of chromosome fragments between homologous chromosomes during meiosis, provides a measure of the relative distance between two genes. Whereas closely linked genes will rarely be separated by crossing over, genes that are far apart will be frequently separated. Therefore, the lower the crossover frequency between two genes, the closer together they are on the chromosome.

It was not until 1931, however, that Barbara McClintock and Harriet Creighton, of Wellesley College, obtained cytological proof that the inheritance of novel gene combinations during crossing over is due to the physical exchange of specific chromosome segments. They crossed corn strains whose 9th chromosome contained distinctive features that could be distinguished under the light microscope. This allowed them to determine which chromosome exchanges accounted for each phenotype in the F_1 hybrid plants. In the same year, Curt Stern, at the University of Berlin, reported the use of a similar approach to study the X chromosome of *Drosophila*.

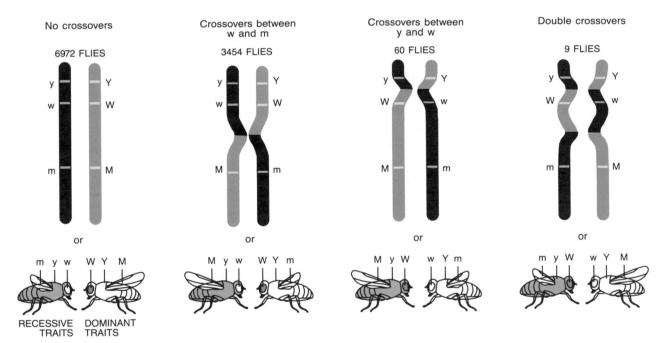

Sturtevant's Linkage Experiment in *Drosophila*, 1913
Sturtevant examined the X-linked inheritance of three recessive traits in yellow body (*y*), white eyes (*w*), and miniature wings (*m*). He crossed recessive males (*y,w,m*) with heterozygous females having recessive genes on one X chromosome (*y,w,m*) and dominant genes on the other (*Y,X,M*). Because a male parent *only* can pass on recessive genes on its single X chromosome, the phenotypes of both male and female offspring are due entirely to the inheritance of the maternal X chromosomes. Mendelian analysis predicts that all of the 10,495 offspring in Sturtevant's experiment would show either a purely dominant phenotype, normal body/eye color/wings, *or* a purely recessive phenotype, yellow body/red eyes/miniature wings. However, offspring inherited various mixtures of dominant and recessive traits. Sturtevant deduced that the mixed phenotypes were caused by genetic exchange between a female's two X chromosomes during gamete formation. The frequency of exchange is a measure of the distance between two genes located on the same chromosome.

What Is the Job of the Gene?

Archibald Garrod, about 1910
(From *Genetics: A periodical record of investigations bearing on heredity and variation* 56: frontispiece, 1967.)

Edward Tatum, 1965
(Courtesy of The Rockefeller Archive Center.)

George W. Beadle, about 1940
(Courtesy of Stanford University.)

Throughout the first 40 years of genetics, the gene was merely a concept whose true identity was shrouded in mystery. Genes were known only by their outward manifestation as visible traits. Nothing was known about their function in the biochemical life of organisms.

The British physician Sir Archibald Garrod proposed, in 1908, that some human diseases are "inborn errors of metabolism" that result from the lack of a specific enzyme needed to perform a biochemical reaction. He speculated that lack of enzyme function resulted from a defective gene inherited at birth. It would take more than 30 years to prove this prophetic hypothesis, for it could not be seriously tested by the geneticists of the day. The experimental models they used—primarily fruit flies, mice, and domestic plants and animals—are all highly evolved, multicellular organisms. They were far too complex, as models, to shed light on the connection between genes and cellular biochemistry.

It was George Wells Beadle and Edward Tatum, at Stanford University, who finally introduced a fungus as the first genetic model in which it was possible to study metabolism. The red bread mold *Neurospora* can thrive on minimal medium containing only sucrose, inorganic salts, and the vitamin biotin. Beadle and Tatum reasoned that it must therefore possess a number of specific enzymes that metabolically convert these simple nutrients (plus water and oxygen) into amino acids, vitamins, and all the other complex molecules necessary for life.

They hypothesized that if they could induce a mutation in a gene that produces one of these enzymes, it would be lethal to the mold. They employed the powerful genetic tool that had been introduced in 1927, when Hermann Muller (then at the University of Texas) showed that mutations can be generated at will by irradiating organisms with X-rays. Beadle and Tatum's ingenious experimental design allowed them to select for X-ray-induced mutations that would, under normal circumstances, be lethal.

First, they irradiated *Neurospora* with X-rays and allowed it to produce spores; then they examined the spores' growth on minimal medium. Spores that failed to grow on minimal medium were, in turn, tried on several types of media—each containing a single nutritional supplement. The 299th spore they examined would grow only when supplemented with vitamin B_6. Thus, irradiation had mutated a gene that produces an enzyme necessary for the synthesis of vitamin B_6. Numerous metabolic mutants for other essential vitamins and amino acids were found. Moreover, multiple mutant strains were isolated that failed to produce the amino acid arginine. Beadle and Tatum found that each mutant strain lacked a different enzyme needed at different points along the arginine synthesis pathway.

Beadle and Tatum's work, published in 1941, introduced the concept that genes mediate cellular chemistry through the production of specific enzymes. Garrod's essential hypothesis, now directly stated as "one gene/one enzyme," was confirmed. (It is most useful to broaden that adage to read "one gene/one protein.") Their experiments also pointed

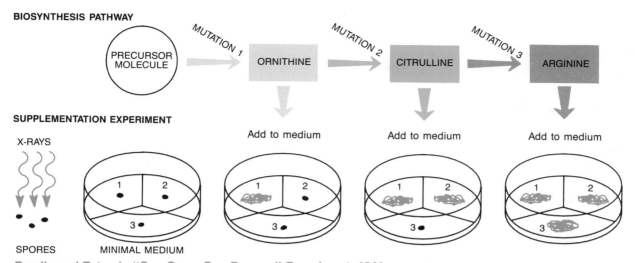

Beadle and Tatum's "One Gene–One Enzyme" Experiment, 1941
Following irradiation, *Neurospora* spores that failed to germinate on minimal medium were sequentially screened for growth on media supplemented with three nutrients in the arginine biosynthesis pathway. The pattern of supplementation for spores 1, 2, and 3 indicates mutations in genes for enzymes at corresponding points in the pathway.

to the advantages of microorganisms as model systems. Using their method of plating onto selective media, it became possible to identify extremely rare genetic events, such as one-in-a-million biochemical mutants.

What Molecule Is the Genetic Material?

In a series of experiments with *Diplococcus pneumoniae*, the ball-shaped bacterium that causes pneumonia, English microbiologist Fred Griffith discovered the model system that provided the key to answer the next question in the development of molecular biology. Two naturally occurring strains of the pneumococcus bacterium have markedly different properties. The virulent smooth (S) strain possesses a smooth polysaccharide capsule that is essential for infection. The nonvirulent, rough (R) strain lacks this outer capsule, giving its surface a rough appearance.

Following injection with the S strain, mice succumb in several days to pneumonia. Although neither living R strain nor heat-killed S strain causes illness when injected alone, Griffith was surprised to find that coinjection of the two produces a lethal infection. Furthermore, he was able to retrieve virulent S strain bacteria from the infected mice. Through some process, the innocuous R strain had been transformed into the infective S strain. Griffith presented his hypothesis in 1928: Some "principle," transferred from the killed S strain, converted the R strain to virulence by enabling it to synthesize a new polysaccharide coat.

Over the next decade and a half, the "transforming principle" was earnestly pursued by a research team headed by Oswald T. Avery at the

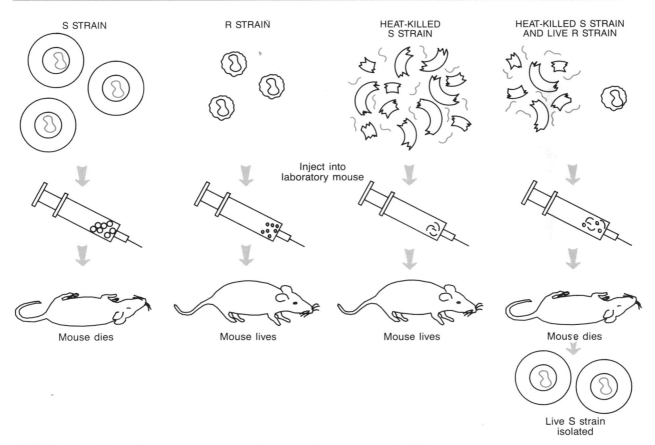

Griffith's Transformation Experiment with Smooth (S) and Rough (R) Strains of Pneumococcus, 1928

Rockefeller Institute. Between 1930 and 1933, they achieved transformation outside the body of a living mouse, observing microscopically the formation of polysaccharide coats in R pneumococci when the cultured bacteria were treated with extracts purified from the heat-killed S strain. A newcomer to Avery's laboratory, Colin MacLeod, showed that an extract containing only the polysaccharide coat could not transform R strains, disproving the belief that the coat itself was passed between strains during transformation.

In 1944, Avery, MacLeod, and Maclyn McCarty reported that they had purified the transforming principle. Analysis of molecular composition and weight indicated that their highly active fraction was primarily DNA. Several enzyme tests were conclusive. Transforming activity was unaffected by treatment with trypsin and chymotrypsin (which digest protein) and ribonuclease (RNase, which digests RNA). However, deoxyribonuclease (DNase) destroyed all transforming activity. Their work was meticulous, and Avery's interpretation was clear: "The inducing substance has been likened to a gene, and the capsular antigen which is produced in response to it has been regarded as a gene product." Therefore, the gene is composed of DNA. (In retrospect, we now know that synthesis of the capsular antigen requires several genes, not a single gene as originally implied by Avery.)

MOLECULAR "INTELLIGENCE"

Avery's work should have immediately focused attention on DNA as the molecule of heredity, but it did not. Like many scientific advances, his conclusions did not easily mesh with prevailing dogma—in this case, beliefs about the relative "intelligence" of protein versus DNA molecules. It was agreed that the molecule that functioned as the carrier of heredity must have the capacity to store, presumably in its molecular structure, huge amounts of genetic information. When compared to protein, DNA did not seem to be a very intelligent molecule. If genetic information was likened to a language, then the DNA language seemed impoverished compared to the protein language.

At a basic level, the protein alphabet has 20 letters (the amino acids), whereas DNA has only 4 letters (the nucleotides). On the face of it, DNA had a more limited combinatorial power—its language had fewer possible words. (Of course, this was before it was known how well a computer can function with a two-letter language.) If the vocabulary of DNA seemed limited, its syntax (sentence structure) was worse. Amino acids were known to articulate into an incredible diversity of large and complex protein molecules. The "tetranucleotide hypothesis," ardently propounded by noted chemist Phoebus Levene at the Rockefeller Institute, depicted DNA as a monotonously regular polymer, where the four nucleotides followed one another in an unchanging pattern. Furthermore, proteins were known to perform numerous important cellular functions, both as enzymes and structural elements. No obvious function for DNA had yet been discovered.

It is thus not hard to see why, for the better part of a decade, many scientists persistently argued against Avery's conclusion. They could not believe that his purified transforming principle had been completely cleansed of protein. DNA might merely be the scaffold to which were attached traces of the true transforming principle—a protein that had escaped enzyme digestion.

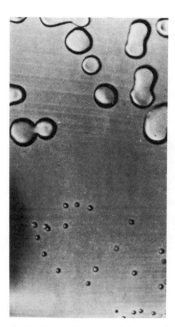

Colonies of Rough (Bottom) and Smooth Pneumococci (From Avery et al. 1944. Studies on the chemical nature of the substance inducing transformation of pneumococcal types. *J. Exp. Med.* 79: 137–158.)

Oswald Avery (Center Foreground) and Associates, 1932 (Courtesy of The Rockefeller Archive Center.)

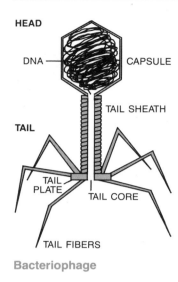

Bacteriophage

THE PHAGE GROUP

As World War II drew to a close, a remarkable group of scientists coalesced around three men who were primarily responsible for taking the search for the physical basis of heredity to the very fringe of life and nonlife. Max Delbrück, a German working at Vanderbilt University, and Salvador Luria, an Italian working at Indiana University, had fled to the United States from Nazi Europe; Alfred Hershey was an American working at the Carnegie Institution's Department of Genetics at Cold Spring Harbor, New York.

The object of their research was a group of tiny bacterial viruses, called bacteriophages (or simply, phages). It was known that during infection, phage particles reproduce inside the bacterial cell, which ruptures to release a new generation of viruses. Delbrück, Luria, and Hershey reasoned that the genetic interaction between bacteria and their virus parasites might provide an ideal model system to study the mechanism of heredity.

Although bacteriophages had been discovered in 1917 by a Canadian, Felix d'Hérelle, Delbrück and Luria were the first to work out quantitative methods for studying the phage life cycle. Key among these was "one-step growth"—ensuring that all viruses begin the infection cycle simultaneously. This synchrony was essential to study organisms so minute that they could not be studied individually.

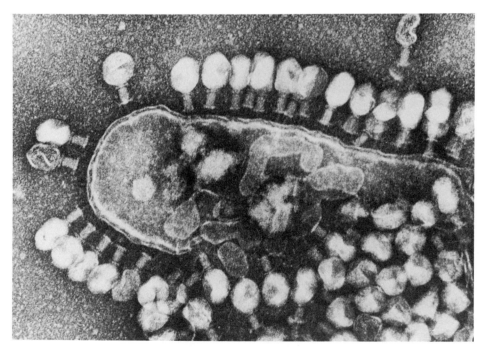

Bacteriophage T4 Infecting a Single *E. coli* Cell
The reproduction of bacterial viruses within *E. coli* provided a simple model to study the molecular basis of heredity. The phage particle is essentially a protein capsule surrounding a core of DNA. (Courtesy of Lee D. Simon, Photo Researchers Inc.)

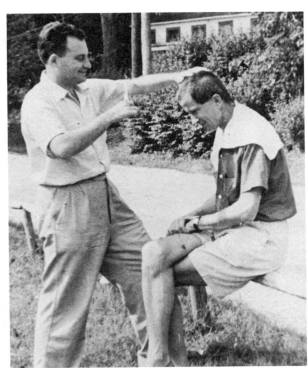

Max Delbrück (Left) and Salvador Luria, 1953

Seymour Benzer Gives Max Delbrück a Haircut, 1953

Camaraderie ranked high with research among members of the American Phage Group, which convened each summer at Cold Spring Harbor, New York. (Courtesy of Cold Spring Harbor Laboratory Archives.)

In 1945, Delbrück organized at Cold Spring Harbor Laboratory a course to introduce researchers to the methods that he and Luria had developed. Many historians regard the phage course as the intellectual watershed of molecular biology, for it drew into contact an ever-widening group of scientists, from both the biological and physical sciences, who would hammer out the molecular mechanics of heredity. The phage course, which was taught at Cold Spring Harbor for 26 consecutive years, was a training ground for the first two generations of molecular biologists.

THE BLENDER EXPERIMENT

Out of the research on phages came final proof that DNA is the molecule of heredity. The blender experiment, performed by Alfred Hershey and his assistant Martha Chase in 1952, focused attention on DNA in a way that Avery's experiments had not. Their experiment, although regarded as biochemically sloppy, more clearly made the connection between DNA and heredity.

The experimental design took advantage of the unique structure of the bacteriophage, in which an outer capsule of protein surrounds an inner core of DNA. No other organism yet discovered was so perfectly designed to settle the protein versus DNA debate. Also key to the

The Waring Blendor Used in the Hershey-Chase Experiment

experiment were radioactive isotopes, which had just become available following World War II. Radioactive "labels" allowed them to follow the fate of protein and DNA during bacteriophage replication.

Hershey and Chase ran parallel experiments with two populations of phages—one in which the protein capsules had been labeled with radioactive sulfur (^{35}S), and the other in which DNA was labeled with radioactive phosphorus (^{32}P). The radioactively labeled phages were allowed to infect cultures of bacteria, giving enough time for the viruses to attach to the bacterial cells. The cultures were then chilled to arrest growth and agitated for several minutes in a Waring Blendor, which detached the phage particles from the bacterial cells. The cultures were centrifuged at a speed fast enough to pellet the bacterial cells at the bottom of the tube, leaving the smaller phage particles in the supernatant. The cell pellet and supernatant were then analyzed for the presence of radioactively labeled phage protein or DNA. Hershey gave the following synopsis of the results in a later publication:

1. Most of the phage DNA remains with the bacterial cells.
2. Most of the phage protein is found in the supernatant fluid.
3. Most of the initially infected bacteria (in the cell pellet) remain competent to produce phage.
4. If the mechanical stirring is omitted, both protein and DNA sediment with the bacteria.
5. The phage protein removed from the cells by stirring consists of more or less intact, empty phage coats, which may therefore be thought of as passive vehicles for the transport of DNA from cell to cell and which, having performed that task, play no further role in phage growth.

Martha Chase and Alfred Hershey, 1953
(Courtesy of Cold Spring Harbor Laboratory Archives.)

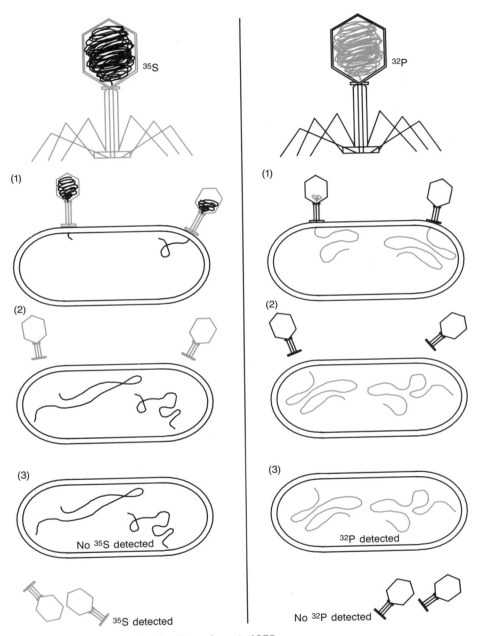

The Hershey-Chase Blender Experiment, 1952
Side-by-side experiments were performed with separate phage cultures in which either the protein capsule was labeled with radioactive sulfur (^{35}S) or the DNA core was labeled with radioactive phosphorus (^{32}P). (*1*) The radioactively labeled phages were allowed to infect bacteria. (*2*) Agitation in a blender dislodged phage particles from bacterial cells. (*3*) Centrifugation pelleted cells, separating them from the phage particles left in the supernatant. Radioactive sulfur was found predominantly in the supernatant. Radioactive phosphorus was found predominantly in the cell fraction, from which arises a new generation of infective phages.

What Is the Structure of the DNA Molecule?

Solving the structure of DNA is surely the most important biological discovery of the 20th century. James Watson, trained in the Phage Group as a geneticist, and Francis Crick, a physicist schooled in X-ray crystallography, were the embodiment of the confluence of genetics and physics that led to a detailed understanding of the physical basis of heredity. In their 1953 letter to *Nature*, they assembled pieces of a chemical puzzle that had been accumulating for more than 80 years.

Ironically, DNA was discovered in 1869, only 10 years after the publication of Darwin's *Origin of Species* and 4 years after Mendel's "Experiments in Plant-Hybridization." An enterprising German doctor, Friedrich Miescher, isolated a substance he called "nuclein," from the large nuclei of white blood cells. His source of cells was pus from soiled surgical bandages.

By 1900, the basic chemistry of nuclein had been worked out. It was known to be a long molecule composed of three distinct chemical subunits: a five-carbon sugar, acidic phosphate, and five types of nitrogen-rich bases (adenine, thymine, guanine, cytosine, and uracil). By the 1920s, two forms of nucleic acids were differentiated by virtue of their sugar composition: *ribo*nucleic acid (RNA) and *deoxyribo*nucleic acid (DNA). These forms were also found to differ slightly in base composition; thymine is found exclusively in DNA, whereas uracil is found only in RNA.

Knowledge about the physical arrangement of atoms within the DNA molecule was made possible by two achievements that occurred prior to World War II—the codification of physical chemistry and the development of X-ray crystallography. The laws of chemical bonding that govern the arrangement of atoms within molecules were rigorously formulated by Linus Pauling in *The Nature of the Chemical Bond*, a series of monographs he wrote between 1928 and 1935, while working at the California Institute of Technology. By the time Watson and Crick became interested in DNA, the molecular structures of all of its individual subunits were known—deoxyribose sugar, phosphoric acid, and each of the four nucleotides.

Determining the three-dimensional arrangement of the subunits within the huge DNA macromolecule was beyond the ability of Pauling's laws. This required the use of X-ray crystallography. In 1912, the German physicist Max von Laue discovered that X-rays are diffracted by the regularly arranged atoms of a simple crystal. In the same year, Australian physicist William Henry Bragg and his son, William Lawrence, worked out the mathematical equations needed to interpret diffraction patterns and used them to determine the molecular structure of table salt (sodium chloride). However, it was not until 1934 that John Desmond Bernal, at Cambridge University, obtained the first X-ray photograph of a biologically important molecule—the protein pepsin. This showed that the crystalline structures of giant organic molecules, including DNA, can be studied using X-ray diffraction methods.

The final three pieces of the DNA puzzle were uncovered only several

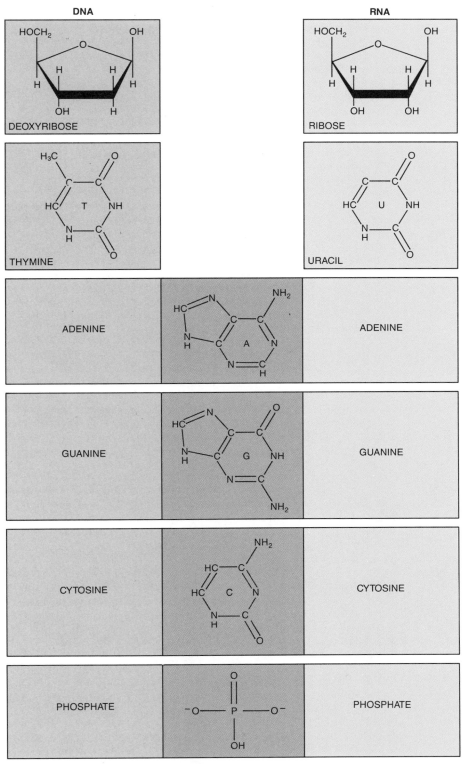

Components of DNA and RNA Molecules
(Art concept developed by Lisa Shoemaker.)

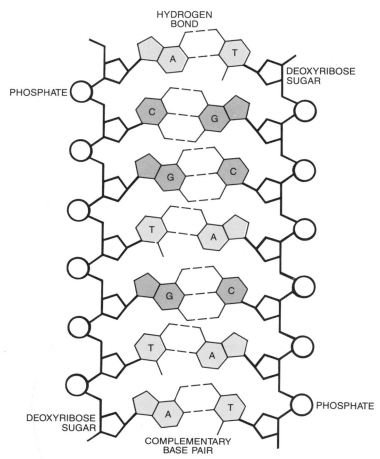

Chemical Structure of DNA

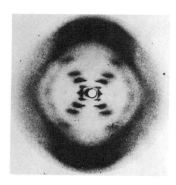

Rosalind Franklin's X-ray Diffraction Photograph of DNA, 1953
(From R.E. Franklin and R.G. Gosling. 1953. Molecular configuration in sodium thymonucleate. *Nature* 171: 740–741. Courtesy of Cold Spring Harbor Laboratory Archives.)

years prior to the publication of the Watson-Crick *Nature* paper. In 1950, Erwin Chargaff of Columbia University discovered a consistent one-to-one ratio of adenine to thymine and guanine to cytosine in DNA samples from a variety of organisms. In 1951, Linus Pauling and R.B. Corey obtained precise atomic measurements of a helical polypeptide structure, the α-helix. At about the same time, Maurice Wilkins and Rosalind Franklin, co-workers of Watson and Crick at Cambridge University, obtained sharp X-ray diffraction photographs of DNA. The diffraction patterns strongly suggested a helical molecule with a repeat of 34 Å and a width of 20 Å.

So Watson and Crick were faced with the problem of trying to get the DNA subunits to fit in a structure that conformed to known biochemical data and the laws of physical chemistry, yet at the same time functioned as the carrier of heredity. The structure they arrived at, by manipulating paper and then metal models, was elegant in its simplicity.

The DNA molecule they proposed is composed of two antiparallel α-helices and resembles a gently twisted ladder. The rails of the ladder, which run in opposite directions, contain alternating units of deoxyribose sugar and phosphate. The planar nucleotides stack tightly on top of one another, forming the rungs of the helical ladder. Each rung is composed of a pair of nucleotides—a base pair—held together by relatively weak hydrogen bonds. Consistent with the 34-Å repeat calculated from the X-ray diffraction data, there are 10 base pairs per turn of the helix, with 3.4 Å between adjacent base pairs. Of key importance is the complementary relationship between the nucleotides in each pair. In agreement with Chargaff's observation, adenine always pairs with thymine and cytosine always pairs with guanine. Thus, the nucleotide alphabet on one half of the DNA helix determines the alphabet of the other half.

Linus Pauling, about 1950

Rosalind Franklin, 1948

Maurice Wilkins, about 1955

Erwin Chargaff, 1947

Key Contributors to the Discovery of the Structure of DNA
Linus Pauling's rules of atomic bonding and elucidation of the α-helix structure provided the intellectual scaffold on which the DNA structure was built. Erwin Chargaff's data on the relative abundance of the four nucleotides in the DNA of a number of organisms suggested the complementary arrangement of adenine-thymine and guanine-cytosine pairs. While working in the laboratory of Maurice Wilkins, Rosalind Franklin took the clearest X-ray diffraction photo of DNA, from which were derived measurements of the key physical parameters of the DNA molecule. (Courtesy of California Institute of Technology (*left*); Anne Sayre (*top right*); Cold Spring Harbor Laboratory Archives (*bottom, far right*).)

Frank Stahl, about 1953

Matthew Meselson, 1958

Meselson and Stahl provided experimental evidence for the semiconservative replication of DNA. (Courtesy of Cold Spring Harbor Laboratory Archives; California Institute of Technology Archives.)

HOW DOES DNA STRUCTURE DESCRIBE REPLICATION?

The genetic implication of the Watson-Crick structure was immediately apparent. This "pretty molecule," as Watson informally described it, embodied the organizing thesis of molecular biology—understanding the structure of a molecule gives clues to its biological function. The Watson-Crick structure could neatly explain how the DNA molecule precisely replicates during cell division, so that each daughter cell receives an identical set of hereditary instructions. The hydrogen bonds between nucleotides could break, allowing the DNA ladder to unzip. Each complementary half would serve as a template for the reconstruction of the other half. The end result would be two identical DNA molecules, one for each daughter cell.

Experimental support for the Watson-Crick hypothesis came in 1958. Matthew Meselson and Frank Stahl, at the California Institute of Technology, devised an ingenious proof of semiconservative replication, during which one parental DNA strand serves as the template for the synthesis of a new, complementary strand.

Their experiment relied on the ability to differentiate the densities of two isotopes of nitrogen, "heavy" ^{15}N and "light" ^{14}N. *Escherichia coli* bacteria were first grown in a nutrient medium where ^{15}N was the only source of nitrogen. During replication, heavy nitrogen was incorporated into the nucleotides of all bacterial DNA. After 14 generations of growth, the bacteria were abruptly switched to a medium containing only ^{14}N and allowed to continue to reproduce. Light nitrogen would be incorporated into all DNA subsequently replicated.

Samples of cells were taken from the generation before the switch in growth media (1) and from the two generations following the switch (2 and 3). DNA extracted from the cells was centrifuged in a solution of cesium chloride, which forms a density gradient when spun for 20 hours

Arthur Kornberg, about 1965 (Courtesy of Stanford University.)

at 40,000 revolutions per minute. DNA molecules of equal density settle during centrifugation to form discrete bands, "floating" at a point where their density exactly equals that of the cesium chloride gradient.

The results were picture perfect. A gradient of DNA from generation 1 contained a single high-density band—all DNA molecules from the parent generation contained two heavy strands. DNA from generation 2 contained a single band of medium density—every daughter molecule had one heavy strand from the parent and one complementary strand constructed of light nitrogen. DNA from generation 3 showed two density bands. The DNA in the intermediate-density band (as in generation 2) was composed of a heavy parental strand and a new, light complementary strand. The DNA in the light-density band consisted of two strands of light DNA, an inherited parental strand and its new complement. Mixing the DNA from generations 1 and 3 showed bands representing all three types of molecules—heavy, intermediate, and light.

In the same year, Arthur Kornberg, at Washington University, elucidated the enzymatic mechanism of DNA replication. He purified an enzyme from *E. coli*, called DNA polymerase I, that synthesizes a polynucleotide chain from trinucleotide phosphate precursors. Consistent with the Watson-Crick model, he found that synthesis only occurs in the presence of a DNA template.

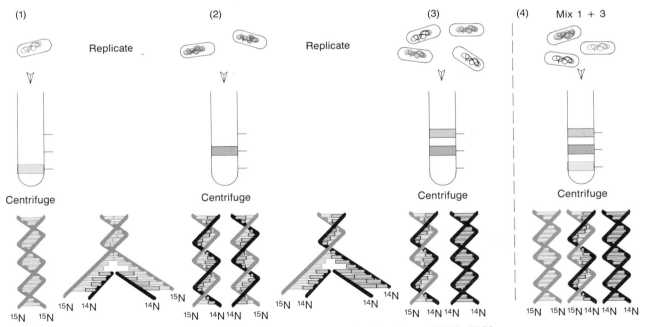

The Meselson-Stahl Experiment Showing Semiconservative Replication of DNA, 1958
(*1*) Following density-gradient centrifugation, DNA isolated from bacteria grown for many generations in a medium containing heavy nitrogen (^{15}N) forms a single band of high density. (*2*) DNA isolated from the generation following a switch to ^{14}N-containing medium shows a single, medium-density band. (*3*) In the subsequent generation, the DNA segregates into a medium-density and a low-density band. (*4*) Mixing DNA from cultures 1 and 3 compares the three density bands.

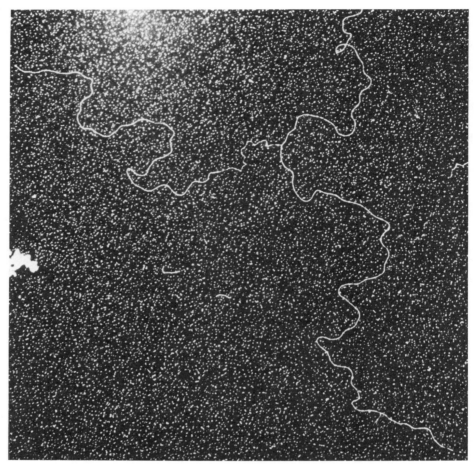

DNA Replication
Parental DNA molecule forks near middle of this electron micrograph to form two daughter molecules. (Courtesy of D. Dressler and J. Wolfson, Harvard Medical School.)

How Does DNA Structure Describe Protein Synthesis?

Replication was the simple half of the DNA structure-function question. More difficult was a molecular explanation of the biochemical basis of heredity—the one gene/one enzyme hypothesis of Garrod, Beadle, and Tatum. How is the genetic language of DNA translated into the synthesis of specific proteins? In 1957, Francis Crick laid the intellectual framework that would guide the cracking of the genetic code:

1. The sequence hypothesis stated that DNA sequence and protein sequence are colinear. Genetic information must therefore be arrayed in a strictly linear fashion along the length of the DNA molecule.

2. The "central dogma" stated that genetic information stored in DNA flows through RNA to proteins. RNA is the intermediate translator of the genetic code.

Proof of the sequence hypothesis came from extremely fine genetic mapping of mutations in *E. coli* by Charles Yanofsky (at Stanford

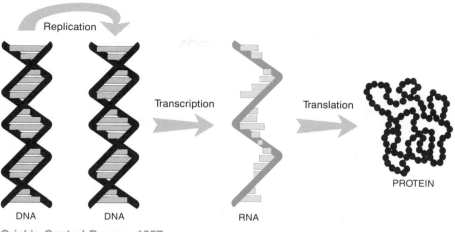

Crick's Central Dogma, 1957

University) and in the bacteriophage T4 by Seymour Benzer (at Purdue University). They confirmed that mutations at various points in a single gene—changes in its linear DNA sequence—produce parallel changes in the linear amino acid sequence of the gene's protein product.

Several types of RNA molecules were found to be involved in the flow of genetic information. In the first step, the DNA code is *transcribed* in the nucleus into messenger RNA (mRNA), which stores a complementary copy of the DNA code. In the second step, the mRNA transcript associates with the ribosome, where with the help of ribosomal RNA (rRNA) and transfer RNA (tRNA), the mRNA transcript is *translated* into amino acids.

Evidence for the transcriptional role of RNA was provided in 1960, when three American research groups independently isolated RNA polymerases that synthesize an RNA strand from a DNA template.

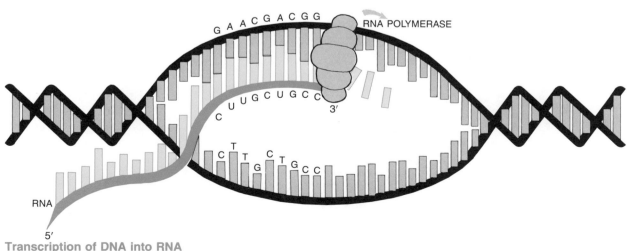

Transcription of DNA into RNA

Sydney Brenner, 1980
(Courtesy of Cold Spring Harbor Laboratory Archives.)

Stronger support was provided a year later by Benjamin Hall and Sol Spiegelman, at the University of Illinois. They purified RNA from bacteriophage T2 and found it would hydrogen bond to T2 DNA that had been denatured into single strands, forming hybrid RNA/DNA molecules. However, the T2 RNA would not form hybrids with the DNA from other organisms, even from the closely related T5 phage. This indicated that a specific DNA code is transcribed into a complementary mRNA sequence.

As early as 1955, Francis Crick had hypothesized that some type of "adaptor" molecule must attach to each amino acid and align it on the mRNA template during translation. This prediction was proved true in the next decade, when it was discovered that a different tRNA is the acceptor for each of the 20 amino acids.

In 1961, Crick and Sydney Brenner performed at Cambridge University experiments suggesting that the genetic code consists of three-letter sequences of nucleotides, called "codons." Led by Robert Holley, at Cornell University, tRNAs were purified and sequenced in the mid-1960s. It was shown that on one side of the "cloverleaf" structure is a three-nucleotide "anticodon" that binds to a complementary mRNA codon; on the other side is the binding site for one of the 20 amino acids. During protein synthesis, tRNAs successively bind to their complementary mRNA codons. The ribosome acts as a scaffold to hold the tRNA molecules in close contact while peptide bonds are formed between adjacent amino acids that form the specified protein.

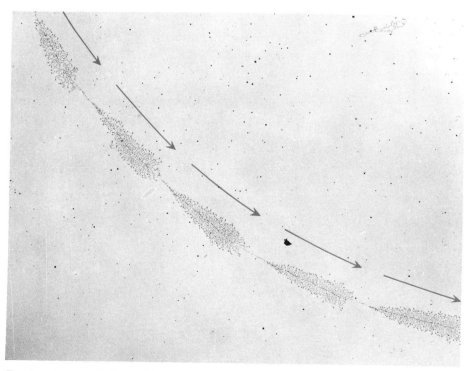

Tandem Transcription of Several Genes
Newly synthesized RNA molecules appear as "bristles" off transcribed regions of a DNA molecule. Arrows indicate direction of transcription, with lengthening RNA chains toward the 5' end of each gene. (Courtesy of O.L. Miller, University of Virginia.)

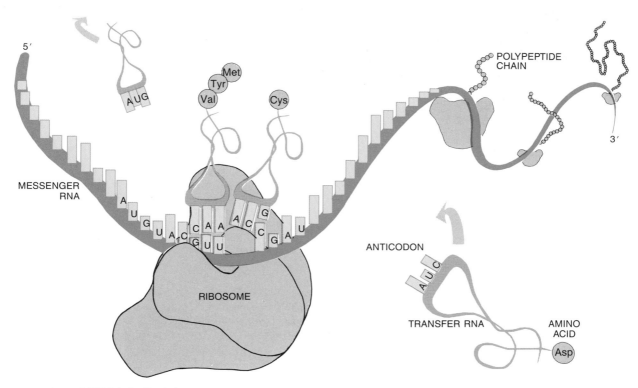

Translation of RNA into Protein

Marshall Nirenberg, about 1983
(Courtesy of Cold Spring Harbor Laboratory Archives.)

The genetic code was cracked by 1966, primarily by teams led by Marshall Nirenberg, at the National Institutes of Health, and H. Gobind Khorana, at the University of Wisconsin. Working independently, they developed cell-free translation systems—consisting of ribosomes, tRNA, and amino acids—that efficiently manufactured polypeptides in vitro. By introducing synthetic mRNAs of known sequence, they confirmed that triplet RNA codons specify each of the 20 amino acids. For example, in 1961, Nirenberg and his colleague Heinrich Matthei first found that a synthetic mRNA consisting only of uracil (making codons UUU-UUU-UUU...) produced polypeptides consisting of long chains of the amino acid phenylalanine.

All possible mRNA combinations were tried, yielding a complete genetic "dictionary" for the translation of mRNA into amino acids. Since nearly all proteins begin with the amino acid methionine, its codon (AUG) represents the "start" signal for protein synthesis. Three codons for which there are no naturally occurring tRNAs—UAA, UAG, and UGA—are stop signals that terminate translation.

Interestingly, only methionine and tryptophan are specified by a single codon; all other amino acids are specified by two or more different codons. Because of this redundancy, single-base changes in RNA often do not change the amino acid coded. For example, any codon beginning GG, regardless of the nucleotide in the third position (GGU, GGC, GGA, or GGG), specifies the amino acid glycine.

AMENDMENTS TO THE CENTRAL DOGMA

Significant amendments and refinements have been made to Crick's central dogma since the cracking of the genetic code. It is now known that the flow of genetic information is not one-way. The molecular genetic system (DNA-RNA-protein) is dynamic; elements exert feedback regulation in ways that were not originally conceived. Amendments have been so extensive that it is perhaps better to think that there is no longer a central dogma of DNA.

Although DNA is, by and large, a very stable molecule, the classic concept of a gene as a unique entity is becoming less and less useful. The genetic material is plastic, so that a single gene may, through rearrangement and biochemical editing, be responsible for the production of several different proteins. The amendments listed below will be discussed in subsequent chapters.

1. Some organisms do not use DNA as the storage molecule for the genetic code. RNA viruses, or retroviruses, store genetic information as RNA. Prior to replication and protein synthesis, their genetic information is converted to DNA by the enzyme reverse transcriptase. This represents a backflow of genetic information.
2. Genes are not immutably fixed on the chromosomes. Transposable genetic elements move about from one chromosomal spot to another and may act as molecular switches to regulate gene expression. Physical rearrangement of genes permits a vast array of proteins to be produced from a relatively limited amount of DNA code.
3. DNA sequence and protein sequence are not entirely colinear. The RNA transcript is often extensively "edited" prior to protein synthesis. Interspersed among eukaryotic genes are DNA sequences with no apparent function. These are spliced out of the primary RNA transcript to yield coherent directions for protein synthesis. Differential splicing can produce several different proteins from the same "gene."

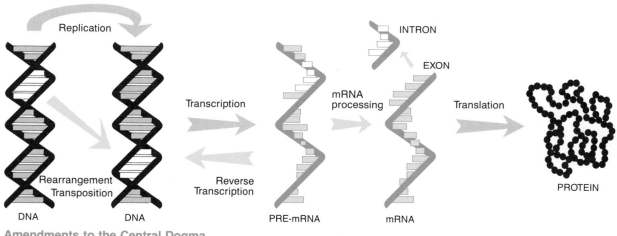

Amendments to the Central Dogma

4. RNA has catalytic ability previously attributed only to proteins. In some organisms, RNA carries on self-splicing reactions, suggesting that self-replicating RNA may have preceded DNA in molecular evolution.
5. Depending on precisely where transcription begins, a single DNA sequence can be read in several overlapping "reading frames" of codons, which may be translated into different polypeptides.

For Further Reading

Crick, Francis. 1988. *What mad pursuit.* Basic Books Inc., New York.

Dickerson, Richard E. 1983. The DNA helix and how it is read. *Sci. Am.* 249/6: 94–111.

Judson, Horace Freeland. 1979. *The eighth day of creation.* Simon and Schuster, New York.

Korey, Kenneth, ed. 1984. *The essential Darwin.* Little, Brown and Company, Boston.

McCarty, Maclyn. 1985. *The transforming principle: Discovering that genes are made of DNA.* W.W. Norton and Co., New York.

The molecules of life: Readings from Scientific American. (Special issue.) 1985. 253/4.

Moore, John A., ed. 1986. *Science as a way of knowing. III—Genetics.* American Society of Zoologists, Washington, D.C.

Sayre, Anne. 1975. *Rosalind Franklin and DNA.* W.W. Norton and Co., New York.

Watson, James D. 1980. *The double helix: A personal account of the discovery of the structure of DNA.* W.W. Norton and Co., New York.

CHAPTER 3
Basic Tools and Techniques of DNA Science

The comedy movie "Sleeper," which tells of a totalitarian society of the 21st century, contains a cloning scene that is hard to forget. "The Leader" has been assassinated, but his nose has been kept alive through the wonders of futuristic science. The bungling Miles Monroe (Woody Allen) and his sidekick Luna Schlosser (Diane Keaton) masquerade as surgeons who know the procedure to reproduce a new Leader from the saved nose cells. Asked if he minds performing the operation while numerous doctors observe in the surgical theater, Miles replies, "Never clone alone." Setting the nose in its proper position on a person-shaped cloning table, he confirms his mission: "What you want, basically, is a whole entire person connected to that nose, right? Do you want me to leave room for a mustache?"

This scene satirizes the hype and misconception of the word cloning. It conjures up images of regenerating an entire human being, or other complex organism, from a small sample of body tissue. In theory, this is possible. Since every cell carries in its chromosomes the same basic set of genetic instructions as in the fertilized ovum, somatic cells that make up body tissues could provide the raw genetic information needed to recreate "from scratch" an entire organism.

Woody Allen and Diane Keaton as Cloners in "Sleeper", 1973
(Courtesy of the Museum of Modern Art Film Stills Archive.)

There are real-life examples of this sort of cloning. The mature cells from certain plants *can* be stimulated to produce a new adult plant. All animals have some ability to regenerate injured cells and, in some cases, even entire body parts. Amphibians and reptiles, for example, can regenerate new tails. In practice, however, fully differentiated cells taken from most adult organisms lack the regenerative ability of plant cells and cannot be stimulated to "clone" a complete organism. In the vast majority of higher organisms, only the fertilized ovum is able to progress through the complex developmental stages that culminate in a fully functional adult. Technology for cloning mammals from single cells does not now exist, nor does it seem very close at hand.

Taken in its pure sense, the term cloning refers to the reproduction of daughter cells by fission or mitotic division. During these processes, DNA from a parent cell is replicated, and identical sets of genetic information are passed on to daughter cells. Successive generations of cells in turn divide, giving rise to a population of genetically identical *clones*, all derived from a single ancestral cell.

Gene cloning uses the natural replicating ability of cells to isolate and duplicate an individual gene. First, the gene of interest is inserted into a carrier DNA molecule, termed a vector, that can self-replicate within a host cell. Then the vector, with its gene insert, is introduced ("transformed") into an appropriate host cell. Subsequent mitosis of the host cell creates a population of clones, each containing the gene of interest.

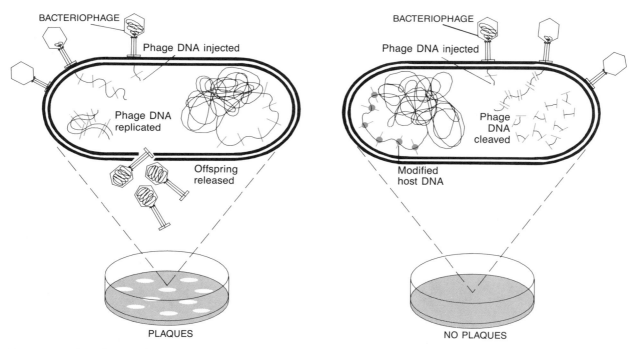

Host-controlled Restriction
Bacteriophage infects strain at left, producing clear plaques in bacterial lawn. Resistant strain at right possesses a restriction enzyme that cleaves incoming phage DNA, as well as a modifying enzyme that methylates its own DNA to protect it from cleavage. (Art concept developed by Lisa Shoemaker.)

Restriction-Modification

In the early 1950s, Salvador Luria and Giuseppe Bertani (University of Illinois) and Jean Weigle (California Institute of Technology) found evidence for a sort of primitive immune system in bacteria. They observed that certain strains of the bacterium *Escherichia coli* are resistant to infection by various bacteriophages. The phenomenon seemed to be a property of the bacterial cell, which is able to *restrict* the growth and replication of phages. In 1962, Werner Arber, at the University of Geneva, provided the first evidence that the resistant bacterium possesses an enzyme system that selectively recognizes and destroys foreign phage DNA within the bacterial membrane and that also *modifies* the chromosomal DNA of the bacterium to prevent self-destruction.

Several years later, Arber and his associate Stuart Linn, as well as Matthew Meselson and Robert Yuan at Harvard University, isolated *E. coli* extracts that efficiently cleave phage DNA. These extracts contained the first known *restriction endonucleases*, enzymes that attack and digest internal regions of the DNA of an invading bacteriophage but not that of the host. This was shown to occur because, in addition to a restriction (cutting) activity, these enzymes also possess a modification (protecting) activity.

It was later found that the modifying enzyme protects host DNA from digestion by adding methyl groups to a nucleotide within the sequence recognized by the restriction enzyme. This modification of the sequence blocks the restriction enzyme from digesting the DNA. Although typically both strands of host DNA are methylated, bacterial DNA is protected from digestion even when one strand is methylated. Therefore, during DNA replication, the methylated parental strand protects the molecule until the newly synthesized daughter strand can also be methylated.

Because their cutting activity is not precise, the enzymes isolated by Arber, Linn, Meselson, and Yuan were of no practical value as tools for manipulating DNA. Although these enzymes recognize specific nucleotide sequences, they cut the DNA molecule randomly at positions that may be thousands of nucleotides distant from the recognition site.

The phenomenon of restriction-modification thus remained of purely academic interest until 1970, when Hamilton Smith and his student Kent Wilcox, at Johns Hopkins University, isolated a new restriction endonuclease from *Haemophilus influenzae*. The restriction activity of this enzyme, named *Hin*dII, differs from those previously described in two important ways. First, the restriction activity is separate from the modification activity. Second, it cleaves DNA predictably, cutting *within* its recognition sequence.

Daniel Nathans, a colleague of Smith's at Johns Hopkins, then showed the broad applicability of restriction endonucleases. He used the *Hin*dII enzyme to cut the purified DNA of a small virus that infects monkeys (simian virus 40, or SV40) and separated the resulting *restriction fragments* by size in an electrical field. He deduced the order of the fragments (and corresponding restriction sites) in the 5000-nucleotide circular chromosome, creating a *restriction map* that was then related to the existing genetic map of SV40.

The First Recombinant DNA Molecules

In the early 1970s, the first recombinant molecules were produced independently using two methods. One method was discovered at Stanford University under the direction of Paul Berg; the other was developed jointly by the laboratories of Stanley Cohen, also at Stanford, and Herb Boyer at the University of California, San Francisco.

In 1972, Berg's group worked out the "tailing" method of joining DNA molecules, which was modeled after the "sticky ends" found at the chromosome ends of the bacteriophage λ. These complementary, single-stranded regions of DNA join spontaneously to form a circular DNA molecule. Using the restriction enzyme *Eco*RI that had been recently isolated from *E. coli*, they cut the circular SV40 chromosome and a small, circular chromosome (termed a plasmid) from *E. coli*. The restriction enzyme cuts each molecule in only a single place, opening the DNA loops to produce linear strands.

To produce sticky ends, a single-stranded "tail" of 50–100 adenine residues was added to the ends of the SV40 DNA using the enzyme terminal transferase. A tail of thymine residues was added to the plasmid DNA by the same method. When the two DNAs were mixed together, the complementary adenine and thymine tails base-paired to form a

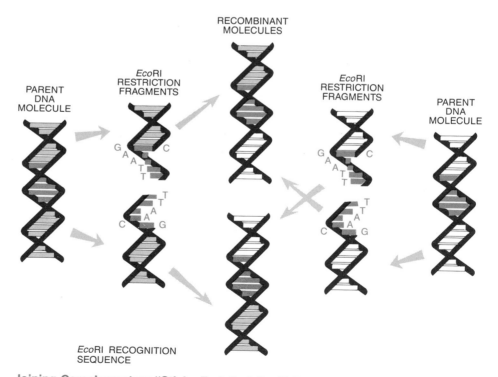

Joining Complementary "Sticky Ends" of *Eco*RI Restriction Fragments

circular, recombinant DNA molecule. Two enzymes completed the job: DNA polymerase filled in any single-stranded gaps, and DNA ligase sealed the junction points between the SV40 and plasmid fragments.

Berg had planned to introduce his recombinant plasmid into animal cells and then look for expression of the bacterial genes in the infected animal cells. We know in hindsight that this would have failed, because expression of a foreign gene requires a detailed understanding of its regulation (see Chapter 4). Regardless, Berg never got the opportunity to test his new DNA molecule. Concern in the scientific community over the potential danger of the experiment halted research on gene transfer into mammalian cells for several years.

While not completed, Berg's experiment provided two key pieces of the recombinant DNA puzzle. He showed that a restriction enzyme can be used to cut DNA in a predictable manner and that DNA fragments from different organisms can be joined together. In 1973, Stanley Cohen and Annie Chang carried the Berg experiment a step further, adding the final piece to the puzzle. They showed that a recombinant DNA molecule can be maintained and replicated within *E. coli*.

Cohen and Chang purified from *E. coli* a plasmid that incorporated several important features: (1) a unique restriction recognition site, to allow the molecule to be cut at a single location by the restriction enzyme *Eco*RI; (2) a nucleotide sequence, the origin of replication, to allow the plasmid to be replicated within a host bacterial cell; and (3) a gene coding for resistance to the antibiotic tetracycline, to allow for selection of bacteria incorporating the molecule. The new plasmid was named pSC101 (SC for Stanley Cohen).

The first step was to devise a method to introduce this plasmid into *E. coli* efficiently. Using a modification of a transformation protocol described 3 years earlier by Morton Mandel and A. Higa (at the University of Hawaii), they inserted pSC101 into *E. coli* cells. The transformed cells were then spread on nutrient agar plates containing tetracycline; the appearance of colonies of bacteria with resistance to the antibiotic showed that the plasmid had been taken up and expressed.

The next step was to construct and insert a recombinant plasmid. For this, a second plasmid, pSC102, was isolated that contained an antibiotic resistance gene for kanamycin. Herb Boyer's laboratory had recently demonstrated that *Eco*RI creates its own sticky ends when it cuts, making Berg's tailing method unnecessary. So, pSC101 and pSC102 were simply cut with *Eco*RI, mixed together, and rejoined with DNA ligase. The resulting recombinant plasmid was transformed into *E. coli* cells, which were plated on media containing both tetracycline and kanamycin. The appearance of colonies with *dual* resistance confirmed that a recombinant DNA molecule had been introduced successfully into living bacterial cells.

Shortly thereafter, Cohen and Boyer teamed up to produce a recombinant molecule containing DNA from two different species. To accomplish this, they spliced a gene encoding a ribosomal RNA from the toad *Xenopus laevis* into the plasmid pSC101 and transformed the recombined DNA into *E. coli*. Some of the colonies resistant to tetracycline were found to contain the toad ribosomal RNA gene.

The Boyer-Cohen-Chang Experiment, 1973

The recombinant DNA techniques introduced by Berg, Chang, Boyer, and Cohen enabled biologists to change the genetic constitution of a living thing in a controlled manner and to transcend established species barriers. Each novel DNA combination creates a new biological entity with altered genetic and biochemical characteristics.

Restriction Endonucleases

Restriction endonucleases, or restriction enzymes, are used as molecular scalpels to cut DNA in a precise and predictable manner. They are members of the class of nucleases, which display the general property of breaking the phosphodiester bonds that link adjacent nucleotides in DNA and RNA molecules. *Endo*nucleases cleave nucleic acids at internal positions, while *exo*nucleases progressively digest from the ends of nucleic acid molecules.

There are three major classes of restriction endonucleases. Type I and type III enzymes have both restriction (cutting) and modification (meth-

Hamilton Smith (Left) and Daniel Nathans, 1978

Paul Berg, 1980

Herbert Boyer, 1977

Stanley Cohen and Annie Chang, 1973

Key Figures in the Construction of the First Recombinant DNA Molecules
(Courtesy of (*Top*) The Baltimore Sun; News and Publication Service, Stanford University; (*Bottom*) Paul Conklin for Academy Forum; News and Publications Service, Stanford University.)

ylating) activity. Both types cut at sites some distance from their recognition sequences; ATP is required to provide energy for movement of the enzyme along the DNA molecule from recognition site to cleavage site. Type I enzymes, such as *Eco*K isolated by Meselson and Yuan, cut at random sites 1000 nucleotides or more away from the recognition sequence. Type III enzymes cut at specific sites quite near the recognition sequence, but these may be difficult to predict.

The restriction enzymes used in DNA science are invariably type II. They are most useful for several reasons: (1) Each has only restriction activity; modification activity is carried by a separate enzyme. (2) Each cuts in a predictable and consistent manner, at a site within or adjacent to the recognition sequence. (3) They require only magnesium ion (Mg^{++}) as a cofactor; ATP is not needed.

Today, more than 1200 type II enzymes have been isolated from a variety of prokaryotic organisms. Enzymes have been identified that recognize more than 130 different nucleotide sequences; more than 70 types are commercially available. To avoid confusion, restriction endonucleases are named according to the following nomenclature:

1. The first letter is the initial letter of the genus name of the organism from which the enzyme is isolated.
2. The second and third letters are usually the initial letters of the organism's species name. Since they are derived from scientific names, the first three letters of the endonuclease name are italicized.
3. A fourth letter, if any, indicates a particular strain of organism.
4. Originally, Roman numerals were meant to indicate the order in which enzymes, isolated from the same organism and strain, are eluted from a chromatography column. More often, though, the Roman numerals indicate the order of discovery.

*Eco*RI E = genus *Escherichia*
       co = species *coli*
       R = strain RY13
       I = first endonuclease isolated

*Bam*HI B = genus *Bacillus*
       am = species *amyloliquefaciens*
       H = strain H
       I = first endonuclease isolated

*Hin*dIII H = genus *Haemophilus*
       in = species *influenzae*
       d = strain Rd
       III = third endonuclease isolated

By some unknown mechanism, a type II restriction endonuclease scans a DNA molecule, stopping only when it recognizes a specific sequence of nucleotides. Most restriction enzymes recognize a four- or six-nucleotide sequence. Assuming that the four component nucleotides (A,C,T,G) are distributed randomly within a DNA molecule, then any four nucleotides

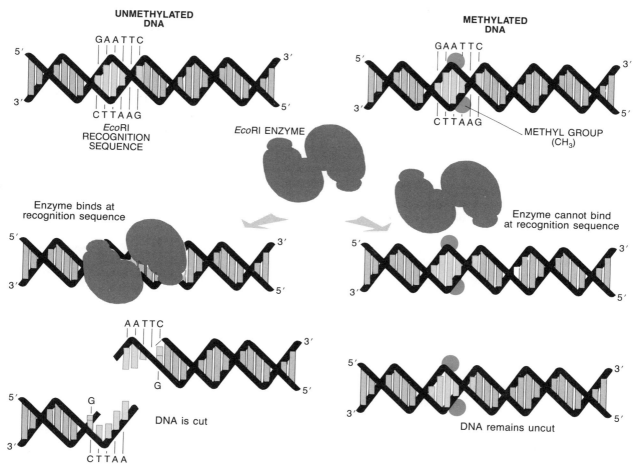

Molecular Detail of *Eco*RI Restriction-Modification
(Art concept developed by Lisa Shoemaker.)

will occur, on average, every 256 nucleotides ($4 \times 4 \times 4 \times 4$), and a six-nucleotide recognition site will occur every 4096 nucleotides ($4 \times 4 \times 4 \times 4 \times 4 \times 4$).

Many restriction enzymes have recognition sites that are composed of symmetrical, or palindromic, nucleotide sequences. This means that the recognition sequence read forward on one DNA strand is identical to the sequence read in the opposite direction on the complementary strand. Put another way, the 5'→3' sequence is identical on each DNA strand.

In a general sense, the terms 5' and 3' refer to either end of a single DNA strand. Specifically, they designate carbon atoms on opposite sides of the deoxyribose ring that are joined to form the DNA polymer. The 3' carbon is linked, through phosphoester bonds with an intervening phosphate, to the 5' carbon of the adjacent nucleotide. By convention, the nucleotide sequence is "read" from the 5' end to the 3' end. In duplex DNA, it is usual to show only one of the strands, since the complementary strand can be deduced from the base-pairing rules. Note that in duplex DNA, the strands are antiparallel; that is, 5'→3' reads in opposite directions on the two complementary strands.

Within or very near the recognition site, the restriction enzyme catalyzes a hydrolysis reaction that uses water to break a specific phosphodiester linkage on each strand of the DNA helix. Two DNA fragments are produced, each with a phosphate group at the 5' end and a hydroxyl group at the 3' end.

For each type II endonuclease, there is a corresponding modifying enzyme that blocks restriction activity by methylating specific nucleotides within the recognition sequence. The protruding methyl group presumably prevents catalysis by interfering with the close molecular interaction between the restriction enzyme and its recognition site. *Eco*RI methylase, for example, adds a methyl group to the second adenine residue within the *Eco*RI recognition site.

Some endonucleases, such as *Hin*dII, cut cleanly through the DNA helix by cleaving both complementary strands at the same nucleotide position, typically in the center of the recognition site. These enzymes leave flush- or blunt-ended fragments. Other endonucleases cleave each strand off-center in the recognition site, at positions two to four nucleotides apart. This creates fragments with exposed ends of short, single-stranded sequences. Various enzymes leave single-stranded "overhangs" on either the 5' or 3' ends of the DNA fragments. *Eco*RI, *Bam*HI, and *Hin*dIII, for example, each leave 5' overhangs of four nucleotides.

Single-stranded overhangs, also called cohesive or "sticky" ends, are extremely useful in making recombinant DNA molecules. These exposed nucleotides serve as a template for realignment, allowing the complementary nucleotides of two like restriction fragments to hydrogen bond to one another. A given restriction enzyme cuts all DNA in exactly the same fashion, regardless of whether the source is a bacterium, a plant, or a human being. Thus, any sticky-ended fragment can be recombined with any other fragment generated by the same restriction enzyme.

Specificities of Some Typical Restriction Endonucleases

Source	Enzyme	Recognition sequence	Number of cleavage sites		
			λ	Adenovirus-2	SV40
Bacillus amyloliquefaciens H	*Bam*HI	G↓GATCC	5	3	1
Bacillus globigii	*Bgl*II	A↓G$\overset{*}{A}$TCT	6	12	0
Escherichia coli RY13	*Eco*RI	G↓$\overset{*}{A}$ATTC	5	5	1
Escherichia coli R245	*Eco*RII	↓CCTGG	>35	>35	16
Haemophilus aegyptius	*Hae*III	GG↓CC$_*$	>50	>50	19
Haemophilus influenzae R_d	*Hin*dII	G$\overset{*}{T}$Py↓PuAC	34	>20	7
Haemophilus influenzae R_d	*Hin*dIII	$\overset{*}{A}$↓AGCTT	6	11	6
Haemophilus parainfluenzae	*Hpa*II	C↓CGG	>50	>50	1
Nocardia otitidis-caviarum	*Not*I	GC↓GGCCGC	0	7	0
Providencia stuartii 164	*Pst*I	CTGCA↓G	18	25	2
Serratia marcescens S_b	*Sma*I	CCC↓GGG	3	12	0

Recognition sequences are written 5' to 3'. Only one strand is represented. The arrows indicate cleavage sites. Pu (purine) denotes that either A or G will be recognized. Py (pyrimidine) denotes that either C or T will be recognized. Asterisks represent positions where bases can be methylated.

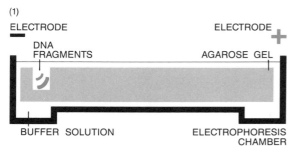 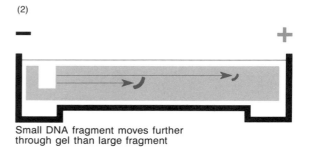

Agarose Gel Electrophoresis of DNA Fragments
(Art concept developed by Lisa Shoemaker.)

Agarose Gel Electrophoresis

Polyacrylamide gel electrophoresis had been used for many years to separate RNA and protein molecules by molecular weight. In 1970, Daniel Nathans used polyacrylamide gel electrophoresis as a simple and rapid means to separate DNA restriction fragments. Prior to this, DNAs of different sizes were separated by velocity-sedimentation ultracentrifugation, a laborious method that determines only the relative sizes of DNA restriction fragments. Whereas centrifugation uses gravitational force to separate molecules, electrophoresis means literally *to carry with electricity*.

Gel electrophoresis takes advantage of the fact that, as an organic acid, DNA is negatively charged. DNA owes its acidity to phosphate groups that alternate with deoxyribose to form the rails of the double helix ladder. In solution, at neutral pH, negatively charged oxygens radiate from phosphates on the outside of the DNA molecule. When placed in an electric field, DNA molecules are attracted toward the positive pole (anode) and repelled from the negative pole (cathode).

During electrophoresis, DNA fragments sort by size in the polyacrylamide gel. The porous gel matrix acts as a molecular sieve through which smaller molecules can move more easily than larger ones; thus, the distance moved by a DNA fragment is inversely proportional to its molecular weight. In a given period of time, smaller restriction fragments migrate relatively far from the origin compared to larger fragments. Because of its small pore size, polyacrylamide efficiently separates small DNA fragments of up to 1000 nucleotides. However, this level of resolution was inappropriate for isolating gene-sized fragments of several thousands of nucleotides.

Although less time consuming than centrifugation, early polyacrylamide gel electrophoresis was still labor intensive and required the use of radioactively labeled DNA fragments. Following electrophoresis, the polyacrylamide gel was cut into many bands, and the amount of radioactivity in each slice was determined in a scintillation counter. The pattern of radioactivity was used to reconstruct the pattern of DNA bands in the gel.

A research team at Cold Spring Harbor Laboratory, led by Joseph

Sambrook, introduced two important refinements to DNA electrophoresis that made possible rapid analysis of DNA restriction patterns. First, they replaced polyacrylamide with agarose, a highly purified form of agar. An agarose matrix can efficiently separate larger DNA fragments ranging in size from 100 nucleotides to more than 50,000 nucleotides. DNA fragments in different size ranges can be separated by adjusting the agarose concentration. A low concentration (down to 0.3%) produces a loose gel that separates larger fragments, whereas a high concentration (up to 2%) produces a stiff gel that resolves small fragments.

Second, they used a fluorescent dye, ethidium bromide, to stain DNA bands in agarose gels. Following a brief staining step, the fragment pattern is viewed directly under ultraviolet (UV) light. This technique is extremely sensitive; as little as 5 ng (0.005 μg) of DNA can be detected. Thus, it is not difficult to understand why ethidium bromide staining quickly replaced radioactive labeling for the routine analysis of DNA restriction patterns.

Currently used methods are identical to those published by the Cold Spring Harbor team in 1973. Molten agarose is poured into a casting tray in which a plastic or Plexiglas comb is suspended. As it cools, the agarose hardens to form a Jell-O-like substance consisting of a dense network of cross-linked molecules. The solidified gel slab is immersed in a chamber filled with buffer solution, which contains ions needed to conduct electricity. When the gel solidifies, the comb is removed, leaving a number of wells into which DNA samples are loaded. Just prior to loading, the digested DNA is mixed with a loading solution that consists of sucrose and one or more visible dyes. The dense sucrose solution weights the DNA sample, helping it to sink when loaded into a well.

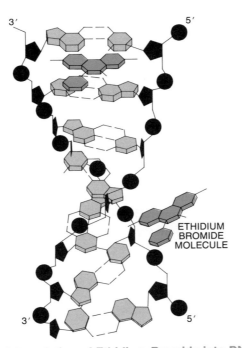

Intercalation of Ethidium Bromide into DNA Helix

Current supplied through electrodes at either end of the chamber creates an electric field across the gel. The negatively charged DNA fragments move from the wells into the gel, migrating through the pores in the matrix toward the positive pole. The negatively charged dye molecules do not interact with the DNA, but migrate independently toward the positive pole. For example, the commonly used marker bromophenol blue migrates at a rate equivalent to a DNA fragment of approximately 300 nucleotides (in a 1% gel). The visible movement of the dye allows one to monitor the relative migration of the unseen DNA bands.

Following electrophoresis, the gel is soaked in a dilute solution of ethidium bromide. The stain diffuses throughout the gel, becoming highly concentrated in regions where it binds to DNA fragments. (Alternately, ethidium bromide is incorporated into the gel and buffer prior to beginning electrophoresis.) A planar molecule, the ethidium bromide intercalates between the stacked nucleotides of the DNA helix, staining DNA bands in the gel.

The stained gel is then exposed to medium-wavelength UV light. The DNA/ethidium bromide complex strongly absorbs UV light at 300 nm, retains some of the energy, and reemits visible light in the orange range at 590 nm. Under UV illumination, the stained restriction fragments appear as fluorescent orange bands in the gel. It is important to understand that a band of DNA seen in a gel is not a single DNA molecule. Rather, the band is a collection of millions of identical DNA molecules, all of the same nucleotide length.

DNA Recombination and Ligation

A key step of gene cloning is to recombine a gene of interest into a plasmid vector. Ideally, a gene is isolated on a restriction fragment created by two different endonucleases that cleave on either side of the gene and that generate distinctive single-stranded ends. The sticky ends of the restriction fragment are then rejoined to the complementary ends of a plasmid vector that has been opened up with the same two enzymes. Such "directional cloning" using two different enzymes produces restriction fragments that have noncomplementary overhangs at each end. This prevents any fragment from rejoining its own ends and encourages recombination between different fragments.

The double digestion of the source DNA produces two or more restriction fragments. The entire mixture of fragments can be ligated directly into plasmid vectors. Alternately, if one begins with a small DNA, such as a plasmid or virus, gel electrophoresis can be used to "gel-purify" a particular restriction fragment of interest. The cut DNA is electrophoresed, and the band containing the desired DNA is cut from the agarose gel. The DNA fragment is eluted from the gel and can be ligated into an appropriate vector.

The single-stranded overhang of a sticky end can form hydrogen bonds with the complementary nucleotides in the overhang of another fragment generated by the same restriction enzyme. Hydrogen bonding of several

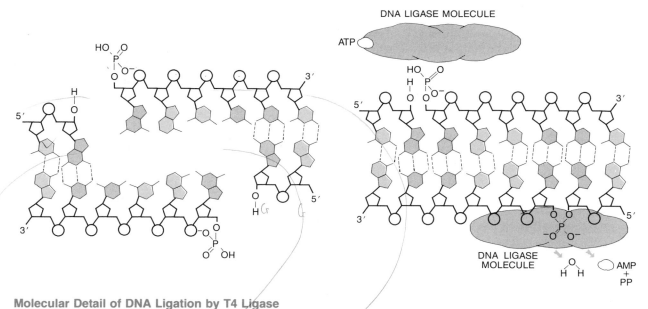

Molecular Detail of DNA Ligation by T4 Ligase
Hydrogen bonding between complementary nucleotides aligns *Bam*HI fragments while ligase reforms phosphodiester bonds on each side of the DNA molecule.

nucleotides is not sufficient to form a stable molecule, so associations between complementary ends constantly form and break. This transient interaction, however, does hold the two restriction fragments together long enough for DNA ligase to re-form phosphodiester bonds between adjacent nucleotides. This covalently links the deoxyribose-phosphate rails of the two fragments into a stable double helix. During the ligation reaction, an ester linkage is formed between the terminal phosphate of the 5′ overhang of one fragment and the adjacent deoxyribose ring at the 3′ end of the second fragment. This is accompanied by the loss of one molecule of water, making ligation an example of a condensation reaction.

The Host Cell: The Bacterium *Escherichia coli*

The manipulation of DNA and creation of recombinant DNA molecules described so far takes place in the test tube. Ultimately, the propagation of a DNA sequence must take place inside a living cell. Thus, transformation—the cellular uptake and expression of DNA in a bacterium—is crucial to the research process. The following elements are required to make the transformation process efficient and controllable enough to be of general use for introducing foreign genes into living cells: (1) a suitable host organism in which to insert the gene, (2) a self-replicating vector to carry the gene into the host organism, and (3) a means of selecting for host cells that have taken up the gene.

The bacterium *E. coli* has become the most widely used organism in molecular biology because it provides a relatively simple and well-understood genetic environment in which to isolate foreign DNA. Its primary genetic complement is contained on a single chromosome of approximately 5 million base pairs, making it only 1/600th the size of a haploid set of human chromosomes (the human genome). The chromosomal locations and sequences of a large number of its genes are known.

Because the genetic code is nearly universal, *E. coli* can accept foreign DNA derived from any organism. The DNA of a bacterium, a human, a corn plant, or a fruit fly is constructed of the same four nucleotides (adenosine, cytosine, guanosine, and thymidine), is assembled in the same structure, and is replicated by the same basic mechanism. Each organism transcribes DNA into messenger RNA, which is in turn translated into proteins according to the genetic code. A foreign gene inside *E. coli* is replicated, and in some cases translated, in exactly the same manner as the native bacterial DNA. *E. coli* "sees" foreign DNA as its own.

Under the best of circumstances, the uptake of a specific foreign gene is a relatively rare occurrence and is thus most easily accomplished in a large population of organisms that are reproducing rapidly. *E. coli* is an ideal genetic organism in this regard.

A recombinant plasmid is biologically amplified when a transformed bacterium replicates by binary fission to create a clone of identical daughter cells. Under favorable conditions, *E. coli* replicates once every 22 minutes, giving rise to 30 generations and more than 1 billion cells in 11 hours. This number of cells can be contained in a single milliliter of culture solution. Moreover, since each bacterium can carry up to several hundred copies of a cloned gene, the foreign DNA sequence is potentially amplified by a factor of several hundred billion.

E. coli is a constituent of the normal bacterial fauna that inhabits the human colon, where it absorbs digested food materials. Thus, it grows best with incubation at 37°C in a culture medium that approximates the nutrients available in the human digestive tract. An example of such a medium is LB broth, which contains carbohydrates, amino acids, nucleotide phosphates, salts, and vitamins derived from yeast extract and milk protein.

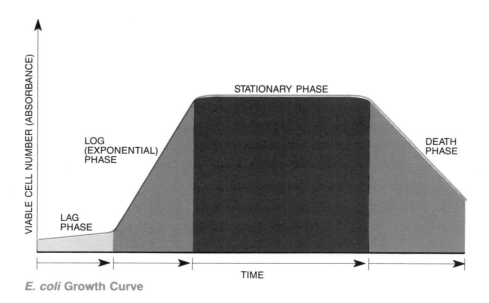

E. coli Growth Curve

Bacterial growth falls into several distinct phases. During *lag phase*, cells adjust to the nutrient environment and gear up for rapid proliferation; little or no cellular replication takes place. During *logarithmic (log) phase*, the culture grows exponentially and the cell number doubles every 22 minutes. During *stationary phase*, the cell number remains constant as new cells are produced at the same rate as old cells die. After an extended period, the culture enters *death phase;* the number of viable cells decreases as nutrients deplete and wastes accumulate.

Masses of bacterial cells are grown in a suspension culture; shaking provides aeration and keeps cells suspended in the medium. To isolate individual colonies, cells are spread on the surface of LB agar plates. Although the individual cells are invisible to the naked eye, after plating onto solid medium, each cell divides to form a visible colony of identical daughter cells in 12–24 hours.

The Plasmid Vector

In medical terminology, a vector is an organism that carries a pathogen from one host organism to another. In molecular biology, a vector is a DNA molecule that is used as a vehicle to carry foreign DNA sequences into *E. coli* or another host cell.

Plasmids are the simplest bacterial vectors. Ranging in length from 1000 to 200,000 base pairs, they are circular DNA molecules that exist separate from the main bacterial chromosome. To be propagated through successive bacterial generations, the plasmid vector must contain specific DNA sequences that allow it to be replicated within the host cell. DNA polymerase and other proteins required to initiate DNA synthesis bind to this region, which is called the origin of replication.

Plasmids can be divided into two broad groups, according to how tightly their replication is regulated. Plasmids that are under *stringent* control only replicate along with the main bacterial chromosome and so exist as a single copy, or at most several copies, within the cell. *Relaxed* plasmids, on the other hand, replicate autonomously of the main chromosome and have copy numbers of 10–500 per cell.

Generally, only those plasmids that confer some selective advantage are maintained in a given bacterial population. A particularly important

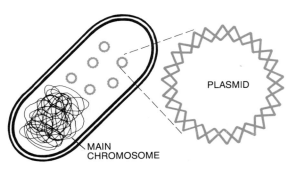

E. coli with Plasmids

selective advantage offered by many plasmids is antibiotic resistance genes that code for proteins that disable antibiotics secreted by microorganisms with which bacteria compete. Plasmids, and antibiotic resistance, can be passed from one bacterial strain to another during conjugation.

Antibiotics function by several different mechanisms. For example, members of the penicillin family (including ampicillin) interfere with cell wall biosynthesis. Kanamycin, tetracycline, and chloramphenicol arrest bacterial cell growth by blocking various steps in protein synthesis. Likewise, there are various mechanisms of antibiotic resistance. The

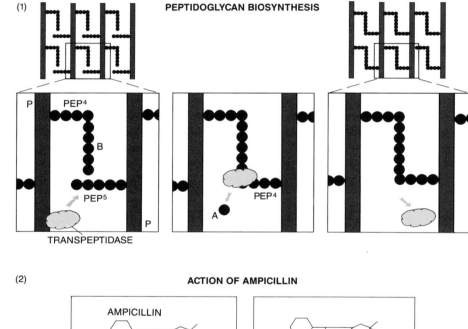

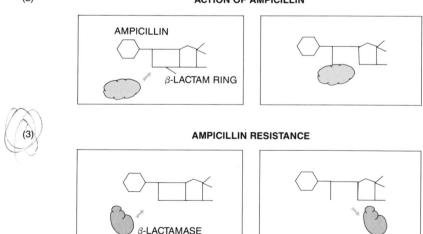

Peptidoglycan Biosynthesis, Ampicillin Action, and Ampicillin Resistance
(*1*) A transpeptidase removes an alanine residue (A) from a pentapeptide (PEP^5). The resulting tetrapeptide (PEP^4) is joined to the peptide bridge (B) to cross-link two adjacent polysaccharide chains (P). (*2*) The β-lactam ring of ampicillin structurally mimics the peptide bridge and irreversibly binds the transpeptidase, making it unavailable for peptidoglycan synthesis. (*3*) The pAMP resistance protein, β-lactamase, cleaves the β-lactam ring of ampicillin, making it unable to bind the transpeptidase.

ampicillin, kanamycin, and chloramphenicol resistance genes produce proteins that inactivate their target antibiotics through chemical modification. The tetracycline resistance gene specifies an enzyme that prevents transport of the antibiotic through the cell membrane.

Induced Transformation

The phenomenon of transformation, which provided a key clue to understanding the molecular basis of the gene, also provided a tool for manipulating the genetic makeup of living things. The natural transformation described by Griffith and Avery is an exceedingly rare event. However, in 1970, Mandel and Higa found that *E. coli* becomes markedly *competent* for transformation by foreign DNA when cells are suspended in cold calcium chloride solution and subjected to a brief heat shock at 42°C. They also found that cells arrested in early- to mid-log growth can be rendered more competent than can cells in other stages of growth.

Their calcium chloride procedure, which is still in wide use, yields transformation efficiencies of 10^5 to 10^7 transformants per microgram of plasmid DNA. (Transformation efficiency is generally expressed as the number of transformed cells that would be obtained from a microgram of intact plasmid DNA.) However, there is nothing particularly magical about calcium (Ca^{++}) ions.

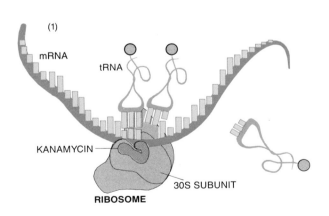
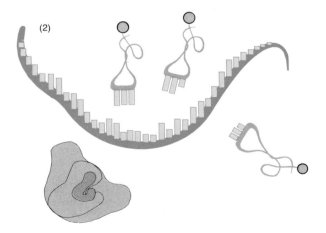

Kanamycin Action

Kanamycin poisons protein synthesis by irreversibly binding the 30S subunit of the ribosome. (*1*) The kanamycin/ribosome complex initiates protein synthesis by binding mRNA and the first tRNA. (*2*) However, the second tRNA is not bound, and the ribosome-mRNA complex dissociates. The pKAN resistance protein, from the aminoglycoside family, is a phosphotransferase that adds phosphate groups to the kanamycin molecule, thus blocking its ability to bind the ribosome. Other resistance proteins from this family can be acetyltransferases, which add acetyl groups to the antibiotic. Three forms of the phosphotransferase (A, B, or C) vary by the group present at R and R', as indicated in the figure.

Subsequent research showed that treatment with other divalent cations, such as magnesium (Mg^{++}), manganese (Mn^{++}), and barium (Ba^{++}), produces comparable or even greater transformation efficiencies. A protocol using a reducing agent and a complex transformation buffer composed of a mixture of positive ions (Ca^{++}, Mn^{++}, K^+, and Co^{+++}) achieves efficiencies of up to 10^9 transformants per microgram of plasmid DNA.

Under conditions that yield high-efficiency transformation (10^9 transformants per microgram of plasmid), approximately 1/10th of all viable cells are rendered competent. At an efficiency of 10^5 transformants per microgram of plasmid, only 1/100,000th of cells become competent. Both size and conformation of the DNA molecule affect transformation efficiency. Small plasmids are more readily taken up than larger ones, although no preferred size cutoff is evident. Linear DNA fragments transform at a negligible rate, in part because they are susceptible to degradation by exonucleases present in the periplasm between the inner and outer cell membrane of *E. coli*.

Because transformation is limited to a subset of cells that are competent, increasing the number of available plasmids does not increase the probability that a cell will be transformed. A suspension of competent cells becomes saturated at a very low concentration of DNA, roughly 0.1–0.2 µg per milliliter of mid-log cells (10^8 cells). Increasing plasmid concentration beyond this point decreases transformation efficiency, because the excess DNA does not transform additional cells. Competent cells will, however, readily take up more than one plasmid under saturating conditions. Experiments have shown that when equal amounts of two different plasmids are added under saturating conditions, 70–90% of all transformed cells contain both plasmids.

Mechanism of DNA Uptake

The mechanism of plasmid DNA uptake by competent *E. coli* cells is unknown. Unlike salts and small organic molecules such as glucose, DNA molecules are too large to diffuse or be readily transported through the cell membrane. Some bacteria possess membrane proteins that recognize DNA and facilitate the absorption of short DNA sequences derived from related species. However, *E. coli* appears not to have evolved such an uptake mechanism.

One hypothesis is that DNA molecules pass through any of several hundred channels formed at *adhesion zones*, where the outer and inner cell membranes are fused to pores in the bacterial cell wall. The fact that adhesion zones are only present in growing cells is consistent with the observation that cells in logarithmic growth can be rendered most competent for plasmid uptake. However, acidic phosphates of the DNA helix are negatively charged, as are a proportion of the phospholipids composing the cell membranes and lining the membrane pore. Thus, electrostatic repulsion between anions may effectively block the movement of DNA through the adhesion zones.

Basic Tools and Techniques of DNA Science 57

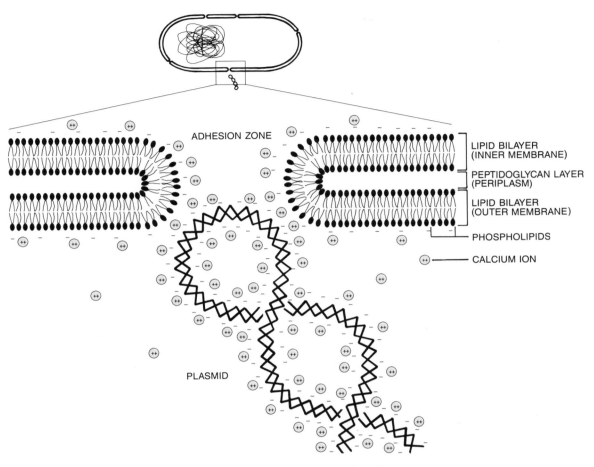

Proposed Molecular Mechanism of DNA Transformation of *E. coli*
Calcium ions (+ +) complex with negatively charged oxygens (−) to shield DNA phosphates from phospholipids at the adhesion zone.

Analysis of this ionic interaction produces a plausible hypothesis for DNA uptake in bacteria. Treatment of the cells at 0°C crystallizes the fluid cell membrane, stabilizing the distribution of charged phosphates. The cations in a transformation solution (Ca^{++}, Mn^{++}, K^{+}, Co^{+++}) form complexes with exposed phosphate groups, shielding the negative charges. With this ionic shield in place, a plasmid molecule can then move through the adhesion zone. Heat shock complements this chemical process, perhaps by creating a thermal imbalance on either side of the *E. coli* membrane that physically helps to pump DNA through the adhesion zone.

Selection of Transformants

Following uptake, the plasmid must become stably established inside the *E. coli* cell. During this period, the plasmid replicates and expresses antibiotic resistance genes that allow selection of a transformed phenotype. Transformed *E. coli* are plated onto culture medium containing one or more antibiotics. Only cells that take up a plasmid and express its

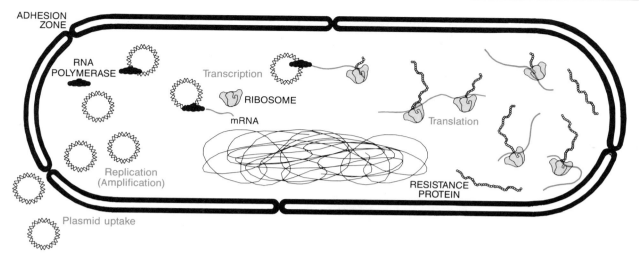

Expression of Antibiotic Resistance by Transformed *E. coli*

antibiotic resistance proteins can thrive on the antibiotic medium; untransformed cells fail to grow. Thus, antibiotic resistance is a selectable marker that allows one to positively identify cells that have been induced to take up plasmid DNA— including recombinant plasmids containing a gene of interest.

This type of culturing strategy, which selects for metabolic phenotypes that cannot be observed directly, was introduced by Beadle and Tatum in 1941. They screened for a *loss of gene function* by supplementing a biochemical mutant with a needed nutrient. Antibiotic selection, on the other hand, screens for a *gain in gene function* by challenging a transformant with an antibiotic.

Isolation and Analysis of Recombinant Plasmids

Growth of colonies on antibiotic medium gives *phenotypic* evidence that the cells have been transformed with a recombinant plasmid. To confirm this at the *genotypic* level, plasmid DNA is isolated from transformants. Restriction analysis of the purified plasmid DNA, along with the parental DNAs used to form the recombinant, provides a "fingerprint" of its genetic identity. Their small size and ring structure make plasmids relatively easy to separate from chromosomal DNA. A rapid method for making a small preparation of purified plasmid DNA, known as a miniprep, is outlined below.

Transformed cells taken from an antibiotic-resistant colony are grown to stationary phase in an overnight suspension culture. The cells are collected by centrifugation and resuspended in a buffered solution of glucose and ethylenediaminetetraacetic acid (EDTA), which binds divalent cations (such as Mg^{++} and Ca^{++}) necessary for cell membrane stability.

The resuspended cells are then treated with a mixture of sodium dodecyl sulfate (SDS) and sodium hydroxide. SDS, an ionic detergent, dissolves the phospholipid and protein components of the cell membrane. This lyses the membrane, releasing the cell contents. Sodium hydroxide denatures both plasmid and chromosomal DNAs into single strands. The chromosomal DNA separates completely into individual strands; however, the single-stranded plasmid loops remain linked together like interlocked rings.

Subsequent treatment with potassium acetate and acetic acid forms an insoluble precipitate of SDS/lipid/protein and neutralizes the sodium hydroxide from the previous step. At neutral pH, the DNAs renature. However, the long strands of chromosomal DNA only partially renature and become trapped in the SDS/lipid/protein precipitate. The plasmid DNA completely renatures into double-stranded molecules that remain in solution and largely escape entrapment in the precipitate.

The precipitate is pelleted by centrifugation and discarded, leaving the plasmid DNA (as well as small RNA molecules) in the supernatant. The supernatant is collected and isopropanol is added to precipitate the plasmid DNA, which is pelleted by centrifugation. The pellet is washed with ethanol and resuspended in a small volume of buffer. Treatment with RNase destroys RNA, leaving relatively clean plasmid DNA.

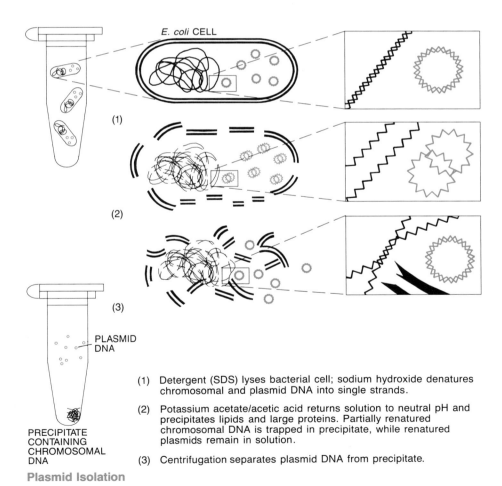

(1) Detergent (SDS) lyses bacterial cell; sodium hydroxide denatures chromosomal and plasmid DNA into single strands.

(2) Potassium acetate/acetic acid returns solution to neutral pH and precipitates lipids and large proteins. Partially renatured chromosomal DNA is trapped in precipitate, while renatured plasmids remain in solution.

(3) Centrifugation separates plasmid DNA from precipitate.

Plasmid Isolation

Samples of the miniprep plasmid and the parental plasmids are cut with the same restriction enzymes used to make the recombinant constructions. The digested DNAs are run on an agarose gel and the fragment patterns are compared. The bands containing the cloned gene and the vector backbone should line up perfectly with the corresponding fragments in the parental DNA. Each fragment of the same size will have migrated precisely the same distance.

For Further Reading

Cherfas, Jeremy. 1982. *Man-made life*. Pantheon Books, New York.

Sylvester, Edward J. and Lynn C. Klotz. 1987. *The gene age*. Charles Scribner's Sons, New York.

Kornberg, Arthur. 1989. *For the love of enzymes*. Harvard University Press, Cambridge, Massachusetts.

Lewin, Benjamin. 1987. *Genes III*. John Wiley and Sons, New York.

Neidhardt, Frederick C., ed. 1987. Escherichia coli *and* Salmonella typhimurium: *Cellular and molecular biology,* volumes I and II. American Society for Microbiology, Washington, D.C.

Novick, Richard P. 1980. Plasmids. *Sci. Am. 243/6:* 103–127.

Shapiro, James A. 1988. Bacteria as multicellular organisms. *Sci. Am. 258/6:* 82–89.

Watson, James D., John Tooze, and David T. Kurtz. 1983. *Recombinant DNA: A short course*. Scientific American Books, New York.

CHAPTER 4
Advanced Techniques of DNA Science

The basic techniques introduced by Paul Berg, Herbert Boyer, Stanley Cohen, and Annie Chang made reality of the science fiction of recombining genes from different species of living things. Their methods allowed new genes to be added to the genetic constitution of bacterial cells. Extension of these techniques has now made it possible to add individual genes to the cultured cells of many eukaryotic organisms, including yeasts, mammals, and dicot plants.

The simple addition of foreign genes or DNA fragments, however, does not necessarily change the character of the host cell. For example, if the chlorophyll biosynthesis genes are cloned into *Escherichia coli*, the bacterium will not produce chlorophyll and turn green. As described in Chapter 5, the expression of a gene is regulated at each step in the molecular genetic pathway, including transcribing a gene into RNA, processing the resulting RNA transcript, translating the messenger RNA (mRNA) transcript into protein, and modifying the protein for cellular transport and secretion. The signaling and modifying proteins that interact in each step of this pathway differ from species to species. Even if the chlorophyll biosynthesis genes are expressed in *E. coli*, the bacterium lacks the specific accessory proteins needed to orchestrate chlorophyll biosynthesis.

Despite this limitation, moving individual DNA sequences into simple genetic systems offers an opportunity to deal with genes as discrete units of cellular behavior. The advent of gene cloning made it possible to address questions that had proved difficult to answer using classic biochemical methods:

- What is the structural organization of DNA?
- Which genes are responsible for specific biological functions?
- How are genes turned on and off?
- What sequences within the DNA molecule regulate the expression of a gene or set of genes?

The Biochemist's Problem

The sequences coding for the major blood proteins were among the first mammalian genes to be isolated using recombinant DNA technology. The research design used by Philip Leder and his co-workers at Harvard

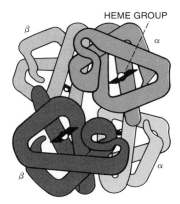

Hemoglobin
(From J. David Rawn. 1989. *Biochemistry*, p. 125. Neil Patterson Publishers, Burlington, North Carolina.)

University to isolate the mouse β globin gene from genomic DNA illustrates how molecular cloning elegantly solves two critical problems of biochemical analysis: (1) separating the gene of interest from the mass of genomic DNA and (2) purifying enough DNA for analysis. The human β globin gene was isolated shortly after by a research team headed by Tom Maniatis, also at Harvard, using a slightly different protocol.

Globin is the protein fraction of hemoglobin, the oxygen-carrying molecule of red blood cells. A hemoglobin molecule is composed of four globin chains—two each of the α and β types—each surrounding a central heme group. Although the two globin proteins are similar in structure, each is encoded by a different gene. The β globin gene contains approximately 2000 base pairs (bp). Since a base pair has an average molecular mass of 635 daltons, the β globin gene has a total mass of about 1.27×10^6 daltons. Multiplying this figure by the number of copies of the gene per cell (2) and the number of cells in the adult human body (10 trillion, or 1×10^{13}), we calculate a total mass of 0.000042 g (0.042 mg or 42 μg) of β globin DNA per individual.

Biochemical analysis might require on the order of 1 mg of DNA, which seems small considering that one teaspoon of sugar weighs approximately 4500 mg. However, it would take the equivalent cell mass of 24 human cadavers to extract 1 mg of β globin DNA. Even if this difficult and onerous task was successfully completed, the problem of separating the globin gene from the rest of the DNA would be nearly impossible.

The β globin gene represents 1/1,500,000 of the estimated 3 billion (3×10^9) bp of DNA that compose a haploid set of human chromosomes (the haploid human genome). Thus, the isolation procedure above would yield a total mass of 15.1 kg of genomic DNA, of which only 1 mg would be β globin DNA. Because DNA is a polymer of only four repeating subunits, for all intents and purposes, any 2000-bp fragment appears biochemically identical to any other 2000-bp piece. The problem is even worse than searching for the proverbial needle in a haystack, where at least the needle looks different from the surrounding hay.

The Molecular Biologist's Solution

Ligating the β globin DNA sequence into a vector molecule, and propagating the recombinant molecule in *E. coli*, provides a single solution for both of the biochemist's problems. First, a sample of human genomic DNA is cleaved with a restriction endonuclease, which cuts on average once every 4000 nucleotides and generates 750,000 DNA fragments. The genomic restriction fragments are mixed with plasmid molecules that have been digested with the same restriction enzyme. The plasmid vector and the genomic DNA fragments are covalently joined with DNA ligase. The collection of 750,000 recombinant plasmids created is called a genomic library. The recombinant plasmids are then transformed into competent *E. coli* cells. In theory, each cell takes up a different plasmid carrying a different fragment of human DNA. The transformed bacteria are spread onto plates as single cells, and each reproduces itself until a visible colony forms. Each colony contains millions of cells, each with many copies of its unique recombinant plasmid.

The problem of separating the gene of interest from all of the other genomic DNA has now been solved. On average, 1 of every 750,000 bacterial colonies should contain an intact β globin gene. Methods for screening the colonies to identify those containing the globin gene are discussed below.

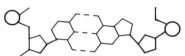

The average atomic mass of one base pair is 635 daltons (a dalton is 1/12 the mass of a carbon atom)	The average atomic mass of one base pair is 635 daltons (a dalton is 1/12 the mass of a carbon atom)
The β globin gene is approximately 2000 bp in length	The β globin gene is approximately 2000 bp in length
So, the atomic mass of the β globin gene is:	So, the atomic mass of the β globin gene is:
2000 bp × 635 daltons/bp = 1.27×10^6 daltons	2000 bp × 635 daltons/bp = 1.27×10^6 daltons

Mass of β globin gene in an adult human

There are two copies of the β globin gene per cell

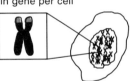

There are 10^{13} cells per individual

So, the total atomic mass of β globin DNA per individual is:

1.27×10^6 daltons/gene
×
2 genes/cell
×
10^{13} cells/individual
= 2.54×10^{19} daltons

If there are 6.02×10^{23} daltons per gram, then:

$$\frac{2.54 \times 10^{19} \text{ daltons}}{6.02 \times 10^{23} \text{ daltons/gram}}$$

= 0.000042 grams

= 0.042 mg

= 42 μg β globin DNA

Mass of β globin gene in a liter of *E. coli*

There are 500 copies of the β globin gene per cell

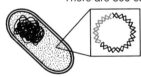

There are 5×10^{11} cells per liter

So, the total atomic mass of β globin DNA per liter is:

1.27×10^6 daltons/gene
×
500 genes/cell
×
5×10^{11} cells/liter
= 3.175×10^{20} daltons

If there are 6.02×10^{23} daltons per gram, then:

$$\frac{3.175 \times 10^{20} \text{ daltons}}{6.02 \times 10^{23} \text{ daltons/gram}}$$

= 0.000527 grams

= 0.527 mg

= 527 μg β globin DNA

Mass of β Globin DNA in Adult Human vs. 1-liter Culture of *E. coli* Carrying β Globin Gene on Plasmid

The problem of purifying enough DNA has also been solved. Only about 50 μg of genomic DNA initially is needed to construct the library, and this amount can be isolated from less than 1 g of tissue or cells growing in culture. Once the bacterial colony containing the β globin gene has been identified, large amounts of plasmid can be prepared from liquid cultures. One liter of *E. coli* culture at stationary phase contains approximately 5×10^{11} cells, each containing up to 500 copies of the recombinant plasmid. Multiplying the total number of plasmids per liter times the mass of the β globin gene (1.27×10^6 daltons), we calculate that 527 μg of β globin DNA can be extracted from each liter of bacterial culture.

Vectors Used in Making Genomic Libraries

Creating a genomic library is often the starting point of a gene-isolation experiment. In the example given here, the goal is to represent the entire human genome, as a set of DNA inserts, in self-replicating vector molecules that can be propagated in an appropriate host. In practice, the library contains the equivalent of several genomes. This "oversampling" helps to ensure that every part of the genome is represented at least once.

Although plasmids can theoretically accept DNA inserts of any size, they are used primarily for cloning small fragments on the order of 4000 bp. Modified forms of the bacteriophage λ are the most commonly used vectors for constructing genomic libraries. Two features of λ vectors increase the probability of creating a genomic library that represents the entire genome. First, because λ vectors typically can accommodate inserts averaging 20,000 bp, one-fifth as many recombinant molecules are required to represent the entire human genome. This makes the job of screening the library much more manageable. Second, introducing recombinant DNA into *E. coli* by phage infection (transfection) is many times more efficient than artificial transformation of *E. coli* with plasmids.

Two other types of vectors can be employed to make genomic libraries. Cosmids, which are hybrid plasmid/λ vectors, can be used to clone fragments averaging 40,000 bp. Recently, artificial chromosomes have been developed that allow fragments on the order of 400,000 bp to be cloned into yeast cells. Obviously, the larger the insert size, the fewer the clones theoretically needed to entirely represent the 3 billion bp of the human genome.

Cloning Vectors and Minimum Number of Clones Needed to Represent the Haploid Human Genome

Vector	Average insert (bp)	Clones to represent genome
Plasmid	4,000	750,000
Lambda	20,000	150,000
Cosmid	40,000	75,000
Yeast artificial chromosome (YAC)	400,000	7,500

The λ Genomic Library

To be useful as a genomic library vector, part of the DNA sequence of wild-type λ (~48,000 bp) must be deleted to make room for large DNA inserts. Charon 4A is among the many λ vectors that have been created to accommodate DNA inserts of 12,000 to 22,000 bp. Digestion of Charon 4A with *Eco*RI generates four fragments: two large outer fragments (the arms) and two small central fragments (the stuffer). The two arms are isolated from the smaller stuffer fragments using gel electrophoresis or density gradient centrifugation. Since the stuffer region is not essential for lytic growth of phage particles, it can be replaced by DNA from any source.

For genomic DNA fragments to be ligated into the Charon 4A vector, they must have *Eco*RI sites at either end. Thus, genomic DNA, typically purified from a tissue sample or cultured cells, is digested with *Eco*RI. With a six-nucleotide recognition sequence, *Eco*RI can be expected to cut on average once every 4000 bp. However, since the object is to produce inserts near the maximum size that can be accommodated by λ, reaction conditions are adjusted so that the *Eco*RI enzyme only cuts at about one in five of the available restriction sites. This "partial digestion" is accomplished by shortening the incubation period and reducing enzyme concentration. Under appropriate conditions, this procedure yields restriction fragments averaging 5×4000 bp, or 20,000 bp.

After the genomic DNA fragments are ligated into λ vectors, the recombinant molecules are "packaged" inside infective phage particles. The phage system automatically screens for large-size genomic inserts. λ particles will not be packaged if the recombined DNA is too much shorter than the wild-type DNA. The upper limit of insert size is dictated by the volume within the capsule, which cannot accommodate more than 54,000 bp of DNA. When cloning into Charon 4A, recombinant DNA molecules containing DNA inserts of less than 12,000 bp or more than 22,000 bp will not be packaged.

Cell extracts are prepared from two different strains of *E. coli*, each of which carries a copy of λ DNA integrated into its genome (termed a lysogen). The activity of the lysogen is temperature-sensitive (*ts*). At 30°C, the λ sequences remain integrated in the bacterial chromosome; however, when the temperature is shifted to 42°C, the λ DNA sequences pop out of the chromosome and lytic growth begins. The λ DNA carried in each lysogen, however, contains a mutation deleting one of the several genes that encode proteins needed for assembly of the phage capsule. The deleted gene is different for each mutant; thus, each strain lacks an essential packaging protein, and the cells fill with unassembled phage proteins. The *E. coli* cells are subsequently lysed, and the phage proteins from each strain are harvested as "packaging extracts."

The packaging process takes advantage of the ability of the protein components of the λ capsule to assemble spontaneously in vitro. When mixed together with the recombinant vector molecules, the packaging extracts complement the missing proteins of one another, thus supplying all of the proteins needed to package the recombinant molecules into

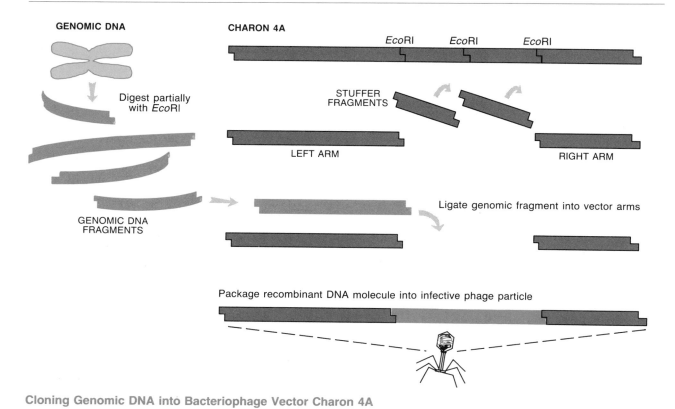

Cloning Genomic DNA into Bacteriophage Vector Charon 4A

viable phage particles. The assembled phage particles are then used to infect *E. coli* cells, which are spread onto plates and incubated. After incubation at 37°C, centers of phage infection appear as relatively clear "plaques" within an opaque "lawn" of uninfected bacteria. Within each plaque are recombinant phages containing millions of copies of a single genomic DNA insert.

To minimize the number of plates to be screened, the cultures are inoculated so that each plate will contain approximately 10,000 plaques. If each plaque represents a different recombinant phage particle with an average 20,000-bp insert, then the entire human genome optimally can be represented on 15 plates. As discussed above, usually the equivalent of three or more genomes worth of recombinant molecules are screened. Hopefully, at least one of the plaques will contain a recombinant vector carrying the gene of interest. But how does one screen the λ library to identify this particular clone out of 450,000 plaques?

Screening a Genomic Library

We have discussed how difficult it is to distinguish biochemically one DNA fragment from another. Because the relative abundance of adenine, thymine, guanine, and cytosine is constant within the DNA polymer, every DNA fragment has more or less the same chemical composition. This was one of the facts that encouraged scientists during the 1940s and 1950s to believe that DNA was incapable of being the genetic material.

However, to work creatively with DNA, one must think in terms of the exact sequence of base pairs along the length of a particular DNA fragment. A genomic library is screened on the basis of two special attributes of the DNA molecule: (1) the uniqueness of even relatively short nucleotide sequences and (2) the complementary binding of nucleotide pairs within the sequence.

THE UNIQUENESS OF DNA SEQUENCES

Consider how long a chain of nucleotides would be required to be to make a unique sequence of human DNA—that is, a sequence that statistically would be found only once in the entire human genome. To be unique, the sequence must have a probability of occurring less than once in 3 billion nucleotides.

For the purpose of this discussion, we will adopt the view that, on the whole, the four bases are randomly distributed in a DNA molecule. Thus, any particular nucleotide—adenine, thymine, cytosine, or guanine—will generally occur once in every four nucleotide positions. (Remember, for simplicity's sake, a DNA sequence usually takes into account only the nucleotide arrangement on one strand, or side, of the double helix.)

The occurrence probability of any DNA sequence can be calculated by taking 4^n, where 4 equals the number of different nucleotides and n equals the total number of nucleotides in that sequence. For example, starting at any nucleotide position along a DNA strand, there is a one in four (1/4) chance that it will be, say, adenine. There is also a one in four (1/4) chance that the adjacent nucleotide will be, say, cytosine. Therefore, the combined chance of encountering the dinucleotide sequence adenine-cytosine (AC) is $1/4 \times 1/4$ or 1/16. This can be represented by the expression 4^2. Every other dinucleotide sequence would also have an occurrence probability of 1 in 16—AA, TT, GG, CC, AT, AG, TA, TG, TC, GA, GT, GC, CA, CT, CG.

Occurrence Probabilities of DNA Sequences of Increasing Numbers of Nucleotides

Nucleotides	Occurrence Probability (1/)
1	4
2	16
3	64
4	256
5	1,024
6	4,096
7	16,384
8	65,536
9	262,144
10	1,048,576
11	4,194,304
12	16,777,216
13	67,108,864
14	268,435,456
15	1,073,741,824
16	4,294,967,296

A simple exercise with a calculator shows that the chance of encountering a specific 10-nucleotide sequence is 4^{10}, or 1 in 1,048,576. Statistically, only 3000 copies of a specific 10-nucleotide sequence should be found in the human genome. A 15-nucleotide sequence has an occurrence probability of 4^{15}, or 1 in 1,073,741,824, whereas a 16-nucleotide sequence has an occurrence probability of 4^{16}, or 1 in 4,294,967,296. A 15-nucleotide sequence should occur in three copies in the human genome, whereas a 16-nucleotide sequence should occur only once.

This calculation suggests that sequences of 16 nucleotides or longer are likely to be unique in the human genome. This works as a general rule of thumb; however, we must bear in mind that repetitive DNA sequences likely compose a significant percentage of the human chromosomes. Obviously, a 16-nucleotide sequence from *within* such a repetitive element would be common to each of the repeated units. Even so, we could be comfortable assuming that nearly all of the 20,000-bp fragments cloned into Charon 4A vectors are unique.

IDENTIFYING SPECIFIC DNA SEQUENCES

In 1961, Spiegelman and Hall discovered that single-stranded DNA will hydrogen bond to its complementary RNA sequence to form a stable, double-stranded (duplex) molecule. This complementary base pairing provides a powerful tool to probe for unique DNA sequences in a genomic library.

Incubation of duplex DNA at a temperature above 90°C, or at a pH of greater than 10.5, or with organic compounds (such as urea or formamide) disrupts the hydrogen bonds between base pairs, causing the complementary strands to disassemble. This process is called denaturation. Under proper conditions of salt, temperature, and pH, the two single-stranded molecules can be renatured to re-form the original duplex DNA molecule. This process of complementary single-stranded molecules aligning and forming double-stranded molecules is known as hybridization.

Under reaction conditions of "high stringency," stable DNA duplexes form only when complementary base pairing is essentially perfect along the entire length of the DNA strands. Under conditions of "low stringency," partial hybridization occurs between strands that have lesser degrees of complementarity.

SCREENING A GENOMIC LIBRARY BY COLONY HYBRIDIZATION

The screening process begins with making replica copies of the phage plaques on each plate in the genomic library. This process takes plaque samples from each plate for testing, while maintaining the original plates as a master copy of the genomic library for future use. A nitrocellulose filter disk, slightly smaller than the culture dish, is placed directly in contact with the agar surface of each plate. Some phage particles in each plaque adhere to the filter, so that the pattern of adsorbed phage particles

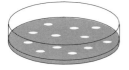
Grow plaques of recombinant phages

Overlay plate with nitrocellulose or nylon filter

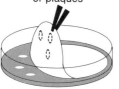
Lift filter containing replica of plaques

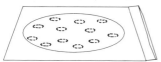
Lyse cells; denature DNA and hybridize to radiolabeled probe in sealed plastic bag

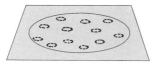
Overlay filter with X-ray film

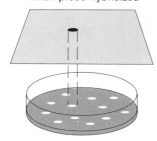
Developed autoradiogram reveals location of plaque to which probe hybridized

on the filter is a replica of the plaque pattern on the original plate. (This is an adaptation of a technique for making replica bacterial colonies published in 1952 by Joshua and Esther Lederberg, at the University of Wisconsin.)

After drying, the filters are soaked in an alkaline solution of sodium hydroxide, which simultaneously lyses the phage capsules and denatures the phage DNA. The filters are then soaked in buffer at neutral pH. The DNA remains denatured, and the single-stranded DNA is fixed in place on the nitrocellulose by baking the filters under vacuum at 80°C. Recently, several types of support membranes have been developed that covalently bind DNA without baking.

The dried filters are then screened with a probe that has a sequence complementary to the gene of interest. The probe is first radioactively "labeled" so that it can be readily detected. This is usually accomplished by incorporating radioactive nucleotides, such as [^{32}P]deoxyadenosine monophosphate into the sequence. The filters are incubated with the probe under conditions that encourage hybridization of complementary sequences: neutral pH, presence of sodium ions (which neutralize the negative charges along the DNA backbone), and elevated temperature (65°C, which untangles the DNA strands). During an incubation of about 24 hours, the probe should hybridize to the DNA from only those plaques containing all or part of the gene of interest.

Following hybridization, the filters are washed to remove excess probe, blotted dry, and covered in plastic wrap. The covered filters are tightly sandwiched against a sheet of X-ray film, which is exposed for several hours. β particles, released during the decay of the radioactive ^{32}P in the probe, expose the film. Following development, the exposed area appears as a dark spot on the X-ray film. Realignment of reference marks on the film, filter, and culture plate matches an exposed spot to its corresponding plaque on one of the master plates. This plaque contains DNA sequences complementary to the probe and therefore contains the gene of interest.

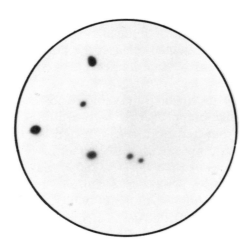
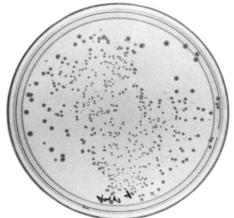

Screening a Genomic Library by Colony Hybridization
Autoradiogram (*left*) indicates locations of colonies on master plate (*right*) to which a radioactive probe hybridized.

MAKING A PROBE FOR THE β GLOBIN GENE

There is a "catch 22" in the protocol that was glossed over in the previous section. To make a probe for a gene of interest requires a purified DNA sequence complementary to that gene. In effect, a purified piece of DNA is needed to identify a cloned piece of DNA. So how does one purify a complementary sequence from which to make the probe? Returning to our original example, we can see how Leder and his co-workers overcame this apparently unsolvable problem in cloning the mouse β globin gene. mRNA is complementary to the DNA sequence of the gene from which it is transcribed, and it will readily hybridize to this DNA sequence. Leder's group reasoned that they might take advantage of hybridization, in combination with unique properties of red blood cells and eukaryotic mRNA, to make a probe for the β globin gene.

Red blood cells can be thought of as bags of hemoglobin because they are filled with this protein and little else. The nucleated precursors of red blood cells, reticulocytes, thus essentially function as hemoglobin factories. Whereas most mammalian cells contain relatively small amounts of thousands of different kinds of mRNA molecules, α and β globin mRNAs account for the major portion of all mRNA molecules in reticulocytes. Furthermore, reticulocytes can be readily harvested from mice in which induced anemia increases the reticulocyte level to more than 50% of circulating blood cells. Thus, blood is collected from the anemic animals, the cells are lysed, and cellular debris is removed by centrifugation. The remaining supernatant is full of globin mRNA.

All but a few eukaryotic mRNAs contain a long stretch of 100–200 adenine residues at their 3' ends. This unique feature, called a poly(A) tail, makes it possible to purify mRNA in a single step using oligo(dT)-cellulose affinity chromatography—a chromatography column filled with cellulose particles, to which are attached short synthetic sequences (10–20 nucleotides) of deoxythymidine (dT). When the reticulocyte lysate is passed through the column in a high-salt buffer containing sodium chloride, the poly(A) tails of the mRNA bind to the complementary deoxythymine residues. mRNA molecules are thus retained on the column while all other molecules pass through. The salt is then rinsed from the column, which releases the bound mRNA. Since α and β globin are the major mRNAs contained in reticulocytes, no further purification is necessary.

A probe is produced by making a radioactive cDNA (c for copy) from the purified globin mRNA. This procedure relies on reverse transcriptase, an enzyme purified from cells infected with retroviruses, which synthesizes a DNA copy from an RNA template. The synthesis of the cDNA probe takes place in a buffered reaction mixture containing magnesium ions (Mg^{++}), reverse transcriptase, and the four deoxynucleotide triphosphates—one or more of which contain ^{32}P in the α position. (Deoxynucleotide triphosphates contain three phosphates: α, β, and γ. During DNA synthesis, the β and γ phosphates are cleaved and only the α phosphate is incorporated into the DNA molecule.)

First, a short "primer" composed of oligo(dT) is hybridized to the 3' poly(A) tail of the purified mRNA. Like all DNA polymerases, reverse

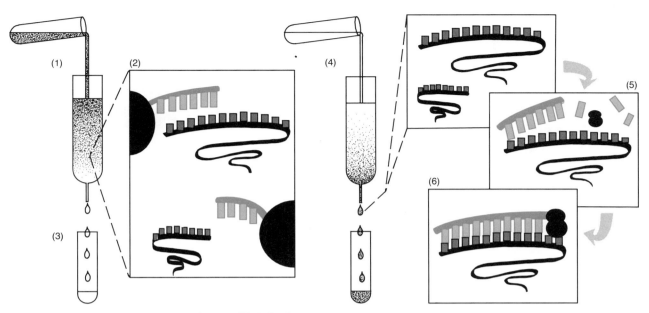

Purifying Globin mRNA and Making a cDNA Probe
(*1*) Lysed reticulocytes in high-salt buffer are poured through column of oligo(dT) cellulose. (*2*) Poly(A) tails of mRNA bind to oligo(dT) molecules on cellulose beads. (*3*) Non-mRNA molecules pass through column. (*4*) Wash with low-salt buffer releases mRNA from column. (*5*) Oligo(dT) primer, radioactively labeled nucleotides, and reverse transcriptase are added to purified mRNA. (*6*) Complementary DNA strand is synthesized on mRNA template.

transcriptase initiates DNA synthesis at the primer; it then builds a DNA chain that incorporates a complementary nucleotide for each position on the mRNA molecule. The reaction produces a radioactively labeled cDNA for both α and β globin and thus will hybridize to λ plaques containing either gene. However, further characterization of the isolated clones allows us to distinguish between the two globin genes.

Restriction Mapping and Subcloning

Typically, a clone isolated from a bacteriophage λ library will contain an insert on the order of 20,000 bp, or 20 kilobase pairs (kb). Cosmids will contain DNA inserts approximately twice as large. Analysis of a cloned DNA sequence usually starts with construction of a restriction map showing the relative positions of endonuclease recognition sites along its length.

Since the globin λ library was made from a partial *Eco*RI digest, it is easiest to map the *Eco*RI sites first. Restriction maps, however, position restriction sites relative to other sites, and thus it is generally necessary to map two enzymes simultaneously. If *Bam*HI is the second restriction enzyme selected, then the cloned DNA is digested in three reactions: with *Eco*RI, with *Bam*HI, and with *Eco*RI and *Bam*HI *together*. The digested DNAs are electrophoresed in separate lanes of an agarose gel, along with a lane containing size markers of a known DNA, such as a *Hin*dIII digest of λ. The size markers allow us to gauge the approximate size of fragments in the three unknown lanes.

The object is to relate the disappearance of bands in each single digest with the appearance of new bands in the double digest. First, we locate a band present in the *Eco*RI digest but missing in the double digest. This means that *Bam*HI has cut *within* the fragment to produce two or more new fragments, whose base-pair size totals the original *Eco*RI fragment. If, for example, a 10-kb *Eco*RI fragment is converted into *Eco*RI/*Bam*HI fragments of 3.5 kb and 6.5 kb, then a *Bam*HI site is located 3.5 kb from

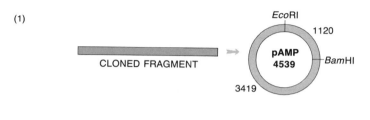

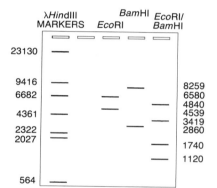

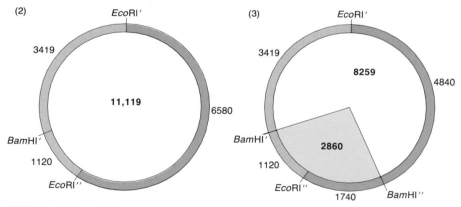

All fragment sizes in bp

Restriction Mapping of an *Eco*RI Fragment Subcloned into pAMP
(*1*) Recombinant plasmid is digested with *Eco*RI, *Bam*HI, and *Eco*RI + *Bam*HI. Digested plasmids are electrophoresed in an agarose gel with λ/*Hin*dIII size markers. (*2*) The *Eco*RI digest separates vector from insert, so the cloned fragment must be 6580 bp. A preliminary map of the vector and cloned fragment can be drawn. (*3*) The *Bam*HI digest yields only two fragments: 8259 bp and 2860 bp. One site, *Bam*HI′, is known to be in the vector, so the other site, *Bam*HI″, must be located in the 6580-bp cloned fragment. To locate this site, determine which *Bam*HI/*Eco*RI fragments combine to equal the length a single *Bam*HI fragment. Since 1120 + 1740 = 2860, the *Bam*HI″ site must be located 1740 bp from the *Eco*RI″ site.

the end of the *Eco*RI fragment. This procedure is repeated until all fragments have been accounted for. Provided there are relatively few sites, it is possible to order the restriction fragments, and hence recognition sites, relative to one another on the cloned DNA.

One goal is to find the smallest DNA fragment (of the cloned insert) that contains the entire gene, including its regulatory information. This portion of the cloned sequence can then be isolated, ligated into a plasmid vector, and transformed into *E. coli*. This process of moving cloned DNA fragments between different vectors is called subcloning. The subcloned fragments can be dealt with more easily in large-scale preparations, sequence analysis, or functional studies. Large genes, encompassing more than 40,000 bp, must be maintained in *E. coli* as several clones.

Constructing a cDNA Library

The abundance and ease of isolation of globin mRNA made it relatively straightforward to make a probe to screen a genomic library for globin sequences. However, mRNAs for most proteins constitute less than 1% of the total mRNA in a cell. Therefore, other screening strategies must be used in cases where the gene of interest produces a low level of mRNA.

One such strategy starts with construction of a cDNA library, which differs in several respects from the genomic library. Whereas a genomic library uses DNA as starting material, a cDNA library begins with mRNA. A cDNA library represents only those genes that are expressed in a particular cell, not the entire genome.

To prepare a cDNA library, cellular mRNA is purified using oligo(dT) affinity chromatography, and cDNA copies are synthesized using reverse transcriptase, as described above. Incubating the resulting mRNA/cDNA duplex at high pH hydrolyzes the mRNA, but leaves the naked cDNA strand intact. For reasons still unknown, reverse transcription produces a cDNA whose 3' end folds back on itself, producing a "hairpin loop" and a short sequence of double-stranded DNA. This double-stranded sequence serves as a primer for synthesis of a DNA strand complementary to the cDNA strand. Second-strand synthesis is carried out with *E. coli* DNA polymerase or reverse transcriptase (which can use either an RNA or a DNA template).

The resulting double-stranded cDNA molecule (dscDNA) must undergo several modifications before it is ready for insertion into a vector. First, the hairpin loop is digested with S1 nuclease, an enzyme that degrades only single-stranded DNA. Next, a restriction endonuclease site is added to either end of the dscDNA using T4 DNA ligase. For example, the recognition site for *Eco*RI is contained in a "linker," a short double-stranded oligonucleotide with the sequence G*GAATTC*C. Prior to the addition of the linkers, the dscDNA is incubated with *Eco*RI methylase, blocking any *Eco*RI sites from cleavage during the next step. (In some strategies, a different linker is added to each end of the dscDNA, creating a different restriction endonuclease site at either end of the sequence to be cloned.)

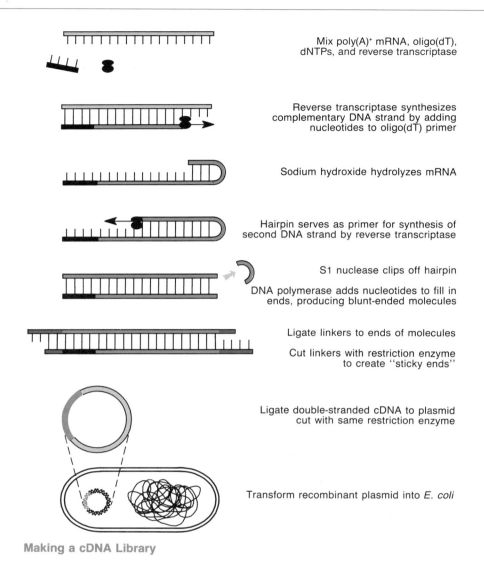

Making a cDNA Library

Plasmid vector and the dscDNA/linker are digested with *Eco*RI, creating complementary ends that are then ligated together. The resulting recombinant plasmids are transformed into *E. coli*, which are spread onto agar plates. Antibiotic in the medium selects for colonies that contain the recombinant plasmid, creating a cDNA library. Alternately, a suitable bacteriophage vector, such as λgt10, can be used. As with a genomic library, the cDNA library allows us to separate a DNA copy of an mRNA of interest from all others in the cell. But again, we face the challenge of identifying the recombinant molecule containing the cDNA of interest.

Constructing and Screening an Expression Library

One very laborious method of screening a cDNA library begins with making suspension cultures that pool numerous colonies from the cDNA library. The plasmid DNA from each culture is bound to a small piece of nitrocellulose, denatured, and then hybridized with the total cellular mRNA. Each different mRNA will hybridize to the filter containing its complementary cDNA sequence. The bound mRNAs are eluted from each separate filter and used to synthesize their encoded proteins in vitro, using a cell-free translation system. The resulting protein products are analyzed to determine if any represent the mRNA of interest, thus identifying the culture containing the desired clone. The pool of colonies used to make this culture is then rescreened, as above, to identify the single clone expressing the gene of interest.

This general scheme was used to identify the cDNA clones of several abundant mRNAs, but it proved infeasible for detecting mRNAs of rarer proteins. Imagine the labor involved in using this procedure to hunt for a message that represents less than 0.1% of total cellular mRNA. A far easier method relies on detecting the protein expressed by the cDNA *inside an E. coli host cell.* An antibody directed against the protein encoded by the mRNA of interest is then used to screen for the colony containing the desired cDNA. It is therefore crucial that the cloned cDNA be faithfully transcribed into mRNA and that the mRNA be translated into protein.

As mentioned at the beginning of this chapter, expression of foreign genes in *E. coli* must overcome the differing gene-regulation mechanisms of each species. Thus, the human cDNA fragments must be cloned into special expression vectors that contain specific regulatory information required by the host bacterial cell. This, in effect, disguises the foreign gene so that the bacterium believes that it is an *E. coli* gene. A number of widely used expression vectors are derived from pUC, a plasmid of about 2700 bp. In addition to a selectable ampicillin resistance marker, pUC contains several key 5' regulatory sequences of the *E. coli* β-galactosidase gene:

1. A strong bacterial promoter contains the binding site where RNA polymerase begins mRNA synthesis (transcription).
2. A ribosome-binding site, which is incorporated into the transcribed mRNA, serves as the point of attachment for the bacterial translation machinery.
3. The first few codons of the β-galactosidase gene include the initiation codon (AUG) that signals the start of translation.

Directly adjacent to these regulatory sequences is a polylinker sequence that contains a cluster of several useful restriction sites, such as *Eco*RI, *Bam*HI, and *Hin*dIII. Following digestion with the appropriate restriction enzyme(s), the expression vector is ligated to the dscDNAs. The recombinant plasmids are transformed into *E. coli*, which are usually plated onto nitrocellulose filters that are placed on top of nutrient agar plates containing ampicillin. This system aids in subsequent replica plating to screen resistant colonies for the cDNA inserts of interest.

Within transformed bacteria, each recombinant pUC vector functions in the following manner. The bacterial RNA polymerase recognizes the β-galactosidase promoter and initiates transcription of the β-galactosidase gene. When the RNA polymerase encounters the cloned cDNA downstream from the β-galactosidase sequence, it simply continues

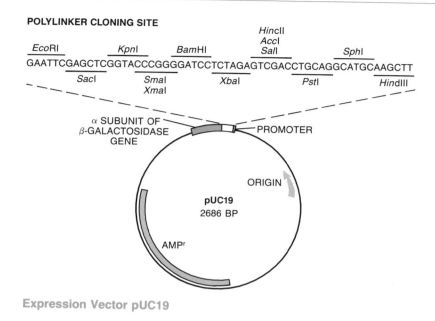

Expression Vector pUC19

synthesis, fusing the cDNA codons to the β-galactosidase codons. The presence of the β-galactosidase ribosome-binding site leads to the translation of this mRNA, producing a hybrid (or fusion) β-galactosidase/cloned cDNA protein. This requires that the cDNA has been cloned in the proper 5' to 3' orientation and that the protein-coding sequence is synchronized with the coding sequence of the β-galactosidase gene.

The cDNA is not usually a complete copy of the mRNA, so the protein translated is likely to contain only part of the information encoded by the cloned mRNA. However, this partial protein is usually sufficient for recognition by a specific antibody, which recognizes and binds to a unique region of the protein composed of only a small number of amino acids.

Assume that we are attempting to locate a cDNA clone of the actin protein. First, replicas are prepared by pressing a clean nitrocellulose filter against each master filter of the cDNA library. A part of each colony is transferred to the new filter, which is placed on a fresh plate where the cells are allowed to grow up to colony size again. The bacteria on the replica filters are then lysed, and the cellular proteins bind to the nitrocellulose. The baked filters are bathed in a solution of anti-actin antibody that will specifically bind to the actin protein. Following incubation at room temperature, excess antibody is washed off. The antibody will have bound only to the spot on the filter where there had been a colony producing the actin protein.

The location of the anti-actin antibody on the filter is detected by using the Staph A protein, which is isolated from the bacterium *Staphylococcus aureus*. The Staph A protein has the unique property of binding specifically to the ends of antibody molecules and will therefore tag the anti-actin antibody. The filters are bathed in a solution of Staph A protein that has been labeled with radioactive ^{125}I. Following incubation, excess Staph A

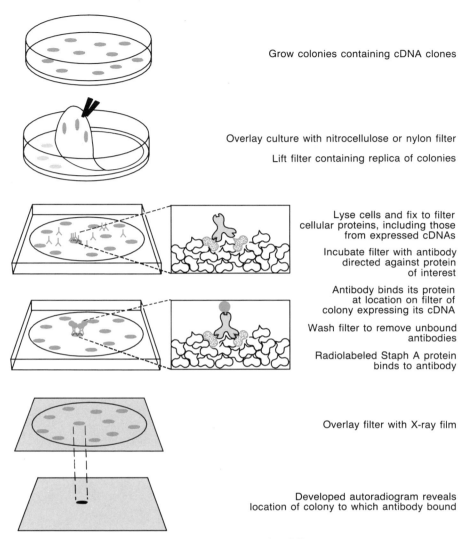

Using Antibodies to Screen a cDNA Expression Library

protein is washed off, and the filters are sandwiched between sheets of plastic wrap. As with genomic libraries, the replica filters are then used to expose X-ray film. After development, a dark spot on the X-ray film indicates the position of the original colony that made the detected actin protein. This spot can be realigned with its colony on the master plate. A sample from this colony can be grown up in a large-volume liquid culture, from which milligram quantities of cDNA/plasmid can be isolated for further analysis.

Monoclonal Antibodies

Antibodies are formed as part of the immune response of animals to the presence of foreign substances (antigens). Exposure to an antigen stimulates B lymphocytes in the bloodstream to produce antibodies (immunoglobulins) that bind to the antigen and facilitate its destruction. Each antibody recognizes a different surface feature—an epitope—of the antigen molecule consisting of as few as only several amino acids. Each type of antibody is secreted by a different clone of lymphocytes.

For years, scientists have produced antibodies directed against specific proteins by injecting them into experimental animals such as mice and rabbits. The antibody-rich serum collected from an immunized animal contains a mixture of *polyclonal* antibodies that are directed against numerous epitopes of the injected protein, as well as against other antigens to which the animal has been exposed. Working at the British Medical Research Council (MRC) Laboratory of Molecular Biology, in 1976, Cesar Milstein developed hybridoma technology that provided a means to make pure preparations of *monoclonal* antibodies with single-epitope specificity.

To make monoclonal antibodies against human actin, for example, the protein is used to immunize a laboratory mouse. The mouse is sacrificed, and B lymphocyte cells from its spleen are mixed with cultured myeloma cells. Derived from a mouse bone cancer, the myeloma cell line is composed of immortalized cells that live essentially forever in culture. The addition of polyethylene glycol (PEG) encourages fusion of the two cell types to form *hybridomas*. The myeloma confers on the hybridoma

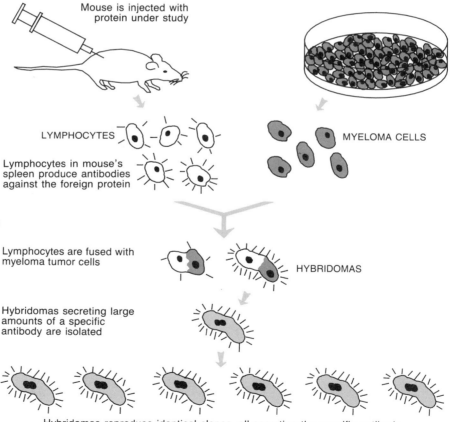

Making Monoclonal Antibodies

the capacity for unlimited reproduction in culture, whereas the B lymphocyte confers the ability to synthesize a specific anti-actin antibody.

The cell mixture is propagated in a selective medium in which only fused cells can survive. The fused cells are then diluted to isolate individual hybridomas, which multiply to produce a pure colony of clones. The progeny clones of the single ancestral hybridoma all produce an antibody of the same specificity, hence the term monoclonal antibody. In this way, antibodies against specific epitopes of an antigen can be purified.

Screening a cDNA Library with an Oligonucleotide Probe

Recombinant DNA technology has made it easier to deduce an amino acid sequence from a cloned gene than from the protein itself. This is largely due to the advent of rapid methods to sequence DNA and of computer controllers to automate the chemical synthesis of short DNA sequences. It is now common practice to determine the amino acid sequence of only a small portion of the protein under study. The DNA-coding sequence derived from these amino acids can be used to synthesize a small DNA molecule—an oligonucleotide—as a probe for the gene sequence encoding the protein. The oligonucleotide is usually radioactively labeled using [γ-^{32}P]ATP and polynucleotide kinase, which transfers the γ phosphate of the ATP to the 5' end of the oligonucleotide. The probe is used to screen a genomic or cDNA library, and the complete amino acid sequence of the protein is predicted from the DNA sequence of the cloned gene.

Since each amino acid is encoded by a triplet codon, a six-amino-acid sequence can be used to derive an 18-nucleotide DNA probe, which should hybridize to a unique sequence in the human genome. (Remember the calculation for the length of a unique sequence?) However, most amino acids are encoded by more than one triplet. Because of this "degeneracy" of the genetic code, the exact DNA sequence cannot be derived precisely from an amino acid sequence.

There are two ways to get around this problem. First, select an amino acid sequence with the least degeneracy—containing amino acids with the fewest codon choices. Then, make a mixed probe from a solution containing all of the possible oligonucleotides that could code for the particular amino acid sequence. (DNA synthesizers can be programed to produce such a mixture automatically.)

Alternately, a single, longer oligonucleotide containing 50–100 residues is synthesized. Although only one of the potential nucleotides can be incorporated at the positions of degenerate codons, the choice of nucleotides is guided by the knowledge that different organisms tend to make preferential use of one of the codon alternatives. The longer probe is hybridized under conditions of lower stringency to ensure that hybridization takes place despite occasional mismatches between the probe and the target sequence.

Determining Sequence of Cloned DNA

Once a gene of interest has been isolated and a sufficient mass purified from *E. coli* or other cultured cells, the next step is often to determine its entire nucleotide sequence. This is accomplished by one of two methods, both of which were discovered in 1977. The chemical cleavage method, of Allan Maxam and Walter Gilbert at Harvard University, involves the specific chemical modification of each different nucleotide. This alteration allows subsequent cleavage at a specific site and identifies the nucleotide present at that position.

The chain termination method, developed by Fred Sanger at the MRC Laboratory of Molecular Biology, is based on two facts of DNA synthesis. First, in the presence of the four deoxynucleotide triphosphates (dNTPs), DNA polymerase will initiate synthesis of a new strand of DNA at the point at which a short DNA primer is hybridized to a single-stranded DNA template. Second, if *di*deoxynucleotide triphosphates (*di*dNTPs) are included in the reaction, DNA chain elongation will cease at whatever position a *di*dNTP is incorporated. This occurs because *di*dNTPs lack a 3' hydroxyl group (OH), which is necessary to form the phosphodiester linkage that joins adjacent nucleotides. In the dideoxy sequencing protocol, four reaction tubes (A,T,C,G) are set up, each containing the following reagents:

POLYACRYLAMIDE GEL ELECTROPHORESIS

Sequencing reactions loaded into denaturing gel

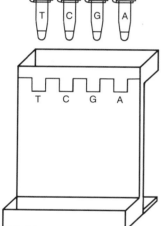

1. A DNA template.
2. A primer sequence.
3. DNA polymerase.
4. The four deoxynucleotide triphosphates (dATP, dTTP, dCTP, and dGTP).
5. One radioactively labeled deoxynucleotide triphosphate, usually [^{32}P]dATP.
6. One *di*deoxynucleotide triphosphate (*di*dNTP) is added to each tube: (Tube A) *di*dATP; (Tube T) *di*dTTP; (Tube C) *di*dCTP; and (Tube G) *di*dGTP.

The ratio of dNTP to *di*dNTP present in the reaction is adjusted so that a *di*dNTP is incorporated into the elongating DNA chain approximately once every 100 nucleotides. Each time a *di*dNTP is incorporated, synthesis stops and a DNA strand of a discrete size is generated.

In Tube A, for example, for every 100 thymine residues encountered on the DNA template, the polymerase will incorporate 99 units of deoxyadenosine. In one instance out of 100, the polymerase will incorporate a *di*deoxyadenosine, halting synthesis of the strand at that point. In other words, the DNA chain will be terminated at each thymine position in 1% of all template molecules. Because billions of DNA molecules are present, the reaction in Tube A results in collections of DNA fragments in which synthesis has been arrested to mirror the position of each thymine residue in the template DNA.

AUTORADIOGRAM OF GEL

Lanes read in sequence from bottom to top of gel

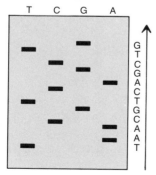

Dideoxy Method for Sequencing DNA

When the reactions are complete, formamide is added to denature the newly synthesized strands from the template DNA. Each reaction is loaded into a different lane of a polyacrylamide gel containing urea, which prevents the DNA strands from renaturing during electrophoresis. This gel system can resolve DNA molecules differing in length by a single nucleotide. Following electrophoresis, the gel is dried and sandwiched with X-ray film. The incorporated radionucleotides expose the X-ray film as a series of bands, indicating the length of fragments generated in Reaction Tubes T, C, G, and A.

Interpreting DNA Sequence

Once a genomic DNA or cDNA sequence is obtained, an obvious first step is to determine the amino acid sequence of its encoded protein. Protein-coding information is generally thought of in terms of mRNA sequence, since the mRNA codons are actually translated into amino acids. By convention, the amino acid codons are expressed as triplets of RNA sequence, where a U replaces a T in the DNA sequence. However, because gene-coding information is generally derived from DNA sequencing gels, it is usually simpler to "read" amino acid codons directly from the DNA sequence.

As reviewed in Chapter 2, protein-coding information is read in triplet codons composed of three adjacent nucleotides. However, three completely different sets of potential codons are generated, depending on where we begin to read the sequence of a DNA strand. For example, the DNA sequence ...ACTCTGAATAG... can be read ...ACT-CTG-AAT... or ...CTC-**TGA**-ATA... or ...TCT-GAA-TAG... Each series of triplet codons is called a "reading frame." Although there are three reading frames per DNA strand—for a total of six reading frames in duplex DNA—generally, only one of these will contain protein-coding information.

To determine the correct reading frame, the cloned DNA sequence is searched for "start" and "stop" codons. Virtually all proteins begin with the start codon ATG (AUG in mRNA), which signals for the incorporation of the amino acid methionine. Of course, an ATG codon can occur elsewhere in the DNA sequence, and the start ATG can only be identified if the very 5' beginning of the gene is present in the cloned DNA.

Of the 64 possible triplet combinations of mRNA nucleotides (4^3), 61 combinations code for incorporation of a specific amino acid by a transfer RNA (tRNA). However, the other three combinations—UAA, UAG, UGA—are stop codons. Because no tRNAs recognize these codons, their presence in an mRNA molecule signals the termination of protein synthesis. The mRNA stop codons are represented by the DNA sequences TAA, TAG, and TGA. A protein-coding sequence will not contain stop codons and is called an "open reading frame." However, stop codons usually will appear in each of the other five reading frames. For example, reading the sequence above as ...ACT-CTG-AAT... codes for ...Thr-Leu-Asn... However, ...CTC-**TGA**-ATA... codes for ...Leu-**STOP**-Ile... and ...TCT-GAA-TAG... codes for ...Ser-Glu-STOP... Thus, the longest open reading frame found is assumed to code for the protein.

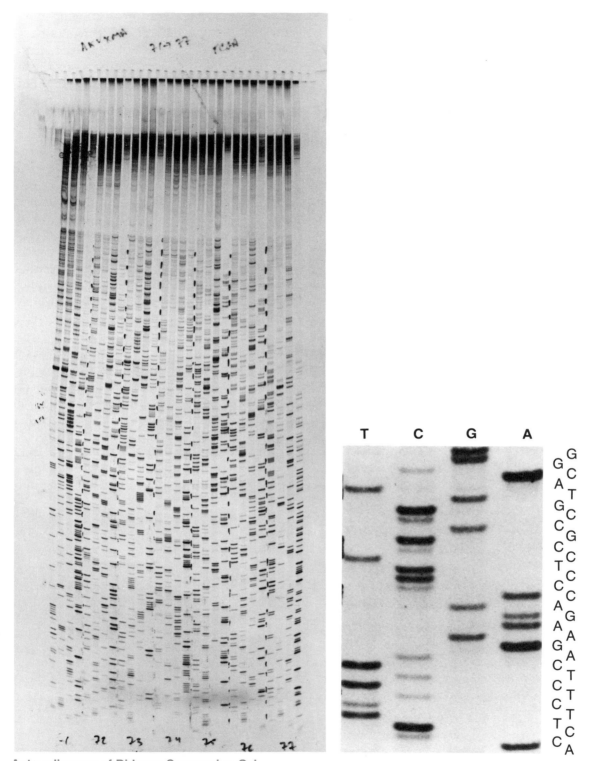

Autoradiogram of Dideoxy Sequencing Gel
Seven separate clones containing restriction fragments were sequenced in side-by-side reactions and coelectrophoresed on a sequencing gel. Each set of four columns (T,C,G,A reactions) on this autoradiogram are the sequence data from a different clone. A portion of the sequence data for clone 5 is shown in detail. Performed in the laboratory of Fred Sanger, who developed the dideoxy sequencing method. (Courtesy of W. Herr, Cold Spring Harbor Laboratory.)

A eukaryotic gene isolated from a genomic library is usually interrupted by large noncoding sequences (introns) of 100 to several thousand nucleotides. Introns, which are excised or "spliced" out of the mature mRNA prior to protein synthesis, can be deduced by comparing the genomic DNA sequence with the corresponding cDNA sequence. (Introns and RNA splicing are discussed in Chapter 5.)

Southern and Northern Hybridization

Southern hybridization (or blotting), named after its discoverer Ed Southern, is used to study the structure of genomic DNA without cloning it. As with library screening, Southern blotting is based on the ability of a DNA strand to seek out and bind to its complementary sequence.

Cellular DNA is first extracted from a tissue of interest and digested in separate reactions with one or several restriction enzymes. The digested DNAs are then electrophoresed on an agarose gel. Following electrophoresis, the gel is soaked in sodium hydroxide solution to denature the duplex DNA. The gel is then neutralized, and capillary action is used to transfer DNAs from the gel onto a nitrocellulose filter, a technique called blotting. Baking fixes the DNA in place on the filter. The filter is then bathed in a solution containing a specific radioactive probe, under hybridization conditions that are essentially identical to those discussed for library screening. Following hybridization, the filter is washed, dried, and sandwiched next to X-ray film. As in plaque or colony screening, the radioactive probe hybridizes to its complementary DNA sequence.

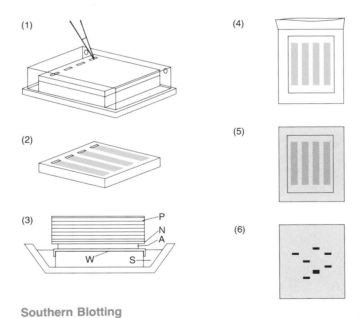

Southern Blotting

(*1*) Load digested DNA into agarose gel, and electrophorese. (*2*) Visualize DNA, and denature in gel. (*3*) Transfer DNA to nitrocelluose filter by capillary action: P = paper towels, N = nitrocellulose, A = agarose gel, W = wick, S = high-salt solution. (*4*) Hybridize radioactive probe to filter, which, if stained, would be a mirror image replica of gel. (*5*) Wash filter, and overlay with X-ray film. (*6*) Developed autoradiogram reveals location of bands to which probe binds. (Art concept developed by Lisa Shoemaker.)

Exposed bands on the film mirror the positions of this sequence on the agarose gel.

Nick translation is one commonly used technique for producing a radioactive probe. A purified phage or plasmid vector containing a cloned genomic or cDNA sequence is treated with a small amount of pancreatic DNase, which hydrolyzes the phosphodiester bonds between nucleotides. At very low concentration, the DNase produces only scattered "nicks" in one or the other strand of the duplex DNA. DNA polymerase and radioactively labeled deoxynucleotides are also added to the DNA sample. Using the unharmed strand as a template and the exposed 3' end at a nick site as a primer, the DNA polymerase synthesizes a new second strand—which then displaces the existing DNA strand from the 5' end of the nick. Radioactive nucleotides are incorporated into the new strand, so a single-stranded probe is created when the duplex DNA is denatured.

Southern blotting is especially useful in detecting structural variations in DNA that result in restriction fragments of differing lengths, known as restriction fragment length polymorphisms (RFLPs). The use of RFLPs as genetic markers to map genes to specific chromosomal locations, in disease diagnosis, and in "DNA fingerprinting" is discussed in Chapter 7.

Whereas in Southern blotting a probe is hybridized to DNA, in Northern blotting, the probe is hybridized to RNA. Northern blotting is the method of choice for detecting specific mRNAs and for monitoring mRNA synthesis and turnover in cells. It can also be used to compare the expression of a particular mRNA at various times during development and in various tissues. An adaptation of this technique, in situ hybridization, detects mRNA in intact tissue or cells in culture.

Northern blotting, for example, can be used to monitor the synthesis of β globin mRNA during reticulocyte development. Total mRNA is isolated from reticulocytes at various stages of maturation and separated by size using agarose gel electrophoresis. However, care must be taken to prevent the mRNAs from hybridizing to themselves, which creates folded secondary structures that alter migration through the gel. Therefore, the samples are electrophoresed in an agarose gel containing a denaturant, such as formaldehyde, which prevents the single-stranded RNA from base pairing.

As in Southern blotting, the separated RNA molecules are transferred to a nitrocellulose filter. The blotted RNA is incubated with the radioactively labeled globin probe that has been prepared by nick translating a cloned β globin gene or by synthesizing a complementary oligonucleotide. The intensities of the bands detected in the autoradiograph of the filter reveal the relative amounts of β globin mRNA at various stages of development.

This chapter has highlighted some of the major techniques that have evolved from the early cloning experiments of Berg, Boyer, Cohen, and Chang; however, any comprehensive survey of recombinant DNA techniques would fill a large volume. Hopefully, this survey has provided a base of knowledge that will allow you to understand better the applications of recombinant DNA described in the following chapters.

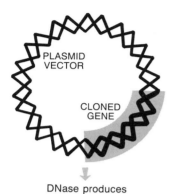
DNase produces nicks in plasmid

DNA polymerase incorporates radiolabeled nucleotides in nicked regions

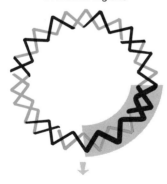
Heat denatures plasmid to single strands

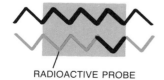
RADIOACTIVE PROBE

Using Nick Translation to Make a Radioactive Probe

For Further Reading

Alberts, Bruce, Dennis Bray, Julian Lewis, Martin Raff, Keith Roberts, and James D. Watson. 1989. *Molecular biology of the cell.* Garland Publishing Inc., New York.

Darnell, James, Harvey Lodish, and David Baltimore. 1986. *Molecular cell biology.* Scientific American Books, New York.

Sambrook, Joseph, Edward F. Fritsch, and Thomas Maniatis. 1989. *Molecular cloning: A laboratory manual,* volumes I, II, and III. Cold Spring Harbor Laboratory Press, New York.

Watson, James D., Nancy H. Hopkins, Jeffrey W. Roberts, Joan Argetsinger Steitz, and Alan M. Weiner. 1987. *Molecular biology of the gene,* volumes I and II. Benjamin/Cummings Publishing Co. Inc., Menlo Park, California.

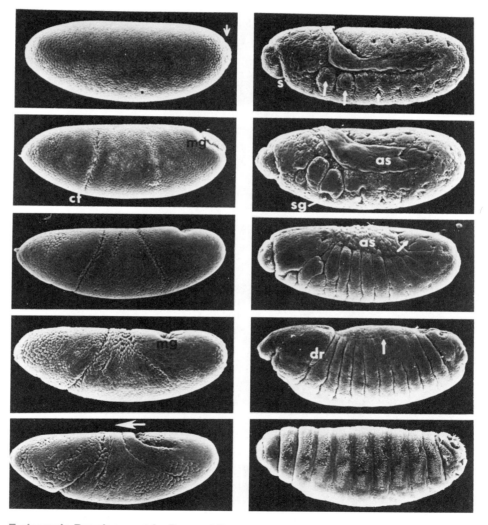

Embryonic Development in *Drosophila*
Scanning electron micrographs show the *Drosophila* embryo 3.5–14 hr after fertilization, during which the body segments become apparent. (Courtesy of R. Turner and A. Mahowald, Case Western Reserve Medical Center.)

CHAPTER 5
Gene Regulation in Development

Two subdisciplines of biology are concerned with time. Evolution deals with changes in the characteristics of a population of organisms over many generations of time. Development deals with cellular and biochemical changes that occur within a single life span. Until relatively recently, development was concerned primarily with embryogenesis—the progression of events, beginning with the fertilized egg, that culminates in a fully functional individual. Armed with the tools of recombinant DNA analysis, biologists are now elaborating the molecular events that govern the differentiation of tissues and the emergence of organs and organ systems.

The concept that every cell carries a virtually identical set of genes was the eventual outcome of the elaboration of the chromosomal theory of heredity in the early 1900s. However, if every cell has the same set of genetic instructions, then how can different types of cells (say a kidney versus a brain cell) have such different structures and biochemical functions? This reasoning led to the inescapable conclusion that cellular specialization results from differential expression of genes in different cell types during development. In this view, gene regulation—the selective turning on and off of genes—is at the core of all developmental questions.

The Molecular Genetic Pathway

In a general way, we understand how DNA works. We know how the information encoded in each gene is transcribed into messenger RNA (mRNA) and then translated into protein. We are beginning to recognize the regulatory sequences in DNA that control gene expression. However, we still have much to learn about how the genetic information stored in the DNA molecule is accessed and utilized in the proper tissue at an appropriate stage in the life of an organism. We know very little about the mechanisms that orchestrate coordinated expression of many genes at the same time.

Proteins are the raw material of cellular differentiation, ultimately determining both the structural characteristics and enzymatic activity of a cell type. However, DNA-binding proteins and "factors" are also key regulators of gene transcription and mRNA translation. Thus, developmental control of gene expression can potentially occur at every step in the molecular genetic pathway that begins with the DNA code and ends with the transport and secretion of a biologically active protein. Working

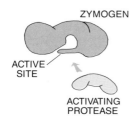
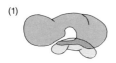
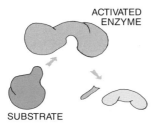
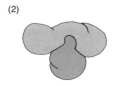

Extracellular Modification of a Zymogen

(*1*) The activating protease cleaves a portion of the zymogen, (*2*) making the active site accessible to bind the substrate in a "lock and key" fashion.

backward along the pathway, we will examine each potential point of regulation, some of which will be discussed in greater detail later in this chapter.

EXTRACELLULAR MODIFICATION

Secretory proteins, including many proteolytic enzymes, are often exported from the cell in nonfunctional forms. These pre-enzymes or zymogens are then modified, as needed, to produce the biologically active form. For example, trypsin is secreted from the pancreas as inactive trypsinogen. Proteolytic cleavage of a segment of the trypsinogen polypeptide occurs after it has been secreted into the digestive system.

PHOSPHORYLATION

Within the cytoplasm, the enzymatic activity of some proteins is regulated by the addition or removal of a phosphate group(s) at a specific serine, threonine, or tyrosine residue(s). Phosphorylation is reversible; protein kinases add phosphate groups, whereas phosphatases remove them. Protein kinases have recently come under intense scrutiny because oncoproteins, the protein products of cancer-causing genes (oncogenes), often exhibit phosphorylating activity (for more details, see Chapter 6).

POSTTRANSLATIONAL MODIFICATIONS

Following release from the ribosome, most polypeptides undergo any of the more than 100 known modifications to amino acid side groups. These modifications regulate the protein's biological activity and may be permanent or reversible. For example, permanent addition of acyl groups (acylation) or of a coenzyme, such as biotin or lipoic acid, is necessary for the action of many enzymes. Within the endoplasmic reticulum (ER) and Golgi apparatus, modifications are made to proteins that enable them to be transported within the cell or externally secreted. Membrane proteins are modified with fatty acid side chains, whereas secreted proteins have added carbohydrate groups (glycosylation).

Proteins destined for modification in the ER contain a signal peptide of 16–30 amino acids at their amino terminus. During protein synthesis, the signal peptide is bound by the signal recognition particle (SRP), which, in turn, binds to an SRP receptor in the ER membrane. This anchors the ribosome in the ER, where the signal peptide conducts the translating protein through the membrane. Within the lumen of the ER, the signal peptide is cleaved by a signal peptidase.

mRNA TURNOVER AND POLY(A) TAIL SHORTENING

The number of protein molecules produced by a single mRNA is dependent on, among other factors, the length of time a transcript exists in the cell before it is degraded or "turned over." The stability of a specific mRNA is determined by measuring the time it takes for half the

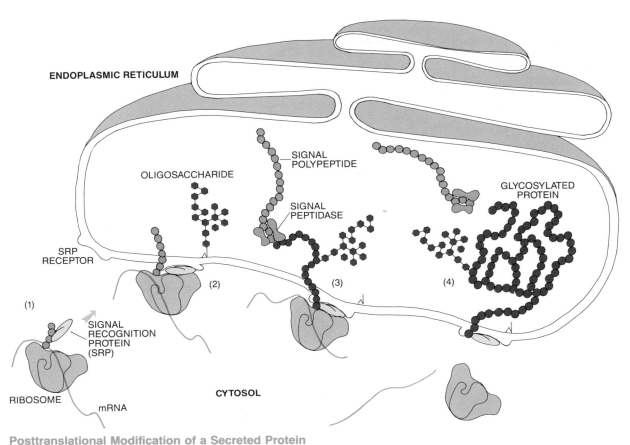

Posttranslational Modification of a Secreted Protein
(1) Signal recognition protein (SRP) binds signal polypeptide, which is translated at a free ribosome following initiation of protein synthesis. (2) SRP binds its receptor on ER membrane. (3) Protein is translated into the ER lumen, where an oligosaccharide side group is added. (4) Signal peptidase cleaves the signal peptide to yield a mature, glycosylated protein.

population of transcripts to be degraded—termed mRNA half-life. For example, the mRNAs that code for α and β globin have very long half-lives, on the order of several days. The mRNAs of some regulatory proteins, on the other hand, have half-lives on the order of seconds. Other factors being equal, the longer the half-life, the more protein molecules synthesized from a given mRNA transcript.

Almost all eukaryotic mRNAs end in a sequence of some 200 adenine residues, called the poly(A) tail. It has been observed that the poly(A) tail gradually shortens (degrades) as the mRNA "ages" in the cytoplasm. This appears to be a contributing factor to mRNA half-life, since mRNA molecules degrade more rapidly as the poly(A) tail shortens.

There is evidence that the 3' noncoding region of mRNA is involved in mRNA stability within the eukaryotic cell. A sequence rich in adenine and uracil residues is frequently found in the 3' noncoding region of mRNAs of genes that are transiently expressed and appears to be a signal for the selective degradation of mRNAs. The AU-rich sequence is recognized by a 3' exonuclease, which degrades an RNA molecule from its 3' end. The effect has been demonstrated by inserting an AT-rich

Turnover Times for Some mRNAs

mRNA	Half-life
Adenovirus E1A	6–10 min
c-*fos*	5–10 min
c-*myc*	10–12 min
β-Interferon	30–60 min
v-*fos*	60–90 min
β-Globin	7–16 hr
Casein	92 hr
Vitellogenin	480 hr

sequence in the 3' noncoding region of the β globin gene. Although the β globin transcript is normally very stable, engineered globin mRNAs with the 3' AU sequence degrade rapidly and accumulate at only 3% of normal levels.

An intriguing experiment with the *fos* oncogene shows that regulation of mRNA half-life can have devastating developmental effects on a cell. The mRNA of the normal cellular gene (c-*fos*) has a very short half-life. The mRNA of the retroviral oncogene (v-*fos*) is considerably more stable and has a much longer half-life. There is convincing evidence that the prolonged presence of the *fos* protein contributes to malignant transformation of a normal cell into a cancerous one. The rapid degradation of the normal c-*fos* mRNA is attributed to the presence of a 67-bp AU-rich sequence in the 3' noncoding region. The mRNA of v-*fos* differs from c-*fos* in several regions, including the absence of the AU-rich sequence. It was found that the normal c-*fos* gene can be made oncogenic by increasing its transcription and deleting the 3' AU-rich sequence, thus making it resemble the 3' region of v-*fos*.

mRNA TRANSLATION: INITIATION, ELONGATION, TERMINATION

Several steps involved in translation of the mRNA transcript determine the rate at which protein is synthesized. Initiation of translation involves binding of the mRNA transcript to the ribosome and its accessory proteins. The cap structure (see facing page) and a ribosome-binding site within the 5' noncoding region position the translation machinery on the mRNA strand so that translation begins at the proper AUG start codon. Transfer RNAs (tRNAs) then must read successive mRNA anticodons and add the proper amino acids to the elongating polypeptide chain. Protein synthesis is terminated when the translation complex encounters one of three stop codons for which no naturally occurring tRNAs exist—UAA, UAG, and UGA.

Mutant tRNAs exist that bind to a stop codon and add an amino acid at that point, rather than terminating translation. These are referred to as suppressor tRNAs, because they suppress termination. Suppressor tRNAs arise from mutations in tRNA genes that change a nucleotide in the anticodon so that it can base pair with a stop codon. For example, three known mutant tRNAs in *E. coli* suppress UAG termination. In one, the anticodon of a tyrosine tRNA has been changed from AUG to AUC, enabling it to base pair with an UAG codon and add a tyrosine residue at that point.

A suppressor tRNA can have survival value for a cell by allowing it to overcome mutations that create a stop codon within the coding sequence of an essential protein. Rather than halting mRNA synthesis at the fallacious stop point, the suppressor tRNA continues translation, producing a full-length protein.

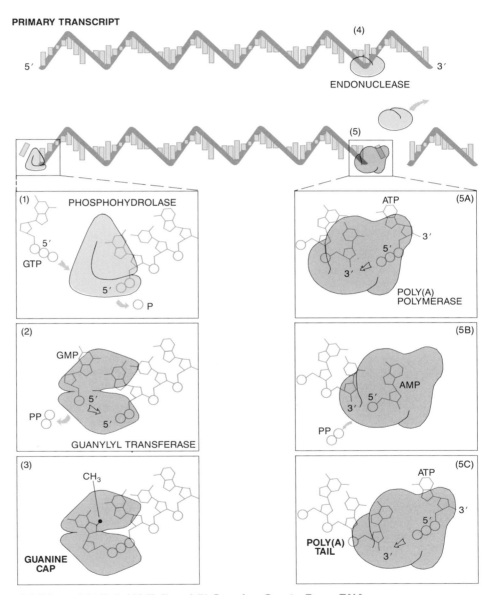

Addition of 3' Poly(A) Tail and 5' Guanine Cap to Pre-mRNA

(1) A phosphohydrolase removes the outermost phosphate from the 5' end of the pre-mRNA. (2) Guanylyl transferase catalyzes the addition of a guanine residue in a reverse orientation, via a 5'-to-5' linkage. (3) The cap is further modified by the addition of a methyl group (CH_3). (4) A specific endonuclease cleaves adjacent to a sequence UAAUUU near 3' end of the pre-mRNA. (5A-C) Poly(A) polymerase catalyzes the addition of some 200 adenine residues to the 3' end.

ADDITION OF 5' CAP AND 3' POLY(A) TAIL

Following transcription, and prior to its transport into the cytoplasm, the mRNA is modified by the addition of nucleotides to its 5' and 3' ends. A cap is made by adding a guanine residue to the 5' end of the mRNA in reversed orientation. A poly(A) tail is created at the 3' end in a two-step process involving cleavage of the mRNA at the proper site and addition of approximately 200 adenine residues. Both of these structures increase mRNA stability and protect it from degradation.

DEGRADATION AND SPLICING OF PRE-mRNA

The mature mRNA that exits the nucleus and serves as a template for protein synthesis is much shorter than the "pre-mRNA" that is transcribed directly from a gene. The longer pre-mRNA molecule contains one or many blocks of sequences, called introns, that are not represented in the mature mRNA. RNA splicing (processing) edits out these introns and joins together the remaining "exons" to form coherent instructions for protein synthesis.

RNA splicing can exert regulatory control in several ways. Research demonstrates that introns are excised at different rates. Therefore, the level of mRNA in the cytoplasm—and, in turn, protein synthesis—potentially can be controlled by slowing down or speeding up the rate of splicing. Different proteins can be produced from the same DNA sequence through the process of alternative splicing. By excising some exons and including others during the splicing process, unique mRNA transcripts are created in different cell types or in cells at different stages of development. Alternative splicing can produce several isoforms of a protein, which, although similar in function, have differing structures.

The degradation of pre-mRNA transcripts is another means of regulating gene expression. Research shows that pre-mRNA transcripts of many different genes are found in the nucleus of various cell types. However, only a fraction of the corresponding mature mRNA transcripts are identified in the cytoplasm. This implies that pre-mRNAs made at the wrong time or in an inappropriate cell are simply degraded.

GENE TRANSCRIPTION

Before any gene can be expressed it must first be transcribed into mRNA. The initial event in this process is the binding of specific proteins to the enhancer (2) and the promoter (3). The binding of these proteins depends on their recognition of specific nucleotide sequences. One of these sequences, the TATA box, positions the RNA polymerase (4) at the start site for mRNA transcription (5). Whereas some DNA-binding proteins are "positive regulators" that stimulate transcription, others are "negative regulators" that block transcription.

DNA REARRANGEMENTS

Although the structures of most genes are stable, DNA rearrangements may function in several ways to regulate the expression of certain genes.

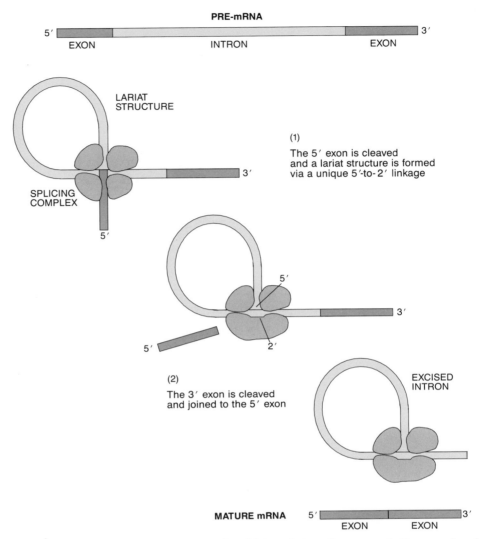

RNA Splicing Is a Two-step Process in which an Intron Sequence Is Removed and Exon Sequences Are Joined

First, the movement of a DNA segment may act as a genetic switch to turn a gene on and off at various stages of development. A DNA rearrangement that brings a gene under the control of a strong promoter activates gene expression, whereas transposition of a DNA fragment into regulatory or protein-coding sequences can interrupt gene expression. Second, gene segments can rearrange prior to transcription to construct novel coding sequences.

Like alternative mRNA splicing, gene rearrangement makes it possible to produce multiple protein products from a single gene or region of DNA. Thus, the nearly infinite diversity of antibody molecules is made possible by recombining a relatively small number of immunoglobulin gene sequences.

Transposable Elements and Gene Rearrangement

MAIZE

From the founding of genetics in the first years of the 20th century, many biologists were intrigued by variegated, or mosaic, pigmentation. In the late 1940s and early 1950s, Barbara McClintock, of the Carnegie Department of Genetics at Cold Spring Harbor, developed a daring hypothesis about the genetic basis of the striking color variations in the leaves and kernels of Indian corn (maize). She proposed that the movement of genetic "controlling elements" from one chromosomal location to another results in unstable mutations in color-forming genes.

One such transposable, or movable, genetic element is Dissociator (*Ds*), which causes mutations when it inserts into an active gene locus. For example, assume that we begin with an embryo (developing seed) in which the *Ds* element has inserted into a color gene, thereby disrupting pigment production. Replication of progenitor cells results in a clone of unpigmented cells, each of which inherits the mutated gene. Subsequent *Ds* transpositions out of the locus restores gene function, giving rise to clones of normally pigmented cells. In this case, the size of pigmented sectors within a nonpigmented background indicates the timing of *Ds* transpositions that restored gene action. Transposition events early in development would give rise to broad stripes or large patches of pigmented tissue, whereas events occurring late in development would give rise to narrow stripes or small speckles. To complicate analysis further, the transposition of *Ds* is controlled by a second element, Activator (*Ac*).

McClintock's theory of transposition contradicted long-held dogma that genes were immutably fixed along the length of the chromosomes, and it was not widely accepted for over 20 years. With the advent of recombinant DNA techniques, her ideas were confirmed. Furthermore, it has been shown that transposition is not merely a peculiarity of maize, but a widespread genetic phenomenon common in both prokaryotic and eukaryotic cells. Transposition is now accepted as a major means to activate and organize gene expression during development. However, McClintock has also suggested that induction of gene instability by transposable elements may provide a mechanism to reorganize the genome rapidly in response to stress and thus may play an important role in generating diversity.

Barbara McClintock, 1951 (Courtesy of Cold Spring Harbor Laboratory Archives.)

SIMPLE SWITCHES IN *SALMONELLA*, TRYPANOSOMES, AND YEAST

An extremely simple transposition event in the *Salmonella* bacterium (which causes food poisoning and other diseases) works exactly like a toggle switch. The inversion of a central 1000-base pair (bp) sequence activates either of two flanking genes that code for surface antigens. This genetic "flip-flop" brings one of the flanking regions under the control of a promoter positioned at one end of the reversible region; each gene is transcribed in the opposite direction. Antigenic switching allows *Salmonella* to evade the immune response of its hosts. When host

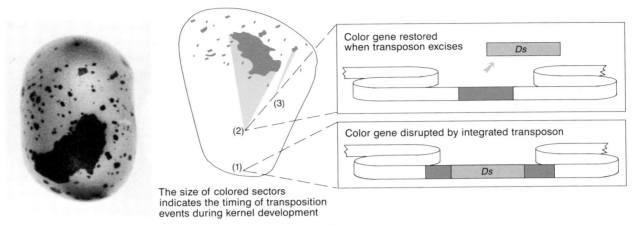

The Phenotypic Effect of *Ds* Transposition Events in a Single Corn Kernel
(*1*) At the beginning of kernel development, the *Ds* transposon is inserted into a gene encoding kernel color, giving rise to a clone of colorless tissue. (*2*) *Ds* transposition in a cell early in development leads to a large clone of colored tissue. (*3*) *Ds* transposition later in development leads to a smaller clone of colored tissue. (Courtesy of B. McClintock, Cold Spring Harbor Laboratory.)

antibodies are produced against one surface antigen, an inversion switch allows a clone to arise that produces the alternate surface antigen. Similar simple switches are found in several bacteriophages.

A more involved sort of antigenic switching is practiced by trypanosomes, the unicellular flagellates that cause African sleeping sickness. Trypanosomes possess a repertoire of up to several thousand different genes encoding variable surface glycoproteins (VSGs). Antigenic switching is accomplished when a duplicate copy of one of the many VSG genes is made and shuttled from its storage area into a single active site, where it displaces a previously active VSG gene. The duplicated VSG gene is then expressed under the control of several promoters resident in the active site. In this way, only one surface antigen is produced at a time.

A switching mechanism that represents a hybrid of the two systems described so far exerts even more fundamental control of development in the brewer's yeast, *Saccharomyces cerevisiae*. Here, transposition is the master switch that sets into motion complex developmental programs that regulate the two most critical aspects of the yeast life cycle—expression of mating type and alteration of haploid and diploid generations. As in *Salmonella*, the yeast mechanism activates one of two alternate genes. As in the trypanosomes, yeast switching involves expression of a duplicate gene at an active locus.

Haploid yeast cells exist as one of two mating types (**a** or α). During the haploid phase of their life cycle, either mating type divides mitotically to give rise to a *mixed population* composed of *both* mating types. This means that a change in the expression of a single (haploid) genome gives rise to a "mating-type switch," which can occur at nearly every mitotic division. Haploid **a** and α cells can fuse to form diploid cells. Under conditions of nutrient depletion, a diploid cell avoids starvation by undergoing meiosis to form four spores.

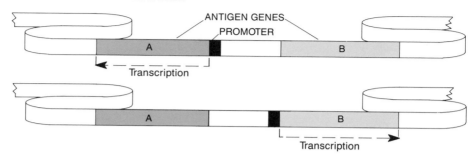

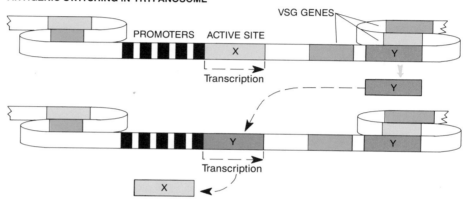

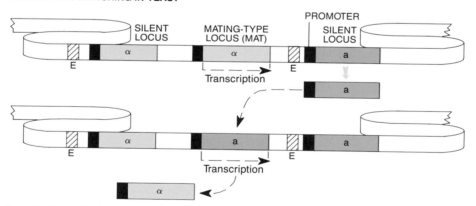

Simple Genetic Switches

In the *Salmonella* bacterium, a promoter lies between genes encoding two different surface antigens. Inversion of a 1000-bp sequence "flip-flops" the promoter to alternately transcribe either gene. The flagellate *Trypanosome* possesses several hundred genes encoding variable surface glycoproteins (VSGs). A transposed VSG gene is expressed at the active site under the control of five resident promoters. The yeast *Saccharomyces* stores unexpressed copies of **a** and α mating-type genes at silent loci, where they are under negative control of the regulatory element E. Transposition of either **a** or α gene releases it from negative control by the E element, allowing it to be expressed at the active mating-type locus (*MAT*).

In the mid-1970s, Jeffrey Strathern, James Hicks, and Ira Herskowitz at the University of Oregon discovered that mating-type switching is controlled by three gene loci, or "cassettes," on yeast chromosome 3. A central mating-type locus (*MAT*) is flanked by unexpressed ("silent") copies of α and **a** cassettes. In the **a** mating-type, a duplicate **a** cassette is inserted, and expressed, at the *MAT* locus. During a mating-type switch, the α cassette is duplicated at its silent locus and then transposes into the *MAT* locus to displace the **a** gene.

Proteins expressed at the *MAT* locus, in turn, control expression of a number of cell-specific genes whose proteins regulate differential metabolic pathways in **a**, α, and diploid (a/α) cells. For example, **a** and α cells prepare for mating by each secreting a pheromone that binds to a specific receptor on the surface of the opposite mating type. Entirely different sets of genes are induced in diploid cells that initiate meiosis and sporulation.

ANTIBODY DIVERSITY

A far more complex series of gene rearrangements enables the human immune system to produce antibodies that recognize and destroy foreign particles and disease-causing pathogens. These and other substances that elicit an immune response are called antigens.

A multitude of antigenic molecules are present on the surfaces of viruses, bacteria, dust, pollen, and other foreign particles that may enter the body. Living pathogens also secrete numerous antigens as toxins or by-products of their metabolism. Although antigens are most often proteins, polysaccharides and nucleic acids can also elicit an immune response. Exposure to an antigen stimulates B lymphocytes in the bloodstream to produce immunoglobulin proteins (antibodies) that bind to the antigen, facilitating its destruction. Moreover, several different immunoglobulins may be produced that recognize different molecular features (epitopes) of a single antigen. For proteins, an epitope consists of only several amino acids.

Obviously, the set of potential antigens (perhaps 1 million) is greater than the total number of genes in the entire genome (~100,000). Because it seemed impossible that the genome could contain pre-existing genes for every potential antibody, controversy long raged about the mechanism through which B lymphocytes can synthesize an almost infinite diversity of antibody molecules. In the late 1970s and early 1980s, Susumu Tonegawa, of the Massachusetts Institute of Technology, was key among researchers who cloned immunoglobulin genes and probed their organization. The controversy was settled by showing that the diversity of functional antibody genes is created by recombining a relatively small number of chromosomal DNA segments.

As elucidated in 1969 by Gerald Edelman at Rockefeller University, an antibody molecule consists of four polypeptide chains: two identical heavy (H) chains of about 440 amino acids each and two identical light (L) chains of about 220 amino acids each. The amino acid sequences of

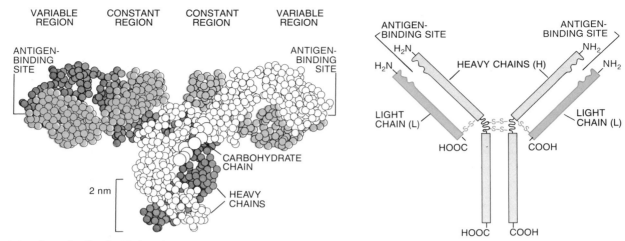

Models of an Antibody Molecule
Each sphere (at left) represents a single amino acid. (After E.W. Selverton et al. 1977. *Proc. Natl. Acad. Sci.* 74: 5140. From B. Alberts et al. 1989. *Molecular biology of the cell*, p. 1013, 1022. Garland Publishing, Inc., New York.)

the carboxyl ends of both heavy and light chains show little variation from antibody to antibody; they are termed constant (C) regions: C_H and C_L. In contrast, the amino ends of both types of chains show great variation in amino acids and are termed variable (V) regions: V_H and V_L. The four chains form a Y-shaped molecule with variable regions all oriented at the tips of the arms, where antigen binding occurs. Moreover, the variable region of each chain contains three complementarity determining regions (CDRs, or hypervariable regions) that form the cavities of the antigen-binding site.

A single locus of genes for immunoglobulin heavy chains is located on chromosome 14; separate light chain loci are found on chromosomes 2 and 22. The approximate number of immunoglobulin genes in these loci, as well as the basic scheme of their rearrangement, was discovered by hybridizing cDNA probes of antibody mRNA to genomic DNA. Constant and variable regions are coded by different sets of genes, widely separated from each other on the chromosome. Cellular DNA contains only about ten different C_L and C_H genes. (Since the constant regions do not interact with the antigen, they are of relatively minor importance in generating antibody diversity.) In contrast, there are 200–300 different V_L genes and a similar number of V_H genes. Taken together, the following molecular genetic events are easily capable of generating a million or more different antibodies:

1. A functional antibody gene is made through DNA rearrangements that randomly join a distant variable gene to any of several small DNA linkers, called joining (J) segments and diversity (D) segments, which are located in relatively close proximity to the constant genes. Light chain genes are linked through a single recombination event between a V_L gene and any of four J segments, which are linked to C_L genes. This is termed V-J joining.

Heavy chain genes (C_H and V_H) are linked through two recombination events. The first links any of 10–20 D segments with any J segment (D-J joining). The second recombination links the D-J segment with any V_H gene (V-D-J joining).

2. Positioning of the V-J and V-D-J junctions, and nucleotide additions at the V-D-J joint, further increase antibody diversity.
3. Point mutations in the fully assembled V_H genes, termed somatic hypermutation, are another source of variation.
4. Alternative RNA splicing of the constant region genes results in antibodies with different properties. For example, membrane-bound antibodies have a C_H exon that codes for transmembrane polypeptides that anchor the antibody to the cell membrane. Differential splicing omits this exon, creating a secreted form of the antibody.

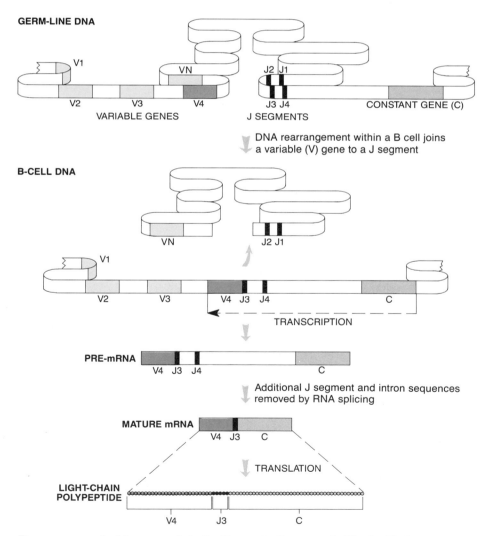

Rearrangement of Immunoglobulin Genes to Form an Antibody Chain

Eukaryotic Promoters

A complete picture of the eukaryotic promoter cannot yet be drawn. However, a basic description of the promoter and the proteins with which it interacts is central to our understanding of gene regulation. The following discussion deals only with the transcription of genes by RNA polymerase II (pol II) to produce mRNAs. Genes coding for ribosomal, transfer, and small nuclear RNAs have different regulatory elements and are transcribed by different RNA polymerases.

As its name implies, the promoter activates transcription of a gene. The promoter is located at the 5' end of the gene, directly adjacent to the protein-coding region that is transcribed into mRNA and extending some 150–300 nucleotides upstream of the start site of transcription. A survey of known promoters has highlighted "consensus sequences" that are conserved among species and are required for the efficient expression of various eukaryotic genes.

Functional studies, where a promoter is linked to a responder gene in a plasmid, have further delineated sequences required for promoter activity. Nucleotides within a suspected consensus sequence are systematically altered or deleted, and transcriptional activity is assayed in one of two ways. One approach is to compare the transcription of a reporter gene in cultured cells that have been transformed with plasmid DNAs carrying either normal or mutated promoter consensus sequences. Another approach is to compare transcription of normal and mutated constructs using an in vitro transcription extract. A difficulty with this system is that tissue-specific genes may not be expressed unless the transcription extract is obtained from cell types containing all of the necessary transcription factors. In addition, necessary transcription factors may be lost during preparation of the in vitro extract.

Consensus sequences are now known to be the binding sites of specific proteins needed for DNA transcription. Sequence-specific binding results from the close interactions between the base pairs of the DNA molecule and amino acid side chains of the protein. X-ray crystallography and comparative studies of amino acid sequences have revealed several hypothetical structures that apparently align regulatory proteins within the major and minor groove of the double helix, where they interact with the sequence of bases that distinguish the binding site. "Zinc fingers" are amino acid loops that are stabilized by the interaction of four cysteine or cysteine and histidine residues with a zinc atom. Another DNA-binding motif is the "helix-turn-helix," which, as its name implies, consists of two α helices oriented at approximately right angles to one another. An intriguing DNA-binding motif, the "leucine zipper," allows for the cooperative binding of several known oncogenes (see Chapter 6 for details).

Although the exact sequences of promoter elements vary slightly from gene to gene, they are present in virtually every eukaryotic promoter. The name of each promoter element is derived from its characteristic nucleotide sequence. The TATA box of alternating thymine and adenine residues is located 25–30 nucleotides upstream of the transcription start site. The TATA sequence is necessary for the proper positioning of RNA polymerase for initiation of transcription.

The CAT box, typically with the sequence CCAAT, lies between 70 and 80 nucleotides upstream of the transcription start site and is the binding site for the CTF protein (CCAAT-binding transcription factor). Recent studies suggest that CTF may be identical to a previously identified nuclear factor (NF1), which is required for the initiation of adenovirus DNA replication. Since not all promoters contain a CAT box, the CTF protein must specifically regulate genes containing this sequence.

Multiple guanine/cytosine (GC)-rich regions are a common feature of the promoters of "housekeeping" genes, whose proteins perform essential metabolic roles in all cells. The GC box is the binding site for the SP1 protein, first identified in simian virus 40 (SV40). Stretches of alternating purines and pyrimidines, such as the GC box, have the potential to flip into a right-handed helix, Z-DNA. (Remember that "normal" B-DNA, with which we are familiar, is a left-handed helix.) The unusual structure of a local region of Z-DNA may make it an easily recognized target for a DNA-binding protein.

Eukaryotic Enhancers

Enhancers generally stimulate the level of mRNA synthesis; removal of the enhancer sequence can reduce transcription by as much as 100-fold. Although their exact function is unknown, it is most commonly believed that enhancers provide an entry site for RNA polymerase II or its subunits, which can then move to the transcription start site. Enhancers are enigmatic because, unlike promoters, they are independent of the position of the gene(s) they regulate. They are not necessarily positioned

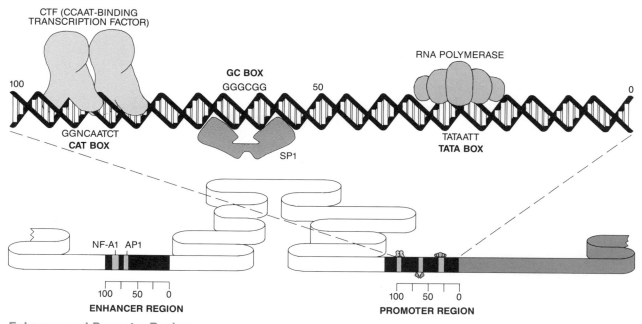

Enhancer and Promoter Regions

upstream of a gene, and thus can be located downstream from, or even within, the transcribed region. They need not be located adjacent to the target gene, but they can also exert influence from positions up to several thousand nucleotides distant. Furthermore, enhancer sequences function even when their 5'→3' orientation is flipped.

Enhancer regions are often composed of several hundred nucleotides and typically contain repeated units that can function independently or coordinately. The SV40 enhancer, for example, is composed of two tandem copies of a 72-bp sequence, only one of which is essential for full transcriptional function. As in promoters, DNA-binding proteins recognize specific nucleotide sequences within the enhancer. For example, the enhancer binding protein AP1 recognizes the sequence TGACTCA.

Enhancers and their binding proteins are both species- and tissue-specific. For example, several DNA-binding proteins have been identified that recognize the sequence ATTGCAT that is found within the enhancer region of nearly all immunoglobulin heavy chain genes. Two nuclear factors, NF-A1 and NF-A2, bind to this conserved sequence. The NF-A1 protein is found in most cell types, whereas NF-A2 is found almost exclusively in B lymphocytes that produce antibodies. Although other factors most certainly are involved, NF-A2 appears to be a cell-specific factor that controls expression of the immunoglobulin heavy chain gene in B lymphocytes.

The β globin gene appears to have two enhancers, one near the 3' end of the protein-coding region and one in the 3'-flanking region. The latter has four consensus sequences that bind several different proteins, including an erythroid-specific nuclear factor (NF-E1).

Metallothionein, a protective protein that binds heavy metals, illustrates that the regulation of eukaryotic genes may often involve numerous transcriptional factors and that others likely remain to be identified. In addition to the enhancer and promoter binding sites described above, the metallothionein gene contains four sites responsive to heavy metals and one site responsive to glucocorticoid (a steroid hormone):

BLE	Basal level enhancers (two AP1-binding sites)
GRE	Glucocorticoid-responsive element
MRE	Metal-responsive elements (4 sites)
GC Box	SP1-binding site
TATA Box	RNA polymerase II start position

The discovery of multiple DNA-binding proteins that regulate DNA synthesis calls to mind still another level of regulation and raises a paradoxical question: What proteins, in turn, regulate the expression of the genes that code for DNA-binding proteins such as CTF, SP1, and AP1?

Alternative RNA Splicing

For years it was known that the nuclei of eukaryotic cells contain large RNA molecules, called heterogeneous nuclear RNA (hnRNA), whose function was obscure. Work on the β globin gene in the late 1970s showed that large nuclear RNAs are, in fact, the precursors of mRNA.

These pre-mRNAs are primary transcripts read directly from DNA and contain extraneous sequences that are removed prior to mRNA transport to the cytoplasm.

Studies on the mRNAs of adenovirus led to the discovery of the phenomenon of RNA splicing. Comparison of a mature, cytoplasmic mRNA transcript with its parental DNA showed that the mRNA sequence is not contiguous with its DNA sequence. The mRNA sequence instead exists as blocks of nucleotides interrupted by DNA sequences that are not represented in the mRNA. Hybridization of an mRNA transcript to a single-stranded parental DNA results in duplex regions of homology, interrupted by loops of single-stranded DNA for which there is no mRNA counterpart.

Thus arose the concept of "split genes." Regions of eukaryotic genes that will ultimately be present in the mRNA (called exons) are interrupted by sequences that are not found in the mRNA (introns). Within the nucleus, introns are excised from the pre-mRNA, and the exons are spliced together to form a mature mRNA transcript.

Intron sequences are common to most eukaryotic genes, and many elaborate gene structures have been discovered. At approximately 2000 nucleotides, the β globin gene is more than three times larger than its functional mRNA transcript. Globin DNA sequences represented in cytoplasmic mRNA exist as three protein-coding blocks (exons) totaling

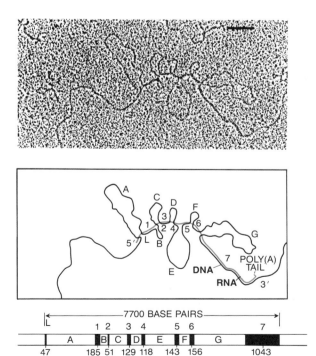

Organization of a Split Gene
Heteroduplex mapping reveals the genetic organization of the ovalbumin gene. Regions where single-stranded ovalbumin DNA hybridizes to ovalbumin mRNA are the gene's eight coding exons (L, 1–7). Loops of DNA (A–G) are the gene's seven intron sequences, which have no corresponding sequences in the mRNA. (Courtesy of P. Chambon, Institut Pasteur. From "Split Genes," May 1981, *Scientific American*.)

approximately 600 nucleotides that are interrupted by intron sequences of 230 and 1100 nucleotides. The rat serum albumin gene is interrupted by 23 introns, the mouse collagen gene contains 50 introns, and the human dystrophin gene contains more than 60 introns.

Many proteins exist in several isoforms that have similar, but not identical, amino acid sequences. Isoforms of a particular protein were presumed to be encoded by unique genes. However, for some of these proteins, it was found that cDNA probes made from cloned mRNAs of various isoforms hybridized to a single fragment of genomic DNA. This indicated that the various protein isoforms are encoded by the same gene.

Experiments with the different isoforms of the muscle protein tropomyosin illustrate how alternative splicing can regulate gene expression at the mRNA level. First, a tropomyosin cDNA was isolated from a cDNA library made from mRNA isolated from skeletal muscle. Radioactively labeled tropomyosin cDNA was then used as a probe to isolate a tropomyosin cDNA from a library made from nonmuscle, cultured fibroblast cells. (Tropomyosin, like many contractile proteins, is found in nearly every cell type.)

Sequence comparison of the cDNAs from the muscle and nonmuscle cells showed that the two mRNAs are identical through the coding region for the first 188 amino acids. From this point, the mRNA sequences differ dramatically for the next 25 amino acids. The sequences are again identical for another 44 amino acids. Then the sequences are divergent for the remainder of the mRNAs.

The two cDNAs were then used to probe a genomic library. In Southern blot analysis, both probes hybridized to identical genomic fragments containing the tropomyosin gene. Sequence comparison of the genomic DNA, the muscle mRNA, and the nonmuscle mRNA revealed alternative splicing of the 11 exons that compose the tropomyosin gene. The first five exons and exons 8 and 9 are common to the mRNAs of both cell types. However, exons 6 and 11 are represented only in the mRNA of fibroblast cells, and exons 7 and 10 are found only in the mRNA of skeletal muscle cells.

It is not yet understood how the alternative splicing of a single pre-mRNA is regulated in different cell types. Presumably, cell-specific factors interact with the splicing apparatus to signal the inclusion of some exons and the exclusion of others. There is, as yet, no evidence of

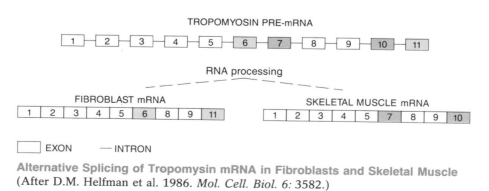

Alternative Splicing of Tropomysin mRNA in Fibroblasts and Skeletal Muscle (After D.M. Helfman et al. 1986. *Mol. Cell. Biol. 6*: 3582.)

conserved intron sequences that would be putative binding sites for such cell-specific factors.

The discovery that a single gene can code for multiple proteins altered forever the "one gene-one enzyme" hypothesis that had held sway for 40 years. Previously, it was believed that the DNA sequence was sole master of the genetic code. However, RNA splicing provides an alternate means to store and access a larger amount of genetic information than can be stored in DNA sequence alone.

Drosophila and the Homeobox

Antennapedia Mutation in *Drosophila*
(Courtesy of Carolina Biological Supply Company.)

Drosophila is an ideal system in which to study development. It is a relatively simple, segmented organism that progresses through distinct larval stages. Since the embryo develops externally, as opposed to in utero, changes can be readily observed. The entire life cycle, from fertilized egg to adult, is accomplished in a matter of approximately 2 weeks.

The discovery over the years of a number of bizarre developmental mutations proved crucial to applying recombinant DNA techniques to *Drosophila* development. These aberrations, which alter the fates of cells during embryogenesis and commit them to new developmental pathways, are known as "homeotic" mutations. For example, the mutation *Antennapedia* causes a pair of normal-looking legs to grow from the head segment, replacing the antenna. *Bithorax* mutations lead to the development of a second pair of partial or complete wings in an inappropriate thoracic segment.

These mutations are clearly mistakes in the spatial arrangement of segments within the overall body plan. Homeotic genes are thus believed to be "master" regulatory genes that orchestrate the coordinated action of a number of genes, which, in turn, determine the development of a large region or body segment. It is hypothesized that homeotic gene products act in one or more of the 19 "imaginal discs" found on mature

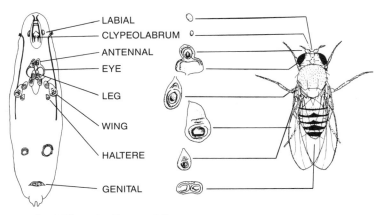

Imaginal Discs in *Drosophila*
Imaginal discs in the *Drosophila* embryo give rise to key structures in the adult fly. (After J.W. Fristom et al. 1969. *Problems in biology: RNA in development*, p. 382. University of Utah Press, Salt Lake City. From J. Watson et al. 1987. *Molecular biology of the gene*, volume II, p. 788. Benjamin/Cummings Publishing Co., Inc., Menlo Park, California.)

larva. (An imago is an adult fly.) Each imaginal disc contains cells that are destined to develop into major body parts, such as antenna, mouth parts, eyes, wings, legs, and genitals of the adult.

Studies of the molecular genetic control of homeotic genes began with cloning the *Antennapedia* (*Antp*) gene. A probe made from a portion of the *Antp* gene was found to hybridize to a second region in the *Drosophila* genome. The sequence homology fell within *fushi tarazu* (*ftz*), which also was known to be involved in regulating the development of embryonic body segments. The homologous region was a highly conserved sequence of 180 bp, within the protein-coding region, now referred to as the homeobox. Subsequently, homology was also found with other known homeotic genes of *Drosophila*. Thus, it appeared that the conserved homeobox sequence might be a general feature of many development regulatory genes.

This was exactly the case. Using the 180-bp sequence as a probe, several previously unidentified homeotic genes of *Drosophila* were discovered. Even more exciting was the discovery that the homeobox probe cross-hybridizes with the genomic DNA of many organisms, including yeast and man. The *Antp* homeobox shows 75–90% homology with sequences from other *Drosophila* genes and greater than 90% homology with two sequences discovered in human genes.

Further clues about the function of the homeobox came from analysis of its encoded polypeptide sequences. First, approximately 30% of the amino acids encoded by homeobox sequences are basic—arginine and

Extensive Homology in Amino Acid Sequences of Five Homeo Domains

	1																			20
Mouse *MO*-10	Ser	Lys	Arg	Gly	Arg	Thr	Ala	Tyr	Thr	Arg	Pro	Gln	Leu	Val	Glu	Leu	Glu	Lys	Glu	Phe
Frog *MM3*	Arg	Lys	Arg	Gly	Arg	Gln	Thr	Tyr	Thr	Arg	Tyr	Gln	Thr	Leu	Glu	Leu	Glu	Lys	Glu	Phe
Antennapedia	Arg	Lys	Arg	Gly	Arg	Gln	Thr	Tyr	Thr	Arg	Tyr	Gln	Thr	Leu	Glu	Leu	Glu	Lys	Glu	Phe
Fushi tarazu	Ser	Lys	Arg	Thr	Arg	Gln	Thr	Tyr	Thr	Arg	Tyr	Gln	Thr	Leu	Glu	Leu	Glu	Lys	Glu	Phe
Ultrabithorax	Arg	Arg	Arg	Gly	Arg	Gln	Thr	Tyr	Thr	Arg	Tyr	Gln	Thr	Leu	Glu	Leu	Glu	Lys	Glu	Phe

	21																			40
Mouse *MO*-10	His	Phe	Asn	Arg	Tyr	Leu	Met	Arg	Pro	Arg	Arg	Val	Glu	Met	Ala	Asn	Leu	Leu	Asn	Leu
Frog *MM3*	His	Phe	Asn	Arg	Tyr	Leu	Thr	Arg	Arg	Arg	Arg	Ile	Glu	Ile	Ala	His	Val	Leu	Cys	Leu
Antennapedia	His	Phe	Asn	Arg	Tyr	Leu	Thr	Arg	Arg	Arg	Arg	Ile	Glu	Ile	Ala	His	Ala	Leu	Cys	Leu
Fushi tarazu	His	Phe	Asn	Arg	Tyr	Ile	Thr	Arg	Arg	Arg	Arg	Ile	Asp	Ile	Ala	Asn	Ala	Leu	Ser	Leu
Ultrabithorax	His	Thr	Asn	His	Tyr	Leu	Thr	Arg	Arg	Arg	Arg	Ile	Glu	Met	Ala	Tyr	Ala	Leu	Cys	Leu

	41																			60
Mouse *MO*-10	Thr	Glu	Arg	Gln	Ile	Lys	Ile	Trp	Phe	Gln	Asn	Arg	Arg	Met	Lys	Tyr	Lys	Lys	Asp	Gln
Frog *MM3*	Thr	Glu	Arg	Gln	Ile	Lys	Ile	Trp	Phe	Gln	Asn	Arg	Arg	Met	Lys	Trp	Lys	Lys	Glu	Asn
Antennapedia	Thr	Glu	Arg	Gln	Ile	Lys	Ile	Trp	Phe	Gln	Asn	Arg	Arg	Met	Lys	Trp	Lys	Lys	Glu	Asn
Fushi tarazu	Ser	Glu	Arg	Gln	Ile	Lys	Ile	Trp	Phe	Gln	Asn	Arg	Arg	Met	Lys	Ser	Lys	Lys	Asp	Arg
Ultrabithorax	Thr	Glu	Arg	Gln	Ile	Lys	Ile	Trp	Phe	Gln	Asn	Arg	Arg	Met	Lys	Leu	Lys	Lys	Glu	Ile

From W.J. Gehring. 1985. *Sci. Am.* 253/4: 159.

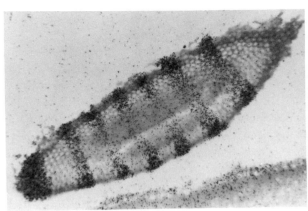

In Situ Hybridization
In situ hybridization of an early *Drosophila* embryo reveals that *ftz* mRNA is expressed in seven stripes, each three to four cells thick. (Courtesy of W.J. Gehring, Biozentrum. From W.J. Gehring. 1985. *Sci. Am.* 253/4: 159.)

lysine. This is a common characteristic of DNA-binding regions of proteins. Second, computer comparison indicates that the amino acid sequence of the homeo domain shares significant homology with the protein products of the yeast mating-type (*MAT*) locus, discussed earlier in this chapter. The homologous regions of the *MAT* proteins bind specific sequences within the promoters of the genes they control. In fact, the homologous sequences are those known to be most critical to the protein/DNA interaction.

The homeo domain and the DNA-binding region of the *MAT* protein also share homology with the DNA-binding regulatory elements (operators and repressors) of prokaryotic genes. All also appear to bind DNA with a closely similar "helix-turn-helix" structure. These observations strengthen the compelling hypothesis that proteins containing homeoboxes regulate gene expression in development by binding to the promoters of their target genes. Thus, homeotic gene products appear to act in much the same way as the SP1 protein discussed earlier in this chapter.

In situ hybridization, employing radioactively labeled sequences from homeotic genes, is now being used to study homeotic gene expression during *Drosophila* development. Embryos and larva at various developmental stages are fixed and cut into thin sections, which are incubated with a homeo-gene probe. The radioactive probe localizes to regions containing the complementary mRNA, thus identifying cells in which the homeotic gene is actively expressed.

These studies revealed that more than one homeotic protein can be localized within the same cell during the same stage in development. Although each homeotic protein is likely to exhibit a degree of specificity in the DNA sequences to which it binds, the homology between homeo domains suggests that several homeotic proteins can potentially bind the same DNA sequence. Thus, the mechanism by which homeotic proteins regulate development is likely to involve competitive binding of promoter sequences.

There is experimental evidence to support this. The *Antennapedia* complex (*ANT-C*) acts primarily on thoracic segment 2 (T2). The *ultrabithorax* (*Ubx*) locus, a subregion of the bithorax region, acts on thoracic segment 3 (T3). Deletion of *Ubx* results in expression of T2-like characteristics in the T3 segment. One conclusion is that during normal development, *Ubx* proteins successfully compete against *Antp* proteins for binding of regulatory sequences in T3 cells and thus control gene expression there. In the absence of *Ubx* proteins, *Antp* proteins can bind *Ubx* regulatory sequences, forcing *Antp*-like expression in T3.

Homeotic proteins may act directly on genes coding for structural proteins, inducing their expression in the correct tissue at the correct stage in development. However, it is more likely that they act in an indirect manner by regulating complex biosynthetic cascades that result in tissue-specific expression of structural proteins. In this scenario, each homeotic protein coordinately activates a number of regulatory genes in several cell types. The regulatory gene products, in turn, target another series of regulatory and/or structural genes, and so on. Various regulatory proteins in the cascade act to promote or repress gene activity in certain cell types. This would explain why the mutation of a single homeotic gene can cause major alterations in body plan that clearly involve a number of genes and gene products.

Transgenic Studies

Development of mammalian cell transfection in the late 1970s made it possible to introduce cloned genes into cultured eukaryotic cells (see Chapter 6). The recent development of gene-transfer techniques has made it possible to study cloned genes in whole organisms, such as mice. A "transgenic" animal is created by inserting a foreign gene into a single-cell embryo, which then ramifies throughout the somatic and germ-line cells during development. The foreign transgene becomes stably integrated into the genome and is inherited in a Mendelian fashion in succeeding generations.

The general scheme for creating transgenic mice is described below; similar protocols have been developed for other laboratory and domestic animals (see Chapter 8), as well as *Drosophila*. Twelve hours after mating, fertilized ova are collected from the oviduct of a female mouse. Embryos harvested at this time are still in the one-cell stage, prior to fusion of the male and female pronuclei. Cloned DNAs are then microinjected into either pronucleus using a capillary pipet whose tip has been drawn to a diameter of several microns (roughly a hundredth the diameter of a human hair). The microinjected embryos are then surgically reimplanted into the oviduct of a female mouse that is rendered "pseudo-pregnant" by mating with a vasectomized male.

Samples of genomic DNA from live-born pups are screened for integration of the transgene using Southern blot analysis. Expression of the foreign gene in various tissues can be detected by Northern hybridization of a probe to mRNA purified from cells from various tissues. Microinjection is fairly efficient—of 100 microinjected embryos, approximately 70 survive microinjection, approximately 20 are carried to term,

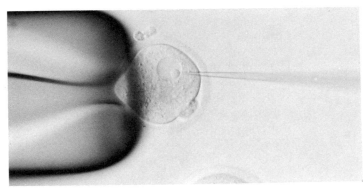

Microinjection of a One-celled Mouse Embryo
(Courtesy of Mark Steinhelper, Cold Spring Harbor Laboratory.)

and approximately 5 offspring stably incorporate the transgene. Transgenic studies, in which a hybrid gene is created by fusing a promoter to a reporter gene, have provided further insight into the 5′ regulatory signals that control tissue-specific gene expression. In one of the first successful experiments of this kind, the 5′ regulatory region was separated from the rat insulin gene, and the protein-coding region of the large T antigen gene of SV40 was isolated from its regulatory sequences. These elements—the rat insulin promoter (RIP) and the large T antigen gene (TAG)—were ligated together to create a fusion gene dubbed RIP-TAG.

The RIP-TAG fusion gene was then microinjected into one-cell embryos to create transgenic animals. Northern analysis of mRNA isolated from various tissues indicated that T antigen mRNA was present only in the β cells of the islets of Langerhans of the endocrine pancreas. This is the specific subset of cells that secrete insulin. Thus, the insulin promoter region carries sufficient regulatory information to drive expression of the T antigen gene in a correct tissue-specific manner.

Attaching a tissue-specific promoter to a toxin gene, such as ricin or diphtheria A (dipA), provides a precise means to study the physiological effects of ablating (destroying) a specific subset of cells. For example, some transgenic mice carrying a fusion pancreatic elastase-dipA gene do not develop a pancreas and die shortly after birth. Fusions of the lens crystallin gene with dipA and ricin have been used to ablate cells of the developing eye.

Developmentally regulated expression of the toxin gene within a cell type can be further controlled by adding an inducible element, such as a metallothionein promoter, to the fusion construct. Females can be treated at various times during pregnancy with an inducing substance—in this case a heavy metal such as cadmium—to trigger expression of the toxin gene within specific tissues of the developing transgenic embryo. In this way, the developmental effects of ablating a cell type at various stages of embryogenesis can be examined.

Another strategy is to interfere with the normal expression of a gene to create a model of a genetic deficiency or of abnormal development. For example, cleft palates develop in transgenic mice expressing a hybrid of the viral *erb*A oncogene and the retinoic acid receptor gene.

For Further Reading

Edelman, Gerald M. 1988. *Topobiology: An introduction to molecular embryology*. Basic Books Inc., New York.

Fedoroff, Nina V. 1984. Transposable genetic elements in maize. *Sci. Am. 250/6:* 85–98.

Fox-Keller, Evelyn. 1983. *A feeling for the organism: The life and work of Barbara McClintock*. W.H. Freeman and Co., New York.

Gehring, Walter J. 1985. The molecular basis of development. *Sci. Am. 253/4:* 152–163.

Ptashne, Mark. 1989. How gene activators work. *Sci. Am. 260/1:* 41–47.

Ross, Jeffrey. 1989. The turnover of messenger RNA. *Sci. Am. 260/4:* 48–55.

Tonegawa, Susumu. 1985. The molecules of the immune system. *Sci. Am. 253/4:* 122–131.

President Richard Nixon Signing the National Cancer Act, December 23, 1971 (Courtesy of the National Cancer Institute.)

CHAPTER 6
Molecular Biology of Cancer

On December 23, 1971, President Richard M. Nixon signed the National Cancer Act, initiating the so-called "war on cancer." It was the culmination of the growth of federal sponsorship of medical research that had been accelerating since World War II. Over the ensuing 10 years, $7.5 billion in research support funneled through the National Cancer Institute stimulated an unprecedented expansion of basic biological research. The increased funding fell on a biological research community enormously excited by its new understanding of the role of DNA in directing the machinery of life. With a basic molecular description of DNA replication and protein synthesis in hand, scientists were now anxious to apply what they had learned to study the most dreaded human disease, cancer.

Research on the molecular genetic basis of cancer came of age precisely a decade after passage of the National Cancer Act. In the fall of 1981, research teams at Cold Spring Harbor Laboratory, the Massachusetts Institute of Technology, and the Sidney Farber Cancer Center of Harvard University announced that they had tracked the origin of human cancer directly to the genetic material. Working independently, each group had employed the tools of recombinant DNA to isolate oncogenes—specific genes that induced cultured cells to become malignant.

The discovery of human oncogenes, the first step in our ultimate understanding of cancer, can be largely attributed to the national research commitment galvanized by the National Cancer Act. The breakthrough stemmed from the confluence of molecular genetics, virology, and animal cell culture that had revolutionized the study of the biochemistry of eukaryotic cells during the 1970s.

By the late 1960s, sophisticated methods had been developed to grow mammalian and human cells lines in culture dishes in the manner of bacteria. It was also found that certain animal viruses, known as tumor viruses, could "transform" cultured cells into cancer cells. It thus became possible to mimic cancer in a petri dish and to perform controlled experiments on the origin and progression of malignancy. Bacteriophages had provided a means to probe the molecular genetics of bacterial cells; now the tumor virus made possible a similar approach to mammalian cells. The potential power of the tumor virus system prompted a migration of former phage biologists into the field of basic cancer research. This discussion will approach the discipline from this historical perspective.

What Is Cancer?

According to the 1988 statistics of the American Cancer Society:

- Approximately 494,000 Americans die each year of cancer—about one person every minute.
- Approximately 75 million (30%) Americans now alive will eventually have cancer.
- Cancer will strike in three out of four American families.
- More than 5 million Americans alive today have a history of cancer. Perhaps 3 million of these can be considered completely cured.
- In 1988, about 985,000 new cases of cancer were diagnosed (not including common, nonmalignant skin cancers). Forty percent of these people will be alive 5 years after diagnosis.

Cancer is perhaps the most perplexing and ubiquitous group of diseases. Forms of cancer strike every part of the body. "Liquid tumors," such as leukemias and lymphomas, affect the white cells of the blood. "Solid tumors," such as sarcomas and carcinomas, are found in various organs and structural tissues. Seemingly different diseases, each of the manifestations of cancer is characterized by the uncontrolled growth and spread of abnormal cells. Cancer cells proliferate without regard for the regulatory mechanisms that restrain cell reproduction, crowding out their normal counterparts.

Scientists have always looked to the cell's replicative machinery for a hint about the common basis of all cancers. This line of reasoning led to DNA—for errors in cell replication are ultimately errors in DNA replication. All cancers can be thought of as genetic diseases, in the sense that they originate with changes to the genetic material, DNA. There is increasing evidence that a number of cancers are also genetic diseases in the sense that a predisposition to the conditions are inherited. There are familial, or inherited, forms of several cancers, including neuroblastoma and colon carcinoma.

To understand the origin of malignancy, however, one must focus on the somatic (body) cells that make up tissues and organs. Most cancers are thought to arise when DNA changes within a *single* somatic cell lead to loss of growth control and rapid proliferation. Thus, the cells that make up a solid tumor are progeny, or clones, derived from a single aberrant ancestor. This is the "somaclonal" theory of cancer formation.

Some Tumor Types and Tissues Involved

Adenoma	gland-like tissue
Blastoma	germ/precursor cells
Carcinoma	epithelial/connective tissues
Hepatoma	liver
Leukemia	leukocytes (white blood cells)
Lymphoma	lymphoid tissue
Melanoma	skin
Myeloma	bone marrow
Myoma	muscle
Sarcoma	connective tissues (bone, cartilage, muscle)

Familial Cancers: Some Cancers That Have a Heritable Basis in Some Families

Breast carcinoma
Colon carcinoma
Endocrine neoplasia
Esophageal carcinoma
Leukemia
Lymphoma
Hepatoma
Neuroblastoma
Sarcoma
Skin carcinoma
Melanoma

The Tumor Virus Model System

Peyton Rous, about 1911
(Courtesy of the Rockefeller University Archive Center.)

Close study of bacteriophage replication within *E. coli*, during the 1950s and 1960s, revealed many basic mechanisms of gene regulation. So, when molecular geneticists moved into basic cancer research, it was logical that they should turn again to a simple viral system for clues to the basic mechanisms of oncogenesis. The viruses that came into use belong to a group known as tumor viruses that cause malignancies in various vertebrate animals. Tumor viruses share striking similarities with bacteriophages:

1. Both types of viruses have small genomes.
2. Both are essentially protein capsules surrounding a core of nucleic acid.
3. Both can be propagated in cultured cells.
4. Both display an infective (lytic) phase, during which virus replication occurs, and a lysogenic phase, during which the viral genome integrates into the host-cell chromosome.
5. Both may transduce host-cell DNA sequences. During lysogeny, the integrated virus may pick up host-cell DNA sequences and subsequently transmit them to other cells during infection.

This discussion will be confined to several of the most intensively studied tumor viruses: Rous sarcoma virus (RSV), adenovirus, polyomavirus, and simian virus 40 (SV40). The closely related polyomavirus and SV40 have ring-shaped genomes the size of bacterial plasmids (5292 and 5243 bp, respectively). At 36,000 bp, adenovirus is smaller than the bacteriophage λ (48,000 bp) and RSV (60,000 bp).

Viruses and Cancer

The relationship between viruses and cancer was first established by Peyton Rous in 1911. He found that a filterable extract from sarcomas (connective tissue tumors) could be injected to induce tumors in healthy chickens. The infective agent was later shown to be a virus whose genome is composed of RNA. RSV thus belongs to the category of RNA viruses or retroviruses.

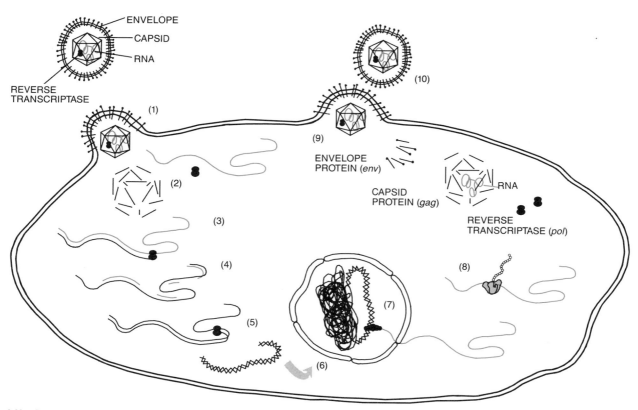

Life Cycle of Rous Sarcoma Virus
(*1*) Viral envelope fuses with host-cell membrane. (*2*) Capsid uncoats. (*3*) Reverse transcriptase synthesizes complementary DNA strand on RNA template. (*4*) RNase H degrades RNA strand. (*5*) Reverse transcriptase synthesizes second DNA strand. (*6*) DNA double helix integrates into host-cell chromosome. (*7*) Host-cell RNA polymerase transcribes viral genes into RNA. (*8*) mRNA translated into viral proteins. (*9*) RNA genome and reverse transcriptase packaged into capsid. (*10*) Viral particle buds from host-cell membrane.

The first DNA tumor viruses—rabbit fibromavirus and papillomavirus—were isolated by Richard E. Shope during the 1930s. He also demonstrated a marked species-specific effect: Papillomavirus, which causes only benign warts in wild rabbits, can induce malignant tumors in domestic rabbits.

Mammalian Cell Culture

The genetic mechanism of tumor induction by DNA and RNA viruses was not amenable to serious study until the development of in vitro systems that allow mammalian cells to be studied in isolation. Increasingly sophisticated techniques were developed in the 1950s and 1960s for growing pure cultures of mammalian cells in defined medium containing amino acids, vitamins, salts, glucose, and animal serum. Particularly important was the isolation of stable cell lines that can be propagated indefinitely in culture.

Most primary cells (those taken directly from a living organism) are difficult to maintain in culture. They have a limited life span, typically undergoing on the order of 50 divisions before dying. However, a number of "immortal" cell types have been discovered that live and reproduce essentially forever in culture. Some cell lines are derived from tumor cells, whereas others arose from mutations that occurred in cultured cells. Some of the common cell lines that have come into common use are the human HeLa cell, the rat embryo fibroblast (REF), and the mouse NIH-3T3 fibroblast.

When grown in culture, noncancerous cells display four important characteristics:

1. Flattened morphology. Cultured cells, such as fibroblasts, have a flattened shape.
2. Anchorage-dependent growth. The cells can only grow when adhered to a solid support; the bottom of the culture plate provides the necessary point of attachment. Noncancerous cells fail to grow when suspended in soft agar.
3. Growth in confluent monolayer. New cells are only reproduced to the point of spreading evenly across the culture dish to form an uninterrupted layer that is one to only several cells thick.
4. Contact inhibition. Replication ceases upon contact with neighboring cells.

The characteristics of 3 and 4 above are analogous to organized tissues, where cells grow only enough to form and maintain specific structures. For example, when a tissue is injured, only enough new cells are reproduced to replace those that were killed in the injured area, then growth ceases.

Viral Transformation

By the mid-1960s, it was established that tumor virus infection of cultured cells can take either of two markedly different courses, which correspond to those previously observed in bacteriophage infection of *E. coli*. During lytic infection, new viral particles are replicated, and host cells are lysed. In contrast, during lysogenic-like infection, the virus integrates into the host's DNA and virtually "disappears"—there is no evidence of virus replication or cell lysis. However, lysogenic-like infection of cultured cells by tumor viruses *does* induce dramatic changes in cell morphology and growth.

Because this result paralleled the transformation process in pneumococcus described by Fred Griffith and Oswald Avery, the term transformation was applied to the cancerous appearance of cultured cells induced by tumor virus infection. Transformation has taken on the expanded meaning of in vitro changes in cultured cells that are the "test tube" equivalent to cancer in a living animal.

Cultured cells transformed by tumor viruses look and act strikingly different from their noncancerous counterparts. They display a rounded, globular morphology. They are not contact-inhibited—cells derived from an original transformant pile up on one another to form a clump called a focus (plural = foci). The dense focus of transformed cells is easily spotted

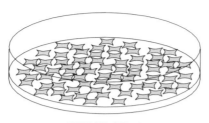
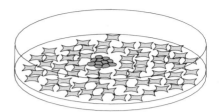

NORMAL CELLS · TRANSFORMED CELLS

Contact-inhibited cells grow in confluent monolayer · Dense foci of transformed cells are not contact inhibited

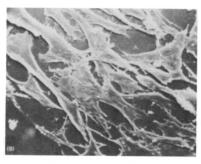
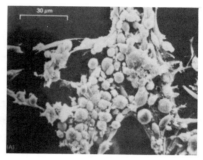

Flattened morphology · Rounded morphology

Characteristics of Normal and Transformed Cells
(Courtesy of L.B. Chen, Dana Farber Cancer Institute.)

against the confluent monolayer of noncancerous cells. Transformed cells are anchorage-independent and will continue growing when transferred to soft agar.

Using the formation of a focus as an assay, the transforming ability of various viruses was tested in a variety of different cell lines. Interestingly enough, as was the case for the Shope papillomavirus, viruses that cause only benign disease in their normal hosts may cause malignant transformation of the cultured cells of other species. For example, adenovirus causes flu-like infections in humans; it is not associated with malignancy of any type. However, adenovirus infection leads to transformation in many rodent cell lines, including 3T3 mouse fibroblasts. This example reminds us that transformation of cultured cells is only a convenient model of oncogenesis. Thus, the events that lead to cancer in vitro may not necessarily mirror in vivo events.

The advent of restriction enzymes and recombinant DNA made possible detailed analysis of viral transformation. Restriction mapping of genomic DNA isolated from transformed cells showed that tumor virus DNA integrates randomly into the host chromosome as a "provirus." The DNA genomes of adenovirus, polyomavirus, and SV40 DNA are directly integrated, whereas the RNA genome of a retrovirus is first reverse-transcribed into a DNA copy. In either case, the proviral DNA is transcribed along with the host-cell DNA to make messenger RNA (mRNA). The host cell provides the ribosomes, transfer RNAs, and amino acids needed to translate the viral mRNA into proteins.

Deletion analysis was coupled with restriction mapping to show the organization of tumor virus DNAs. In this strategy, a portion of the viral genome is deleted, and the mutated virus is tested for replication, capsule formation, and transforming ability. A change in activity can then be correlated to a specific DNA fragment, which can then be mapped in relation to other functional sequences.

The tumor virus genome is typically composed of only a handful of genes that fall into two categories. One set of genes codes for structural proteins that form the virus particle within the host cell. A second set of proteins regulates the expression of both the viral genes and those of the host cell. Some of the genes from this latter class code for proteins that bring about malignant transformation.

Adenovirus contains two oncogenes, E1A and E1B (E is for early), which are located in a region of the genome that is transcribed in the early phase of infection. SV40 and polyomavirus each contain two analogous oncogenes called tumor (T) antigens, confusingly designated large T antigen and small t antigen in SV40, and middle T antigen and large T antigen in polyomavirus. RSV contains a single oncogene, viral (v)-*src*.

The Origin of Viral Oncogenes: Transduction

A detailed understanding of the evolutionary origin of viral oncogenes came in the late 1970s from studies of the v-*src* gene. J. Michael Bishop and Harold Varmus, of the University of California at San Francisco, used a DNA copy of v-*src* to probe the cellular DNA of various animals. A closely related homolog, cellular (c)-*src*, was found in the DNA from normal chickens, fish, mammals, humans, and even *Drosophila*. Subsequently, other viral oncogenes were shown to have related cellular forms that have been conserved from species to species during the course of evolution.

In the early 1980s, Hidesaburo Hanafusa, at the Rockefeller University, found that retroviruses with an incomplete, defective *src* oncogene can still produce tumors in animals. Viruses subsequently isolated from the malignant tumors had reassembled a complete *src* gene, indicating that they had become oncogenic by "capturing" the missing portion of the *src* gene from the host-cell DNA.

The work of Bishop, Varmus, and Hanafusa showed that retroviral oncogenes are modified cellular genes captured from the genomes of their vertebrate hosts. During provirus integration, all or part of the coding region of a cellular gene may be integrated *within* the sequences of the viral genome. Thus, the virus acts like a transducing phage, removing the cellular gene as part of its genome as it excises from the host DNA. The cellular gene is then packaged, along with viral sequences, into an infectious virus particle.

J. Michael Bishop (Left) and Harold Varmus, 1989
(Courtesy of University of California, San Francisco.)

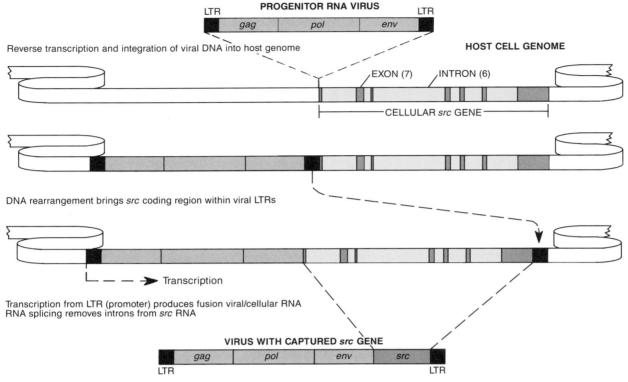

Capture of the Cellular *src* Oncogene by a Progenitor Rous Sarcoma Virus

The Role of Retroviral Genes in Oncogenesis

c-*src* and many other cellular oncogenes encode regulatory proteins that are essential for normal cell proliferation. However, overexpression of the altered v-*src* gene apparently disrupts normal growth control. Consistent with this hypothesis, Bishop and Varmus found that the *src* gene is expressed at low levels in normal cells and at inappropriately high levels in virally transformed cells. These normal cellular genes that have the potential to become oncogenes—through alteration or overexpression—are termed proto-oncogenes.

Overexpression of a viral or cellular oncogene occurs by several mechanisms in cancers caused by retroviruses. Some oncogenic viruses, like RSV, carry a captured cellular oncogene that is under the direct control of a strong viral promoter. The v-*src* gene is under the control of a promoter located in a region of the viral genome called the LTR (long terminal repeat).

Other oncogenic retroviruses carry no oncogene; however, a cellular oncogene adjacent to the site of provirus integration may be overexpressed when it comes under the control of the viral promoter. For example, the proviral DNA of avian leukosis virus (ALV, which causes leukemia in chickens) nearly always inserts upstream of c-*myc*, a known cellular oncogene.

Experiments where the c-*src* gene is overexpressed by placing it in front of the viral LTR show that increased levels of *src* gene product, alone, are not sufficient to cause transformation. Therefore, transformation also appears to require a mutation in the c-*src* gene that arises during the viral infection.

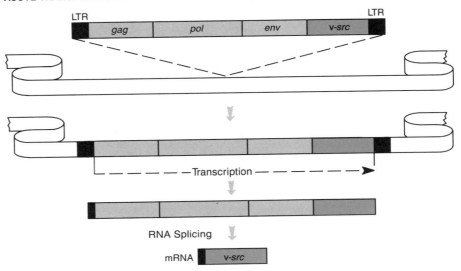

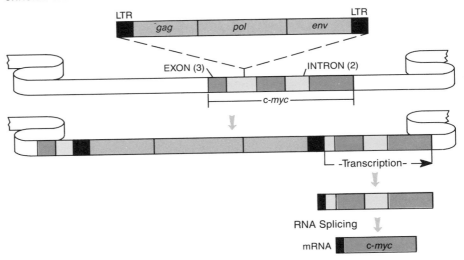

Acute vs. Chronic Transformation by Retroviruses
In acute transformation, a tumor virus carrying a resident oncogene integrates into the host chromosome, where the viral oncogene is transcribed from the viral long terminal repeat (LTR) (promoter). In chronic transformation, a tumor virus lacking a resident oncogene integrates adjacent to the cellular proto-oncogene, which is transcribed from the viral LTR.

There is evidence that transforming viruses, like RSV, integrate preferentially at "fragile sites" on the host chromosome that are prone to breakage. These regions often correspond to "hot spots of recombination," where exchange of DNA sequences occurs at an unusually high frequency. RSV often integrates into a recombinational hot spot located at the 3' end of the c-*src* gene. When RSV excises from the host genome and transduces the cellular gene, the 3' coding sequences are left behind.

Viruses and Human Cancer

Study of the tumor viruses of various vertebrate species revealed much about the basic mechanisms of oncogenesis, both in vivo and in vitro. However, the relationship between viruses and human cancer remained hypothetical until 1980, when Robert Gallo, at the National Institutes of Health, found that a retrovirus is responsible for a rare form of leukemia. Human T-lymphotrophic virus type I (HTLV-I) acts in a manner analogous to that of ALV by integrating upstream of a cellular regulatory gene. The provirus promoter causes overexpression of the cellular gene coding for T-cell growth factor, which stimulates proliferation of T lymphocytes.

There is now compelling evidence to link DNA viruses with several human cancers. Exposure to hepatitis-B virus increases several hundredfold the risk of hepatoma, a type of liver cancer that is the most common malignancy worldwide. In addition to a strong correlation between the worldwide coincidence of hepatitis-B virus and hepatoma, viral DNA has been found integrated in the genomic DNA of hepatic tumor cells. Epstein-Barr virus (EBV), which causes infectious mononucleosis, produces lymphomas in certain primates and immortalizes human lymphocytes grown in culture. Antigens for EBV and copies of viral DNA are present in the malignant cells of two cancers whose worldwide incidence is limited—Burkitt's lymphoma (in Africa) and nasopharyngeal cancer (in China). Human papillomaviruses (HPVs) have been similarly linked to genital cancers, including cervical carcinoma. The PAP smear test, which detects precancerous cell changes of the cervix, has been primarily responsible for a 70% decline in uterine cancer deaths in the United States over the last four decades. However, cervical cancer remains a major cause of female death in underdeveloped nations and among members of lower socioeconomic groups in the developed nations.

Mammalian Cell Transfection

A new approach to understanding the action of oncogenes was opened up by fine tuning techniques for transferring naked DNA into mammalian cells. Prior to the development of these methods, genetically engineered DNA or RNA viruses were the only means to ferry new genes into cultured mammalian cells. In 1973, Alex van der Eb and his associates at

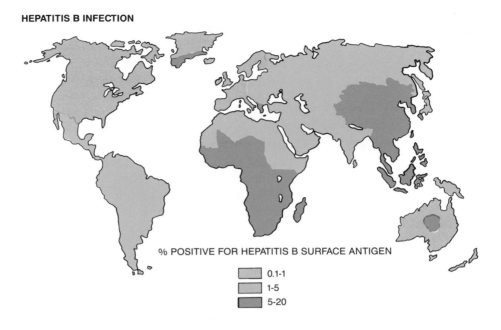

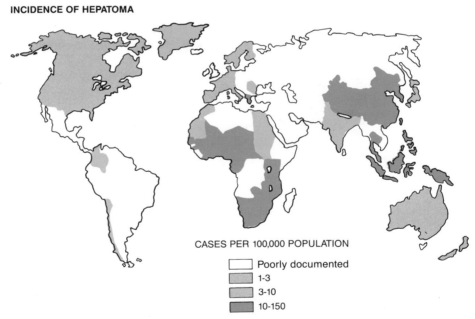

Coincidence of Hepatitis B and Hepatoma (Liver Cancer)
(After P. Maupos and J.L. Melnick. 1981. *Prog. Med. Virol.* 27: 1.)

the University of Leiden, The Netherlands, showed that cultured mammalian cells can be transformed with precipitates of calcium phosphate and purified SV40 or adenovirus DNA. Although the mechanism of uptake, or transfection, is still unknown, it is believed that the cells engulf (phagocytize) the calcium/DNA granules.

Several years later, Richard Axel and Michael Wigler, at Columbia University, found that *co*transfecting a selectable marker *along with* genomic DNA, cDNA, or a cloned gene makes it possible to readily identify cultured cells that have taken up foreign DNA. The first mammalian selectable marker was the thymidine kinase (*tk*) gene, which allows transformants to use an alternative metabolic pathway to incorporate thymidine into DNA. Because of limitations of the *tk* selection system, other dominant-acting markers have been incorporated into cloning vectors.

In one such plasmid vector, SVneo, a bacterial gene for neomycin resistance is fused to the SV40 promoter, allowing it to be expressed in mammalian cells. The foreign DNA need not be ligated directly to the SVneo plasmid; the two DNAs simply can be mixed together and coprecipitated with calcium phosphate. Cells that phagocytize calcium phosphate granules thus receive a mixture of DNAs and are selected by growth in medium containing the neomycin analog G418, which kills nonresistant eukaryotic cells.

The transfected DNA is thought to form long "concatemers" of 800,000 to 1 million base pairs in which multiple copies of the SVneo plasmid are linked together with segments of genomic DNA. The concatemer integrates randomly as a unit into the host-cell chromosome. Since the genomic DNA and the SVneo plasmid are often closely linked within the concatemer, plasmid sequences can be used to probe for a gene of interest.

The First Human Oncogenes

Considering the variety of oncogenes that had been isolated from virally transformed cells and animal tumors, it seemed plausible that cellular oncogenes could be isolated from human cancer cells. Coupling mammalian cell transfection with the focus assay provided the technology for the research teams at Cold Spring Harbor Laboratory, MIT, and the Sidney Farber Cancer Center to test this hypothesis. Their task was to screen the approximately 100,000 genes present in human genomic DNA in the hope of finding a single oncogene.

Although different strategies were employed by each group, the scheme used by Michael Wigler's group at Cold Spring Harbor Laboratory is representative. Genomic DNA was extracted from a cultured cell line (T24) derived from a human bladder carcinoma. The tumor DNA was digested with *Hin*dIII, and the resulting fragments were ligated to a selectable bacterial marker—a suppressor tRNA gene for *E. coli* (discussed in Chapter 5). The recombinant molecules were then transfected into NIH-3T3 mouse fibroblasts. The formation of foci indicated cells harboring restriction fragments containing an intact oncogene.

DNA was isolated from cells from several foci and digested with *Sau*IIA, which left the selectable marker linked to the oncogene. This DNA was religated to a mutant λ vector that *requires* the suppressor tRNA marker for growth. The new recombinant phages that formed

Michael Wigler, 1985
(Courtesy of Cold Spring Harbor Laboratory Archives.)

plaques under selection were reasoned to contain restriction fragments possessing the marker *along with the oncogene.* DNA isolated from these λ plaques was transfected into 3T3 cells; foci formation confirmed which of the λ constructs contained the oncogene.

Prior to the isolation of human oncogenes, it had been feared that the genetic mechanisms of human cancer might be incomprehensible—that each cancer might be the composite result of many different genes. Now,

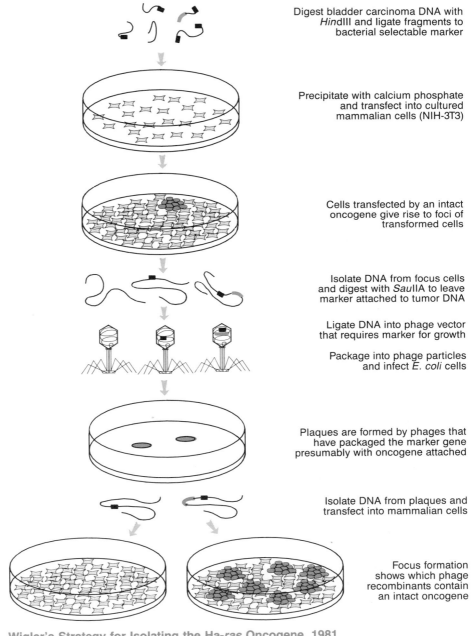

Wigler's Strategy for Isolating the Ha-*ras* Oncogene, 1981
(After R. Weinberg. 1983. *Sci. Am.* 249/5: 138.)

human cancer could be thought of as having discrete causes. Oncogenes might provide insights to cut through the confusion of cancer and provide a common pathway accessible to experimentation. These revolutionary experiments provided a means to analyze human cancer at the level of precision approaching that achieved with viral models.

The next round of experiments compared the DNA sequences of the newly isolated human oncogenes with the known sequences of cellular and viral genes. Not surprisingly, the results showed that, like viral oncogenes, human oncogenes are derived from normal proto-oncogenes with important cellular functions. In the case of the T24 bladder oncogene, sequence comparison revealed the following:

1. The T24 gene is a cellular homolog of the Ha-*ras* oncogene of Harvey sarcoma virus, which causes connective-tissue tumors in rats.
2. Ha-*ras* is related to oncogenes isolated from several different human tumors. An oncogene isolated from both colon and lung carcinoma cell lines is closely homologous to the Ki-*ras* oncogene of Kirsten sarcoma virus of mice. A third member of the *ras* gene family, N-*ras*, was isolated from neuroblastoma cells. Although the three *ras* genes have different arrangements of introns and exons, later sequence analysis showed that they each code for a closely related protein of 189 amino acids.
3. Homologs of the *ras* genes were found in organisms throughout the evolutionary tree, from yeast to *Drosophila* to birds to mammals.
4. Homologs of the *ras* genes were identified in a variety of normal human tissues.

Human Oncogenes: Good Genes Gone Bad?

The results of these homology studies made it obvious that the discovery of human oncogenes would not provide a quick explanation of oncogenesis and raised the unsettling proposition that a cancerous potential exists in the proto-oncogenes in each individual's cells. Oncogenes are not unique to cancerous tissue; they are not simply "bad" genes confined to tumors. Instead, they are essential genes whose normal function has in some way been altered. What mechanisms, then, activate the cancerous potential of human proto-oncogenes?

ACTIVATING MUTATION

Closer examination of the human Ha-*ras* oncogene provided an incredibly precise insight into one mechanism of proto-oncogene activation. The nucleotide sequences of Ha-*ras* genes isolated from carcinoma cells and from normal placental cells were virtually identical. Only a single nucleotide difference was found in the protein-coding region of the two genes! The 12th codon of the normal cellular gene reads G*G*C (which codes for the amino acid glycine), whereas the 12th codon of the T24 carcinoma gene reads G*T*C (which codes for valine). This point mutation

Some Mutations in *ras* Genes from Viruses and Human Tumor Cell Lines and Resulting Amino Acid Changes

	Codons		
	12	59	61
Ha-*ras*			
normal	GGC Gly	GCC Ala	CAG Gln
Harvey sarcoma virus	AGA Arg	ACA Thr	CAA Gln
bladder tumor (T24)	GTC Val	GCC Ala	CAG Gln
lung tumor (HS242)	GGC Gly	GCC Ala	CTG Leu
Ki-*ras*			
normal	GGT Gly	GCA Ala	CAA Gln
Kirsten sarcoma virus	AGT Ser	ACA Thr	CAA Gln
colon tumor (SW480)	GTT Val	GCA Ala	CAA Gln
lung tumor (Calu)	TGT Cys	GCA Ala	CAA Gln
N-*ras*			
normal	GGT Gly	GCT Ala	CAA Gln
neuroblastoma (SK-N-SH)	GGT Gly	GCT Ala	AAA Lys
malignant melanoma (Mel)	GGT Gly	GCT Ala	AAA Lys
lung carcinoma (SW1271)	GGT Gly	GCT Ala	CGA Arg

in the DNA code results in a glycine-to-valine substitution in the *ras* protein. Subsequently, other sites of mutational activation of the *ras* gene family were found notably at codons 59 and 61, which result in a single amino acid change.

Because these experiments were carried out with cultured cells, it was not certain how well they related to cancer formation in a living human being. Mariano Barbacid's research team at the National Cancer Institute was first to show a close relationship between human cancer and mutational activation of the *ras* oncogene. Sequence comparison of the Ki-*ras* gene isolated from malignant and normal tissues taken from a patient with lung carcinoma showed that the 12th codon mutation was present only in tumor cells.

GENE REARRANGEMENT

It is estimated that at least 20% of all human tumors contain *ras* oncogenes with any of several specific "activating" mutations that result in *qualitative* changes in the *ras* oncoprotein. Other types of malignancies result from *quantitative* changes in oncoprotein expression. The *myc* oncogene, first isolated from a chicken myelocytoma, illustrates two molecular mechanisms that can initiate protein overproduction.

Gene translocation may alter the regulation of a cellular oncogene, stimulating its expression to above normal levels. For example, a majority of patients with Burkitt's lymphoma show an exchange of the tip of chromosome 8 (containing the c-*myc* oncogene) and the tip of chromosome 14, 22, or 2. The role of this translocation in a cancer of the immune system makes good intuitive sense. DNA rearrangement is the normal situation for the immunoglobulin loci found at the ends of chromosomes 14, 22, and 2, where several hundred gene segments actively rearrange to generate the huge variety of antibody genes (see Chapter 5). The immunoglobulin loci are also extremely active sites of gene transcription.

There is evidence for two different mechanisms through which translocation into an immunoglobulin locus may activate transcription of the *myc* proto-oncogene, as indicated by increased abundance of mRNA. In some cases, the translocation appears to *de*regulate the c-*myc* gene by destroying a repressor region that normally keeps the gene "silent" in its position on chromosome 8. In other cases, c-*myc* appears to come under the influence of the immunoglobulin enhancer, which stimulates transcription. Similar mechanisms may account for another cancer of the immune system, chronic myelogenous leukemia (CML). Tumor cells from 95% of CML patients show a translocation, where the c-*abl* oncogene on chromosome 9 is moved into the immunoglobulin locus on chromosome 22, known as the Philadelphia (Ph[1]) chromosome.

GENE AMPLIFICATION

Gene amplification, where multiple copies of the proto-oncogene are present, is another means of increasing production of an oncoprotein. Gross chromosome abnormalities, resulting from duplication and recombination of chromosomal fragments, have long been observed as common features of tumor cells. A resident proto-oncogene would, as a matter of course, become amplified during chromosome duplications that result in a large insertion called a homogeneously staining region (HSR) or in numerous small fragments called double minutes (DMs). In situ hybridization, where a radioactively labeled *myc* probe is hybridized to intact chromosomes, has confirmed that numerous copies of the N-*myc* oncogene are present in HSRs and DMs from neuroendocrine tumor cells.

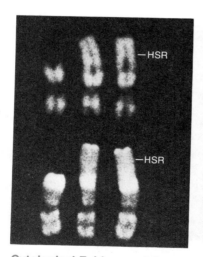
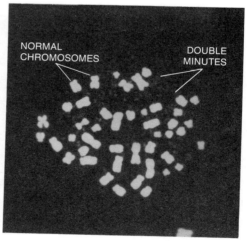

Cytological Evidence of Oncogene Amplification
Amplification of the N-*myc* oncogene results in both homogeneously staining regions (HSRs) and double minutes (DMs) in neuroblastoma cells. (Courtesy of F. Alt, Columbia University.)

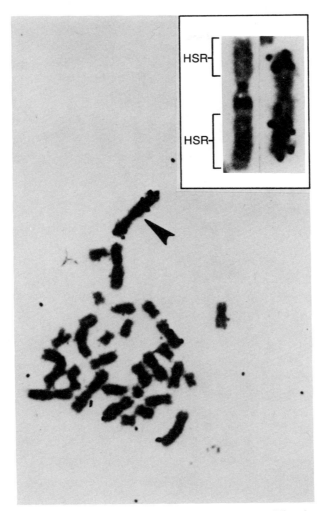

In Situ Hybridization Reveals Oncogene Amplification
Dark grains indicate regions where a radiolabeled probe for the *myc* oncogene hybridizes to homogeneously staining regions (HSRs) of chromosomes from neuroendocrine tumor cells. (*Inset*) Comparison of labeled and unlabeled chromosomes. (Courtesy of J. Bishop and H. Varmus, University of California, San Francisco. From K. Alitalo et al. 1983. *Proc. Natl. Acad. Sci. 80:* 1707.)

Even when chromosomal aberrations are not visible, gene amplification can be detected by measurement of mRNA hybridization to an oncogene probe. This method has identified amplified *myc* oncogenes in numerous human primary tumors and tumor cell lines—including carcinomas of the colon, lung, and breast; leukemia; retinoblastoma; and neuroblastoma. For example, a study of untreated patients with neuroblastoma showed that amplification of the N-*myc* oncogene correlates with advancing stages of the disease. Whereas no amplification was detected in primary tumors taken from patients in early stages of the disease, 3–300 copies of the *myc* gene were present in tumors from half of patients in advanced stages. This suggests that gene amplification is not necessarily a primary cause of cancer, but may be a secondary event associated with tumor progression.

POINT MUTATION (ras)

CHROMOSOMAL TRANSLOCATION (myc)

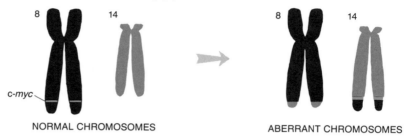

GENE AMPLIFICATION (myc)

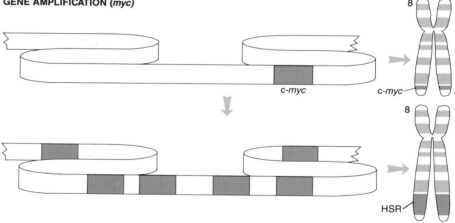

VIRAL INTEGRATION (myc)

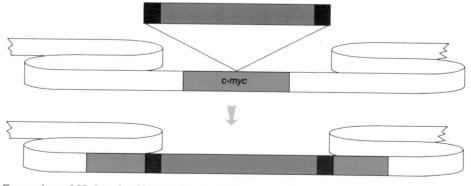

Examples of Molecular Mechanisms of Oncogenesis

Support for "Multi-Hit" Oncogenesis

Epidemiologic studies convincingly show that the incidence of most types of cancers increases dramatically with age. These data strongly support the hypothesis that malignancy results from an accumulation, over time, of several genetic "hits." The transfection experiments just described suggest, however, that changes in a single dominant-acting oncogene are responsible for cancer formation. Many scientists were uncomfortable with this interpretation, in part because they doubted that the 3T3 cells used in the key focus assay are representative of normal cells within the human body. Unlike primary cells taken directly from a living tissue, 3T3 cells are effectively immortal in culture and occasionally even undergo spontaneous transformation.

There was another major inconsistency. Although the *ras* oncogene can efficiently transform 3T3 cells, it fails to transform primary cells. On the other hand, the *myc* oncogene is not detected by the focus assay; it cannot transform 3T3 cells at all!

Again, a lesson was learned from previous work with viral oncogenes. Transfection experiments had shown that the oncogenes of DNA tumor viruses have similar and complementary functions. Adenovirus E1A and E1B genes and polyomavirus large T and middle T genes act in a cooperative fashion to transform primary cells. E1A and large T genes immortalize primary cells, establishing them so that they can grow in culture. E1B and middle T genes convert established cells to the malignant phenotype. However, acting alone, none of the four oncogenes can completely transform primary cells.

Extensions of these experiments worked incredibly well. When cotransfected with E1A or large T, *ras* transforms embryonic rat cells to the malignant phenotype. Primary cells transfected with *myc*, in combination with E1B or middle T, are also transformed. Finally, *myc* and *ras* can cooperate to transform primary cells. The implication was clear—at least two oncogenes are required to transform primary cells in culture. While *ras* is a transforming gene, *myc* is an immortalizing gene. The *ras* oncogene scores positively on the focus assay because the 3T3 cells have already been immortalized—they are already halfway to the transformed phenotype.

The multistep model is supported by other lines of evidence. Several types of tumors appear to contain two independently activated oncogenes. For example, activated forms of both *myc* and *Blym* are found in chicken lymphomas. Chemical carcinogens, such as 3-methylcholanthrene and benzopyrene, immortalize cells and can act cooperatively with transforming oncogenes to transform primary cells. 3-Methylcholanthrene also activates the *ras* oncogene in cultured cells and laboratory animals and thus, in effect, can accomplish two steps of oncogenesis. *N*-nitroso-*N*-methylurea (NMU) induces mammary tumors in rats; 80% of these tumors contain an Ha-*ras* gene with an activating mutation in the 12th codon.

Experiments with transgenic mice suggest that there are other intervening factors in oncogenesis. For example, when the c-*myc* oncogene is

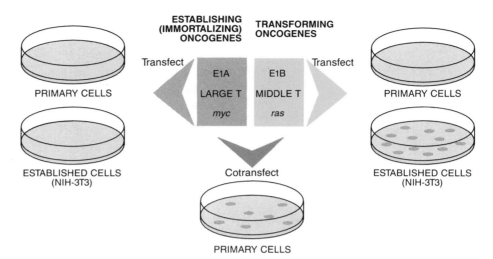

Cooperative Transformation of Primary Cells by Two Oncogenes
Transfected alone, neither immortalizing nor transforming oncogenes transform primary cells. However, cotransfection of both types of oncogenes causes foci of transformed cells. Established cells, such as NIH-3T3, require only a transforming oncogene to form foci.

fused to the promoter/enhancer region of the mouse mammary tumor virus, some transgenic females develop mammary tumors at the time of pregnancy. Hormonal changes associated with lactation are thought to induce sufficient expression of the *myc* protein to induce immortalization of all breast cells. However, localized tumors develop in only a small percentage of mammary cells. This may indicate the relative infrequency of one or more additional steps necessary for malignant transformation.

As mentioned in Chapter 5, transgenic mice carrying a fusion gene of the insulin promoter/large T antigen show correct tissue-specific expression of T antigen in the β cells of the islets of Langerhans. Large T antigen can completely transform primary cells and is thus believed to perform both immortalizing and transforming functions. Although expression of T antigen induces abnormal proliferation of essentially all β cells, tumors develop in only a small percentage of islets. If expression of T antigen is sufficient for oncogenesis, one would expect to observe progression to tumor in every islet. This result suggests that at least a third event is necessary for malignancy.

Rb: A Recessive-acting or Anti-oncogene

The viral and cellular oncogenes discussed above can all be described as "dominant-acting." The *presence* of the gene, in an altered form or in increased copy number, contributes to malignancy. In 1987, a fundamentally different type of oncogene was found to be involved with the rare childhood tumor, retinoblastoma. In this cancer, malignancy results from the *absence* of a functional copy of the retinoblastoma (*Rb*) gene, which is therefore said to be "recessive-acting." *Rb* is an anti-oncogene, because its presence (even in a single copy) inhibits formation of this particular cancer. Whereas the oncogenes described so far encourage cell proliferation, *Rb* suppresses the mitogenic response.

Retinoblastoma also illustrates the genetic hit model of oncogenesis. Children who develop the "familial" form of the disease inherit a normal copy of the *Rb* gene from one parent and a defective copy from the other parent. Subsequently, a spontaneous mutation in a blast cell of the developing retina causes loss of the remaining normal copy and results in tumor cells all lacking functional *Rb* genes. Children who have the "sporadic" form of the disease inherit normal *Rb* genes from each parent and thus are born with two functional *Rb* genes per cell. Two successive somatic mutations *within a single cell* of the developing retina must inactivate both copies of the *Rb* gene.

Not surprisingly, retinoblastoma patients may show deletions and rearrangements at the *Rb* gene site on chromosome 13. Considering that

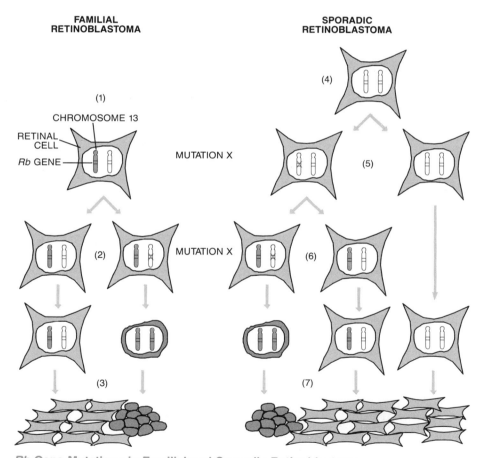

***Rb* Gene Mutations in Familial and Sporadic Retinoblastoma**

In familial retinoblastoma, one normal and one mutated *Rb* gene are inherited (*1*). (*2*) Subsequent mutation in *any* retinal cell inactivates remaining *Rb* gene, (*3*) leading to loss of growth control in a clone of tumor cells. In sporadic retinoblastoma, two normal *Rb* genes are inherited (*4*). (*5*) First mutation inactivates one copy of *Rb* gene; (*6*) subsequent mutation within *same* retinal cell inactivates remaining copy of *Rb* gene, (*7*) leading to loss of growth control in a clone of tumor cells.

chromosome abnormalities are widely observed in cancer cells, the loss of anti-oncogene function may turn out to be a common feature of cancers. Indeed, the *Rb* gene itself has now been implicated in several other human cancers, which indicates that it may play a generalized role in growth suppression in a variety of tissues. For example, patients who are cured of familial retinoblastoma are at increased risk of osteosarcoma (bone cancer), small cell lung cancer, and breast cancer—*Rb* gene deletions or chromosome 13 abnormalities have been found in each of these tumors.

A dramatic gene addition experiment, in principle like those that demonstrated dominant-acting oncogenes, shows that Wilms' tumor of the kidney involves deletion of a possible growth-suppression gene on chromosome 11. If a normal copy of chromosome 11 is introduced into Wilms' tumor cells, they revert to normal and lose the ability to form tumors.

Uniting Oncogenes and Anti-oncogenes

The existence of oncogenes *and* anti-oncogenes suggests an elaborate system of positive and negative controls that maintains cell replication within normal limits. Taken separately, the oncogenes and anti-oncogenes provide alternate pathways of oncogenesis. Mutation or amplification of dominant-acting oncogenes can stimulate mitogenic pathways, resulting in cell proliferation. Mutation of recessive-acting oncogenes results in loss of control of cell proliferation.

One study of colorectal cancer has provided a multistep model of carcinogenesis involving both dominant and recessive oncogenes. In this disease, benign polyps (adenomas) increase in size and may progress to fully malignant carcinomas. An analysis of DNA from adenomas and carcinomas at various clinical stages showed that multiple genetic changes accumulate with progression of malignancy. Observed changes include mutational activation of *ras* oncogenes and several types of deletions at loci on chromosomes 5, 17, and 18. These results suggest that tumor progression may involve mutational activation of an oncogene, *as well as* deletional inactivation of one or more growth-suppression genes.

Another potentially important oncogenesis pathway linking oncogenes and anti-oncogenes was uncovered in 1988, when Edward Harlow and his associates at Cold Spring Harbor Laboratory showed that the adenovirus E1A protein binds to the *Rb* protein. Subsequent discovery of *Rb* binding by SV40 large T antigen reinforced the conclusion that dominant-acting oncogenes may bring about malignancy by interfering with anti-oncogene proteins required to suppress cell proliferation. Presumably, oncoprotein binding to the *Rb* protein has the same cellular effect as deletion of the *Rb* gene—loss of the cancer-protective function. Considering that repression of function is a common theme in biological systems, *Rb* and other negative-acting oncogenes may be the direct regulators of cell growth, whereas the positive-acting oncogenes may merely be "anti-*Rb*" genes.

Signal Transduction: Cancer as a Disease of Communication

In the several years since the cloning of the first human oncogenes, more than 40 others have been isolated. Although much has been learned about the structures and modes of activation of these oncogenes, comparatively little is known about the oncoproteins they encode.

Throughout this chapter, cancer has been framed in terms of cellular *proliferation*. To begin to understand the functions oncoproteins may have in cell biochemistry, it will serve us well to consider cancer as a disease of cellular *communication*. The following sections focus on *signal transduction*—the molecular pathways that relay a command to REPLICATE or TRANSCRIBE from the cell exterior to the DNA molecule in the cell nucleus. Oncoproteins have been found in every step of this pathway.

Cellular communication begins when a messenger molecule, such as a hormone or growth factor, binds to a receptor molecule anchored in the cell membrane. The receptor carries (transduces) the chemical signal through the cell membrane to the internal membrane surface, where it activates a "second messenger," such as cyclic AMP or an inositol lipid. Next, a cascade of reactions, regulated by protein kinases, carries the signal through the cytoplasm and into the nucleus. In the final step, DNA-binding proteins attach to regulatory sequences that initiate DNA replication and/or transcription. Thus, the message DIVIDE or TRANSCRIBE reaches the genetic material.

GROWTH FACTORS, RECEPTORS, AND G PROTEINS

The first direct link between an oncoprotein and biochemical signal transduction came, in 1983, from studies of the amino acid structure of platelet-derived growth factor (PDGF). This protein is released by platelets to stimulate tissue repair at wound sites. Like other growth factors, it is a powerful mitogen that induces replication of cultured cells.

Computerized comparison of the amino acid sequence of PDGF with those of other proteins in a data bank revealed that a 104-amino-acid stretch of PDGF is virtually identical to the protein structure predicted by the nucleotide sequence of the v-*sis* oncogene of simian sarcoma virus. Further experiments showed that the *sis* oncoprotein produced in virus-transformed cells has properties identical to those of PDGF. This result suggested that the simian retrovirus had captured cellular sequences that encode a portion of PDGF or a closely related protein. Virus-transformed cells then secrete a "mimic" PDGF protein that, in turn, stimulates a mitogenic response when it binds to PDGF receptors on the cell surface. Alternatively, internal expression of the *sis* oncoprotein might stimulate replication in cell types lacking PDGF receptors.

Another connection between oncoproteins and growth factors quickly emerged. The amino acid sequence of the cell-surface receptor for epidermal growth factor (EGF) was shown to share remarkable homology with the predicted sequence of the *erb*B oncogene of avian erythroblastosis virus. It appears that *erb*B is an example of viral capture of an

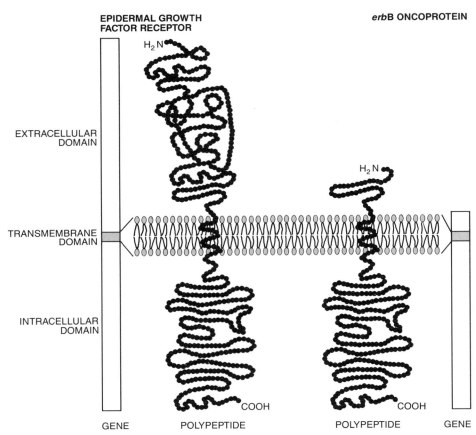

Comparison of Epidermal Growth Factor Receptor and *erb*B Oncoprotein

incomplete cellular gene. The EGF receptor has three domains: (1) An extracellular region projects outside the cell and is the site of EGF binding; (2) a membrane-spanning region anchors the protein in the cell membrane; and (3) an intracellular domain projects into the cytoplasm. The *erb*B protein shares with the EGF receptor common intracellular and membrane-spanning regions, but it is missing all or most of the extracellular binding site. It very well may be that the loss of the binding function leads to deregulation of the receptor and that *erb*B is a growth factor receptor stuck in the "ON" position.

The *ras* oncoproteins appear to be signal transducers that link the primary and second messenger systems. They belong to the group of guanine-nucleotide-binding (G) proteins that localize to the inner surface of the cell membrane, where they receive signals from cell-surface receptors and pass the signal onto second messengers. The hydrolysis of guanosine triphosphate (GTP) to guanosine diphosphate (GDP) is believed to shut off the G protein following transmission of the chemical signal. However, the activated *ras* oncoproteins lack hydrolytic activity. It is hypothesized that this effectively locks the protein in the ON position and causes it to inappropriately activate the second messenger cascade.

PROTEIN KINASES

Protein kinases play key roles in regulating reaction cascades by catalyzing the transfer of phosphate groups to substrate molecules. This phosphorylation may activate (up-regulate) or suppress (down-regulate) the substrate. A number of oncoproteins, including *sis*, *abl*, *neu*, and *erb*B, have kinase activity. A number of cell-surface receptors, including those for EGF, PDGF, and insulin, are protein kinases whose activity is regulated by both autophosphorylation and phosphorylation by other protein kinases.

The vast majority of kinases add phosphate groups to serine or threonine residues. However, a number of oncogenes, growth factors, and growth factor receptors exclusively catalyze the addition of phosphate groups to the amino acid tyrosine. Since phosphates on tyrosine residues account for only a fraction of a percent of total protein phosphorylation, this unusual kinase activity may eventually provide a key to unravel the biochemical pathways that lead to cancer. As mentioned previously, the transduced v-*src* gene is missing 3' c-*src* sequences, which include a codon for tyrosine. The phosphorylation of this position in the c-*src* protein is essential for control of its kinase activity. Thus, the absence of this phosphorylated residue in v-*src* leads to uncontrolled kinase activity in transformed cells.

DNA-BINDING ONCOPROTEINS

As emphasized in Chapter 5, DNA-binding proteins play an important role in turning genes on and off. Immortalizing oncoproteins, such as c-*myc*, N-*myc*, and c-*fos*, localize to the nucleus where they presumably interact with DNA to release the normal constraints on replication. The *fos* oncoprotein is known to interact with several promoters that control transcription of cellular and viral genes. It appears to be a direct

Some Oncogene Proteins Classified by Cellular Location and Function

Location	Protein	Function
Secreted	*sis*	growth factor derived from PDGF
Transmembrane	*erb*B	EGF receptor, tyrosine protein kinase
	*erb*B-2	tyrosine protein kinase
	fms	CSF-1 receptor, tyrosine protein kinase
	ros	tyrosine protein kinase
Plasma membrane	*src*	tyrosine protein kinase
	abl	tyrosine protein kinase
	ras	guanine-nucleotide-binding
Cytoplasm	*fes*	tyrosine protein kinase
	mos	serine/threonine protein kinase
	*erb*A	thyroid hormone receptor
Nucleus	*myc*	DNA-binding protein
	fos	DNA-binding protein
	jun	DNA-binding protein
	Rb	DNA-binding protein?

transducer, accepting the signal of growth factors or second messengers and transposing directly to the nucleus. *Rb* is believed to be a DNA-binding protein, which raises the possibility that it may act as a repressor. In this scenario, the binding of E1A and middle T antigen to *Rb* may release repression.

A new protein structure, the "leucine zipper," has been identified in the DNA-binding domains of the oncoproteins *myc*, *fos*, and *jun*. Repeated leucine residues form an α-helix region in which side chains project, in line, from one side of the protein molecule. The leucine side chains of two such proteins can interdigitate, like teeth of a zipper, to form a stable dimer. This helical pairing allows for a close association of the *fos* and *jun* polypeptides, which is essential to their sequence-specific alignment on the DNA molecule. Many leucine zipper proteins exhibit a conserved region of basic amino acids directly adjacent to the leucine repeats. It has been proposed that two zippered proteins diverge at this point, contacting the major groove of the DNA molecule in a "scissors grip."

Whispers in the Dark

The satisfying discovery of the related viral and cellular oncogenes has not led to a quick understanding of human cancer and has forced biologists to delve more deeply into the complicated protein interactions within the living cell. Although there is only one kind of DNA molecule in all cells, each cell contains several thousand different types of proteins. Teasing out the biochemical pathways of cellular communication that may be disrupted in the cancer cell has proved a challenging task that can be likened to identifying the contents of a sealed black box.

Since oncoproteins have been found at every step along the signal transduction pathway, the number of proto-cancer genes is potentially equal to the number of regulatory proteins that occupy critical positions in *each* biochemical pathway involved in cell proliferation. Thus, an adequate understanding of cancer may only come with a nearly complete understanding of the eukaryotic cell.

For Further Reading

Angier, Natalie. 1988. *Natural obsessions: The search for the oncogene.* Houghton Mifflin Co., Boston.

Bishop, Michael J. 1982. Oncogenes. *Sci. Am.* 246/3: 81–89.

Croce, Carlo M. and George Klein. 1985. Chromosome translocations and human cancer. *Sci. Am.* 252/3: 54–73.

Feldman, Michael and Lea Eisenbach. 1988. What makes a tumor cell metastatic? *Sci. Am.* 259/5: 60–87.

Gallo, Robert C. 1986. The first human retrovirus. *Sci. Am.* 255/6: 88–101.

Gold, Michael. 1986. *A conspiracy of cells.* State University of New York Press, Albany.

Goldberg, Marshall. 1988. *Cell wars: The immune system's newest weapons against cancer.* Farrar-Straus-Giroux, New York.

Kartner, Norbert and Victor Ling. 1989. Multidrug resistance in cancer. *Sci. Am. 260/3:* 44–51.

Sylvester, Edward J. 1987. *Target: Cancer.* Charles Scribner's Sons, New York.

Varmus, Harold. 1987. Reverse transcription. *Sci. Am. 257/3:* 56–65.

Weinberg, Robert A. 1983. A molecular basis of cancer. *Sci. Am. 249/5:* 126–142.

Weinberg, Robert A. 1988. Finding the anti-oncogene. *Sci. Am. 259/3:* 44–53.

CHAPTER 7
Applying DNA Science to Human Genetics

The National Cancer Act of 1971 was the first mobilization of government and public support of basic biological research during the gene-splicing age. In 1988, the second mobilization of biology began: The Human Genome Project. The genome is the genetic complement that defines a given organism; thus, the ultimate goal of the Human Genome Project is to determine the entire nucleotide sequence of DNA composing the human chromosomes. This is the most ambitious experiment in the history of biological science and amounts to deciphering the full set of genetic instructions encoding human life.

The project is expected to cost several billion dollars over a 15-year period, with funding provided primarily by the National Institutes of Health (NIH) and the Department of Energy (DOE). The NIH program is directed by James Watson, who, having played a fundamental role in launching the revolution in our understanding of genetic material, is now steering science toward the frontier of understanding humans as genetic organisms. Watson has set his sights on the year 2003 for the conclusion of the project—the 50th anniversary of his co-discovery of the structure of the DNA molecule.

The human genome is contained in the 23 pairs of chromosomes in the nucleus of each cell. Each chromosome is a complex package containing a single, unbroken strand of DNA. The bulk of the chromosome structure is contributed by protein molecules (histones) that form a scaffold around which the DNA strand is wound, like coils of rope around a capstan. Some 100,000 different genes are carried on the chromosomes, but these account for as little as 5% of the total DNA content of the human genome. Interspersed between functional genes are vast regions of noncoding, and often repetitive, DNA sequences whose functions are unknown.

With the exception of the different X and Y chromosomes (and random mutations), each member of a chromosome pair contains the same linear arrangement of genes. Thus, for simplicity's sake, biologists concern themselves only with the *haploid* genome composed of a single set of 23 chromosomes. The DNA of the haploid human genome contains approximately 3,000,000,000 (3×10^9) nucleotides.

The magnitude of the Human Genome Project becomes apparent when we imagine a "genomic book of life," in which each nucleotide of

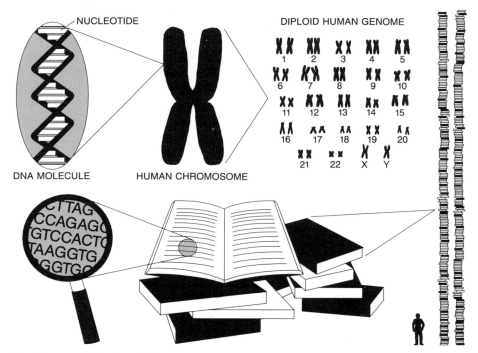

The Genomic Book of Life
If each nucleotide of the genome is equated to a single letter in this textbook and the nucleotide sequence is written *without spaces, punctuation, illustrations, or tables,* then a gene of 4000 nucleotides could be represented on one page of type. At this rate, the DNA sequence of the haploid human genome (23 individual chromosomes) would fill 750,000 pages, 50 meters high. Twice this amount would be needed to represent the genetic information of the diploid genome (23 pairs of chromosomes).

the complement of human DNA is represented by a single letter (A, T, C, or G). Using the typography of this text as an example—without spaces, punctuation, or illustrations—each page would contain approximately 4000 nucleotides. At this rate, the entire sequence of the human genome would fill 750,000 pages. If contained in volumes the length of this text (~300 pages), the genomic book of life would comprise a stack of 2500 volumes. At two centimeters per volume, the stack would measure 50 meters—the height of a 15-story building!

Possession of the entire sequence of the human genome will have many benefits, particularly in the area of health. The causative gene has been identified for only a fraction of the several thousand known genetic diseases of humans. Mapping and sequencing these disease genes will facilitate diagnosis, and detailed biochemical study of their protein products should lead to improved therapies. An understanding of noncoding control regions that regulate coordinated gene expression should give insight into diseases of development and aging. Moreover, among the tens of thousands of unknown genes (and their proteins) there are certainly thousands that have the potential to alleviate the pain and suffering of disease and to improve the quality of life.

Constructing Maps of the Human Genome

Like road maps, human genome maps can be constructed with varying degrees of resolution. *Genetic maps* describe the locations of genes and their patterns of inheritance. *Genetic linkage maps*, like those first constructed for *Drosophila* by Thomas Hunt Morgan and his colleagues in the early 1900s, have been useful in making low-resolution maps of whole chromosomes. *Physical maps*, such as *restriction maps*, describe structural characteristics of the DNA molecule itself and can therefore achieve much higher resolution. The marriage of these two types of maps will yield a molecular genetic description of *Homo sapiens*.

Variable regions of DNA that result in detectable changes in restriction banding patterns—restriction-fragment-length polymorphisms (RFLPs)—provide a bridge between genetic and physical maps. *RFLP maps* thus promise to integrate a wealth of previous research. An immediate goal of the Human Genome Project is to obtain a genetic linkage map of the locations of several thousand RFLPs spaced evenly throughout the genome at 1-centiMorgan (cM = 1 million nucleotides) intervals. Ray White and his associates at the University of Utah have already established a set of RFLP markers that span the chromosomes at approximately 10-cM intervals.

The Human Genome Project really began over 10 years ago, when the first human genes, including the β globin gene, were isolated and sequenced. Some 4000 human genes have been identified, and 1500 have been mapped to chromosomal locations. With a complete genetic linkage map in hand, these genes can be precisely mapped relative to the established set of RFLP markers and to each other. One aspect of the genome project will involve sequencing regions where several genes have already been identified. Sequencing out from a known gene should eventually link it to others located within the same region of a chromosome.

Creating maps of increasingly higher resolution will involve subcloning successively smaller genomic DNA fragments into vector molecules and then determining their sequences. *Contiguous (contig) maps* align the ends of adjacent DNA fragments, thus stringing together continuous sequences. Contig maps orient physical markers on adjacent fragments, including restriction recognition sites and RFLPs, which provide points of reference spaced at regular intervals throughout the genome. As the contig positions of smaller and smaller genomic fragments are established, the resolution of the genomic map will increase.

Sequence-tagged Sites and Polymerase Chain Reaction

Initially, it appeared imperative to construct an ordered set of cloned DNA fragments that span the entire human genome. The prospect of maintaining a central bank of clones and dispersing these among research groups worldwide for detailed mapping and functional analysis loomed like a dark cloud over the genome project. The sheer difficulty of working with large segments of cloned DNA also appeared to favor "big science" and threatened to exclude smaller research groups from meaningful participation.

In 1989, Maynard Olsen (Washington University), Leroy Hood (California Institute of Technology), Charles Cantor (Lawrence Berkeley

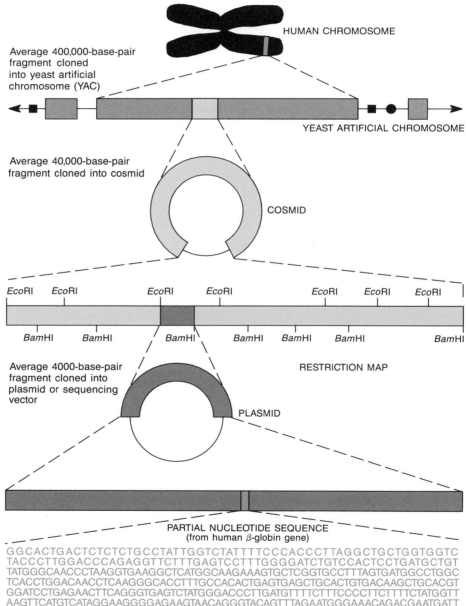

Up to 400 base pairs can be determined from a sequencing reaction

Hierarchy of Ordered Clones
Progressively smaller DNA molecules are subcloned in vectors to ultimately obtain fragments from which a partial sequence can be determined. This information is then compiled to provide a complete sequence across a large chromosomal region.

Laboratory), and David Botstein (Genentech, Inc.) provided a common language of DNA markers with which scientists from large and small institutions alike can easily exchange sequence information and obtain probes for known regions of the genome. They proposed that single-copy sequences of 200–500 nucleotides—sequence-tagged sites (STSs)—be used as physical landmarks along the chromosomes. They suggested that all types of mapping landmarks, including RFLPs and restriction-mapped regions, be "translated" into STSs and set the goal of obtaining some 30,000 STSs to span the genome at intervals of approximately 100 kb.

The clever application of a recently popularized technique, polymerase chain reaction (PCR), provides a simple means to exchange STSs among researchers and makes the routine exchange of cloned DNA unnecessary. Until the advent of PCR, the only way to purify a DNA sequence was through biological amplification in cultured cells. The entire process encompassed a period of several days and required ligating the desired sequence into a vector, transforming the recombinant vector into an appropriate host cell line, culturing the transformed cells, purifying the construct from the transformants, and releasing the fragment from the vector molecule. In contrast, PCR uses enzymatic amplification to increase the copy number of any DNA fragment of up to approximately 6000 bp. Using data available from an STS database, a researcher anywhere in the world can use PCR to assay for a particular STS in as little as 24 hours.

PCR is based on the phenomenon of primer extension by DNA polymerases, which was discovered in the 1960s. First, oligonucleotide (15–20 bp) primers are synthesized that are complementary to the 5' end of each strand of the STS. The two primers are mixed in excess with a DNA sample from which the target sequence (STS) will be amplified, along with a heat-stable *Taq* DNA polymerase from *Thermus aquaticus* (a bacterium that inhabits hot springs). The four deoxyribonucleoside triphosphates are also provided, one or more of which may be radioactively labeled with ^{32}P. The reaction mixture is then taken through multiple synthesis cycles consisting of the following:

1. Denaturing target sequence. Heating to 95°C denatures the target DNA and creates a set of single-stranded templates.
2. Annealing primers. Cooling to 37°C encourages primers to anneal to their complementary sequences on the single-stranded templates.
3. Extending new DNA strand. Heating to 72°C provides optimum temperature for *Taq* polymerase activity. Extending from the oligonucleotide primer, the polymerase synthesizes a second strand complementary to the original template.

During each synthesis cycle of approximately 7 minutes, the number of copies of the target DNA molecule is doubled. Twenty-five rounds of synthesis produce 1,000,000 new copies of the STS in less than 3 hours. If radionucleotides are provided in the reaction mixture, the radioactively labeled STS can be denatured to make a probe to identify clones in a genomic or cDNA library.

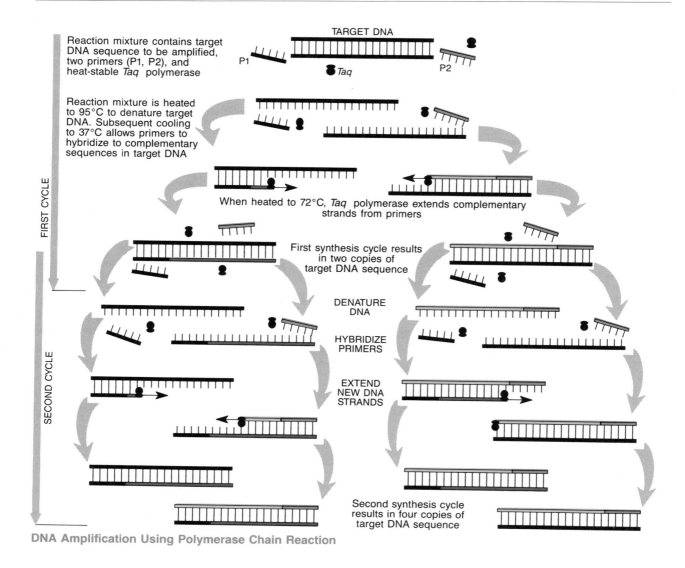

DNA Amplification Using Polymerase Chain Reaction

Pulse-field and Field-inversion Electrophoresis

To simplify the screening and ordering of cloned genomic fragments, it is necessary to manipulate extremely large DNA molecules. Analysis of large DNA molecules (greater than 100 kb) requires special care, since the use of standard extraction techniques would damage large DNA fragments. To minimize manipulations that would break the DNA strands, cell samples are embedded in small agarose blocks, each having a volume of approximately 100 µl. Reactions to release and digest the genomic DNA are then performed by diffusing reagents into the agarose block. First the cells are lysed with a detergent. After the detergent has been washed out, appropriate restriction enzymes are diffused into the block. Following incubation, the entire reaction block is placed into the well of an agarose gel for electrophoresis.

Conventional (fixed-field) electrophoresis can effectively separate

DNA fragments of up to about 20,000 nucleotides in length. However, several newer techniques can separate fragments of up to 10 million nucleotides. In pulse-field electrophoresis (PFE), current is alternated between several pairs of opposing electrodes. In field-inversion or orthogonal-field-alternation gel electrophoresis (OFAGE), a longer forward current is alternated with a shorter reverse current.

The resolving ability of PFE and OFAGE depends on the length of time the two alternating currents are applied. Without field alternation, large DNA molecules migrate through an agarose gel in a size-independent manner. The molecules align parallel to the current and move end-on through the gel, a process called reptation. However, alternation of the electrical field in PFE causes DNA to change its orientation in the gel ("turn"), which is a size-dependent process. Field inversion in OFAGE causes repeated changes in configuration of the DNA molecule, which also causes size-dependent migration.

Cloning into Yeast Artificial Chromosomes

The refinement of techniques for cloning genomic DNA into yeast artificial chromosomes (YACs) makes it possible to analyze very long DNA sequences. First, the genome is cleaved into fragments of 100,000 to 500,000 bp with any of several restriction enzymes that make infrequent cuts in mammalian DNA.

The enzymes used for "megabase cloning" cut 15–75 times less frequently than would be predicted by the random occurrence probabilities of their recognition sequences. For example, NotI recognizes an eight-nucleotide sequence and would be expected to produce restriction fragments averaging 65,536 nucleotides (4^8) in length. In fact, it produces fragments averaging 1 million nucleotides. MluI, NruI, and PvuI each recognize six-nucleotide sequences with random occurrence probabilities of 4096 nucleotides (4^6), yet each produces restriction fragments averaging 300,000 nucleotides. This is because the distribution of certain nucleotide sequences in the human genome is not completely random. The recognition sequences of all megabase-cutting enzymes contain one or more CG dinucleotides, a sequence that turns out to be very rare in mammalian genomes.

Some Megabase Cloning Enzymes and Their Recognition Sequences

Enzyme (bp)	Recognition sequence	Average fragment size
NotI	GCGGCCGC	1,000,000
MluI	ACGCGT	300,000
NruI	TCGCGA	300,000
PvuI	CGATCG	300,000
SfiI	GGCCNNNNNGGCC	200,000
RsrII	CGGWCCG	200,000
SacII	CCGCGG	100,000
SmaI	CCCGGG	100,000
SalI	GTCGAC	100,000
XhoI	CTCGAG	100,000

A YAC genomic library is constructed by ligating the megabase restriction fragments into artificial chromosome vectors. These plasmids include the minimal functional elements needed for accurate chromosome replication and segregation during mitosis in yeast cells. The cloned human DNA fragment, in effect, replaces yeast genes that would constitute the bulk of the wild-type chromosome. The YAC vector is transformed into yeast cells, and transformants are identified by expression of a selectable marker.

The YAC vector includes three types of chromosome-specific DNA sequences: telomere, origin of replication, and centromere. A telomere located at each chromosome end protects the linear DNA from degradation by nucleases. Origins of replication scattered throughout the chromosome are sites of attachment for DNA polymerase that duplicate the chromosomal DNA during mitosis. The centromere is the attachment site for mitotic spindle fibers, which "pull" one copy of each duplicated chromosome into each new daughter cell.

To construct large-order contig maps, the ends of the YAC inserts are sequenced and analyzed by computer to search for overlap with other inserts. In this way, adjacent inserts can be positioned relative to one another in the genome. Locating one or several STSs within each YAC insert will allow researchers to correlate locations on a contig map with the genetic linkage map mentioned earlier.

A higher-resolution map can be made by digesting YAC inserts to produce smaller fragments that are then subcloned into cosmids (fragments averaging 40,000 nucleotides) or λ vectors (fragments averaging

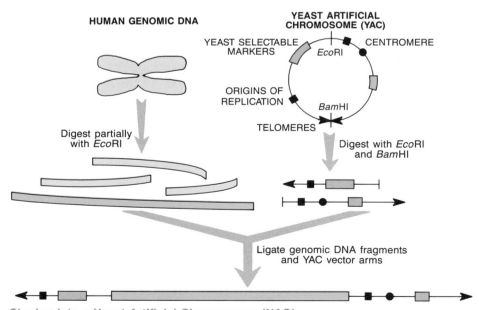

Cloning into a Yeast Artificial Chromosome (YAC)
The YAC vector contains basic yeast chromosomal sequences (centromere, telomeres, and origin of replication), as well as bacterial regulatory information that allows it to be propagated as a plasmid in *E. coli*. (After D.T. Burke et al. 1987. *Science* 236: 807.)

20,000 nucleotides). Aligning the overlapping ends of adjacent cosmid or λ inserts produces a more detailed contig map of the original YAC insert. Linking the restriction sites on adjacent cosmid or λ inserts, and relating these to STSs within YAC clones, will yield a restriction map of the entire human genome.

Collecting and Analyzing DNA Sequences

Restriction fragments from a cosmid or λ library are cloned into specialized sequencing vectors, and their nucleotide sequences are then determined. Relating sequences to specific positions relative to known physical markers will eventually produce the nucleotide sequence of the entire genome. Determining continuous DNA sequences of 1 million or more nucleotides will be a real challenge to sequencing technology; 200,000 nucleotides is the largest continuous sequence determined to date. For example, less than 25% of the sequence of the *E. coli* chromosome (5×10^6 bp) has been published. Although one research group has generated autoradiographs of sequencing gels representing the entire *E. coli* genome, compilation of a complete genomic sequence has been slowed by a lack of automated gel-reading devices and computer programs to organize the data efficiently.

This bottleneck points up the importance of new computer strategies for analyzing and disseminating the huge volume of DNA and protein sequence data to be generated by the Human Genome Project. Most critical are efficient computer programs to search data for biologically important sequences, including promoter, enhancer, and homeo regions that regulate gene expression. Sequence data will likely be stored at several centralized storage facilities, which receive data in a standardized format from sequencing laboratories. Although current computer disk storage systems are capable of storing the amounts of data to be generated, an efficient network must be developed for rapidly distributing sequence data to researchers around the world.

Sequencing by hand is also time-consuming and costly. A well-trained technician using the best available equipment can sequence 50,000 to 100,000 nucleotides per year at a cost of 1 to 2 dollars per nucleotide. At this rate, sequencing the human genome would take 30,000 person-years and cost approximately $3 billion. Thus, another major thrust of the Human Genome Project is to develop automated sequencing technology that will increase sequencing speed to 100,000 or more nucleotides per day at a cost of pennies per nucleotide. A prototype sequencing machine designed by Lee Hood and his co-workers at the California Institute of Technology can theoretically sequence 16,000 nucleotides per day.

Also of concern is the accuracy of DNA sequencing. Even a single-nucleotide mistake can have a major impact on the interpretation of DNA sequence. An insertion or deletion of a nucleotide within a coding sequence (exon) shifts the reading frame and predicts an erroneous amino acid sequence from the point of the error to the end of that coding region. Most authorities believe that it will be difficult routinely to achieve an error rate of less than 0.1%, which translates into 1 error in every 1000 nucleotides. Assuming that the average exon is 200 nucleotides in length, a sequencing error is likely to occur in one exon in five. Although computer programs may detect some of these mistakes, many sequencing errors will only be uncovered by researchers who are actively studying that region of the genome.

Parallel Sequencing Projects

When we speak of sequencing the human genome, are we referring to the genome of a man or a woman; a Caucasian or a Black; an American or a Swede; a Catholic or a Jew; a normal individual or one with muscular dystrophy? And what exactly is normal? Just as every person is different, so too are DNA sequences within each genome. Obviously, a single sequence of 3 billion nucleotides could never account for the variability of human beings. Although it will be impossible to document all the subtle DNA differences between people, biologists will certainly focus on sequence variations in key regions of the genome, especially those containing known disease genes. In reality, the "human genome" assembled over the next decade or so will be a patchwork quilt of sequence data derived from numerous human sources, representing a sort of "everyperson's genome."

The scope of the Human Genome Project will be broadened by parallel sequencing efforts. First, it will be necessary to sequence a library of human cDNAs. Comparison of cDNA and genomic sequences will delineate coding regions from control sequences, intron sequences, and intergenic regions that constitute up to 95% of the human genome. It has been proposed to sequence the genomes of several research organisms, including the bacterium *Escherichia coli*, the yeast *Saccharomyces cerevisiae*, the nematode worm *Caenorhabditis elegans*, the fruit fly *Drosophila melanogaster*, the mouse *Mus musculus*, the simple dicot plant *Arabidopsis thaliana*, and the corn plant *Zea mays*. The function of human genes—and their protein products—can be readily tested in these experimental systems. Conversely, important genes discovered in these organisms can be used to identify corresponding human genes. Comparative studies of conserved gene sequences will yield a molecular taxonomy of how these sequences evolved.

Human Genetics

Victor McKusick, at Johns Hopkins University, has for years compiled information on human genetic diseases. To date, his compendium, entitled *The Morbid Anatomy of the Human Genome*, includes more than 3000 inherited disorders. The causative genes have been identified for approximately 100 disorders, and only a fraction of these have been cloned. So, the immediate challenge to human genetics is to isolate the genes responsible for these illnesses, which will ultimately lead to an understanding of the functions of the normal and mutant proteins made at disease loci. This is a tall order.

Human genetics differs from classical genetics in that the system under study cannot be easily manipulated. The classical approach, used with experimental plants and animals, is to mate parents selectively and observe inheritance patterns in the resulting offspring. Although arranged marriages still take place in some cultures, people, for the most part, choose their own spouses and are generally opposed to being selectively mated. Thus, human geneticists must be content to work with the existing genetic makeup of related family members. Also, they seldom have the luxury of following a single well-defined trait through successive generations.

Humans pose another problem as experimental organisms. The members of most experimental systems, such as mice, are highly inbred. Lineages and traits carried by each individual are known at the outset of a breeding experiment. Furthermore, members of the experimental pool are most often physically and genetically identical, differing from one another only by one or, at most, several traits. This genetic homogeneity allows a specific trait or mutation to be observed against an essentially neutral background. Human populations tend not to be inbred, and they are physically and genetically heterogeneous. (As we will see, biologists have recently turned this disadvantage around and can actually exploit heterogeneity in the cloning of disease genes.)

However, certain human populations, whose members by custom live apart from the general population, have relatively closed gene pools. Isolated groups of individuals belonging to some religious sects, notably the Amish and Mormons, show a degree of genetic homogeneity. The fact that these groups keep meticulous genealogical records and have relatively large families also make them good candidates for genetic analysis. Customs prohibiting alcohol consumption make easier the analysis of mental and behavioral disorders, such as manic depression and schizophrenia, whose symptoms may be masked by alcohol abuse.

Reverse Genetics

Until very recently, the study of genetic diseases relied on discovering a biochemical difference between diseased and normal individuals. This approach is an extension of the prophetic hypothesis of Archibald Garrod (1908) that genetic diseases are "inborn errors of metabolism," where a defect in a gene leads to an abnormal change in its encoded protein. This approach revealed the molecular basis of a number of diseases, such as phenylketonuria, sickle cell anemia, and the thalassemias, whose phenotypes are obvious changes in a single protein or metabolic pathway.

Unfortunately, most diseases are not characterized by a single, well-defined biochemical change. Instead, they result in a complex phenotype of many symptoms (a syndrome) that obscures the primary genetic defect. Thus, a new era of medical genetics was heralded at a scientific meeting at Alta, Utah in 1978, where David Botstein, Ronald Davis, and Mark Scolnick proposed a method that would lead to the identification of the causative mutations in genetic diseases that do not have obvious phenotypes. They suggested that polymorphic DNA regions could be used as genetic markers to "tag" and clone disease genes. This amounted to turning the study of genetic diseases on its ear, by first isolating the disease gene and then working backward to identify the metabolic defect. Although the strategy is complex, its logic is simple: Use the DNA sequence of a cloned disease gene to predict the amino acid sequence of its encoded protein. This approach was dubbed "reverse genetics."

The Botstein/Davis/Scolnick technique combined modern gene cloning with genetic linkage analysis worked out three generations before by Thomas Hunt Morgan and his co-workers (introduced in Chapter 2). During meiosis, paired chromosomes align (synapse) and homologous

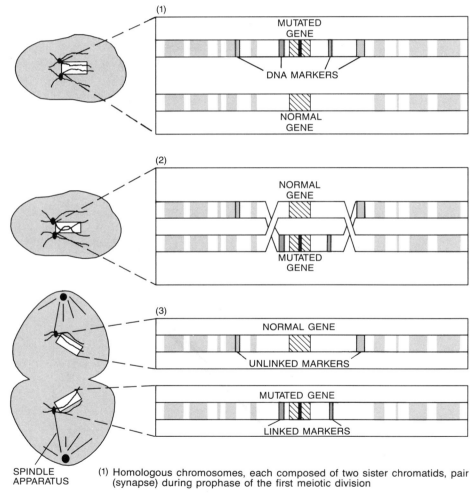

(1) Homologous chromosomes, each composed of two sister chromatids, pair (synapse) during prophase of the first meiotic division

(2) Recombination occurs when chromatids cross over, exchanging DNA fragments. Linked markers remain with the original chromatid, whereas unlinked markers become separated from it

(3) During anaphase, the recombined chromatids separate into two different daughter cells (In a subsequent meiotic division, the sister chromatids will segregate into separate haploid sex cells)

Fate of Linked and Unlinked Markers during Meiotic Recombination

regions of the DNA are exchanged when chromatids "cross over" one another. Usually, large DNA fragments, on the order of millions of nucleotides, are moved between chromatids. Therefore, two loci (genes or markers) lying close together on a chromosome are not likely to become separated during crossing over. Such loci are almost invariably inherited as a single entity and are said to be linked. Conversely, two loci located very far apart are said to be unlinked—they have an equal chance of remaining together or being separated during meiotic recombination.

The frequency of recombination is thus a measure of the genetic distance between two sites on a DNA molecule. If two loci have a 1% recombination frequency, they will become separated once in 100 meiotic recombinations. A recombination frequency of 1% is referred to as 1 cM and is equated to a physical distance of approximately 1,000,000 nucleotides. Thus, the entire human genome of 3×10^9 base pairs is about 3000 cM. At a distance of 50 cM, two chromosomal sites are unlinked. Since the recombination rate is 50%, co-inheritance (or separation) is completely random.

DNA Polymorphisms and RFLPs

DNA polymorphisms (many forms) are differences in DNA sequence that result from point mutations, random deletions or insertions, or the presence of varying numbers of repeated copies of a DNA fragment (tandem repeats). A polymorphic locus exists when two or more variant sequences may occur at that site on the chromosome. A polymorphism in the coding region of a gene may be detected as an alteration in the amino acid sequence of the encoded protein. The mutations causing sickle cell anemia and many of the thalassemias were discovered in this way. Using recombinant DNA techniques, it is now possible to detect polymorphisms in unexpressed regions of DNA as well.

DNA polymorphisms are detected by Southern blot analysis, using a probe that hybridizes to a polymorphic region of the DNA molecule. In one of the earliest studies, the β globin sequence was used to probe the DNA of individuals heterozygous for β-thalassemia. Several individuals were found to have two distinct hybridization patterns: one from the normal allele and one from the mutant allele.

When human DNA is digested with a particular restriction enzyme, a polymorphic locus yields restriction fragments of different sizes, called RFLPs. Insertions, deletions, and tandem repeats of more than 100 nucleotides can be detected as obvious shifts in hybridization patterns in Southern blot analysis. Smaller polymorphisms, even point mutations, can be detected when they result in the loss or addition of a recognition site for a specific restriction endonuclease. Considering that 1 in 200–500 nucleotides differs between paired chromosomes, the loss of a restriction site by point mutation is not a rare event.

For example, the mutation responsible for sickle cell anemia can be specifically detected by RFLP analysis using the restriction enzyme *Mst*II,

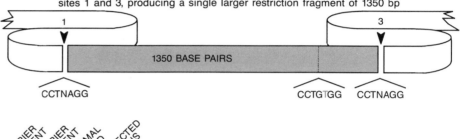
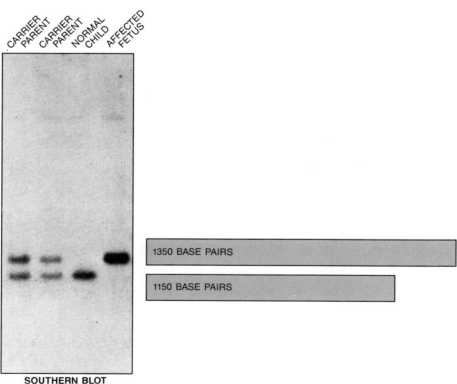

RFLP Diagnosis of Sickle Cell Anemia

The Southern blot shows the RFLP patterns of two carrier parents, a normal offspring, and amnionic fluid from an affected fetus. The carrier parents show a single copy of each RFLP: the 1350-bp fragment associated with the disease allele and 1150-bp fragment associated with the normal allele (the 200-bp fragment is not detected by the probe). The normal child shows the 1150-bp fragment, and the affected fetus shows the 1350-bp fragment. Both normal and affected offspring are homozygous and thus show a single relatively thick band, denoting two chromosomal copies. (Photo reprinted with permission by the *New England Journal of Medicine,* vol. 307, p. 31, 1982.)

which recognizes the sequence CCTNAGG (N equals any nucleotide). The characteristic sickled shape of the red blood cells of affected individuals is due to a single amino acid change in the β globin protein, resulting from a point mutation in the β globin gene that changes the sixth codon from G*A*G to G*T*G. This A-T mutation results in the loss of an *Mst*II recognition site that spans the region of sixth codon. Thus, the DNA from normal homozygous individuals, heterozygous carriers of the sickle cell trait, and homozygous sickle cell patients produces different restriction fragments when cut with *Mst*II. In Southern blot analysis, these RFLPs are detected as characteristic banding patterns, using a radioactive β globin gene probe.

Using Linked RFLP Markers to Map the Huntington's Disease Gene

Huntington's disease is a degenerative nervous system disorder that invariably leads to loss of motor function, mental incapacitation, and early death. It is a rare example of an autosomal dominant disease, meaning that inheritance of a single copy of the defective gene leads to the illness—individuals with two copies of the mutant gene appear to be affected to the same degree as those inheriting only one. Thus, each child born to a family with one parent who is heterozygous for the defective gene has a 50% probability of inheriting Huntington's disease. Although it is caused by a dominant lethal gene, Huntington's disease has been perpetuated in the human gene pool because the onset of symptoms occurs after the child-bearing years (35–45 years of age).

The identification of the chromosomal location of the Huntington's disease gene by James Gusella's group at Massachusetts General Hospital, in 1983, demonstrated the power of RFLP/linkage analysis. Gusella's work was based on studies carried out simultaneously by Nancy Wexler, of Columbia University and the Hereditary Disease Foundation, on the inheritance of Huntington's disease within an interrelated population at

Nancy Wexler and the "Pedigree Wall" at Columbia University, 1987
The pedigree traces the inheritance of Huntington's disease through extended families living at Lake Maracaibo, Venezuela. (Courtesy of S. Uzzell.)

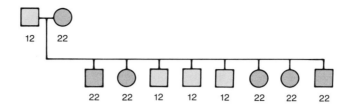

DNA Diagnosis of Huntington's Disease
This pedigree shows the coinheritance of a polymorphic allele and Huntington's disease. Allele 1 is linked to the Huntington's disease gene, and allele 2 is linked to the normal gene. The affected father and three affected sons (red) all carry one copy of each allele. The unaffected mother and offspring (gray) have copies of the normal allele. (Courtesy of T.C. Gilliam, Columbia University.)

Lake Maracaibo, Venezuela. This population made an ideal case study, because it is an extended family composed of 9000 members.

In addition to establishing a pedigree showing the inheritance of the disease, Wexler and her co-workers have collected blood samples from more than 2000 normal and afflicted persons. These samples are returned to the laboratory and used to establish lymphoblastoid cell lines that can be continuously cultured in vitro. This is accomplished by fusing white cells in the blood with immortal cancer cells or by transforming them with an immortalizing oncogene from a retrovirus, such as Epstein-Barr virus (EBV). In either case, the cultured white blood cells provide an infinite source of DNA needed for the next, and most laborious, step.

Gusella radioactively labeled at random a number of cloned DNA fragments from a human genomic library. These were then used to probe Southern blots of DNA from normal and afflicted family members. He was looking for a probe that identifies a polymorphism whose appearance parallels the pattern of inheritance of the disease. The assumption was that all members of the extended Lake Maracaibo family inherited the disease from a single common ancestor. Thus, if a polymorphic marker is tightly linked to the disease gene, it should be co-inherited by all of the afflicted people.

This process is very time-consuming. To pass the threshold for linkage, the polymorphic marker must lie within 20 cM of the disease locus. This means that the two alleles will be separated in 20% of meiotic recombinations—or, conversely, that the polymorphism will be present in 80% of Huntington's disease patients screened. Assuming that the probes are generated in a random fashion and that they are evenly distributed throughout the genome, there is a 1 in 150 probability of any probe being linked to the disease (300 cM per genome/20-cM linkage distance).

Luck was with Gusella: A linked marker was identified with the 12th probe tested. Using this probe, he demonstrated that the Huntington's disease gene is located on chromosome 4. New probes were obtained from a chromosome-4 library and tested for tighter linkage. Gusella's initial good fortune did not continue, and it took almost four years to identify a closer marker. Although the ultimate goal is to identify flanking markers that lie on either side of the disease gene, this has proved difficult, as it has become evident that the Huntington's disease gene lies very near the end (telomere) of chromosome 4.

The power of reverse genetics has been well established with the identification of linked, chromosome-specific markers for several dozen diseases, including Duchenne muscular dystrophy, cystic fibrosis, and Alzheimer's disease.

Cloning the Disease Gene: Chromosome Walking

Once a linked marker has been identified within 1 cM of the disease locus, "chromosome walking" is one of several approaches that can be used to clone the disease gene. First, a genomic library of large DNA fragments (20,000–40,000 bp) is constructed that encompasses the disease locus. The closest linked marker is used as a probe to isolate its corresponding genomic clone. Following restriction mapping of the clone, a restriction fragment is isolated from the end of the clone closest to the disease locus. This fragment is used to reprobe the library to

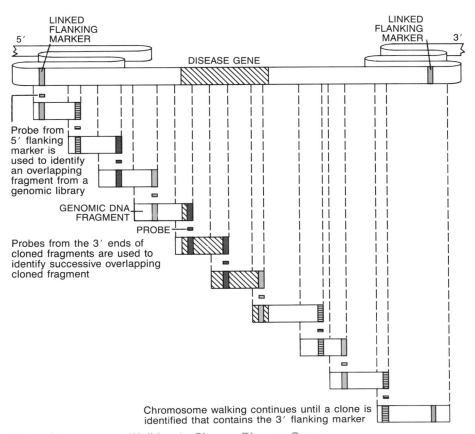

Using Chromosome Walking to Clone a Disease Gene

identify an overlapping clone. The endmost fragment of this clone is then used to reprobe the library, and another overlapping clone is isolated. Through such a succession of overlapping clones, one "walks" along the chromosome region spanning the disease locus, eventually reaching the flanking marker on the other side. The overlapping fragments are then assembled to produce a map of the disease locus.

Disease-causing mutations are identified by comparing the alleles of normal and affected individuals. Obviously, it would be time consuming to sequence and compare a region of cloned DNA spanning a million base pairs. An added difficulty is to distinguish between disease-causing lesions and the normal 0.2–0.5% variation in DNA sequence between individuals. (This amounts to 2000–5000 differences in a 1-million base-pair sequence.) A common strategy is to identify open reading frames within the cloned segment and then search the coding regions of these putative genes for mutations that have an obvious effect on protein structure. If no significant mutations are found within coding regions, an examination of noncoding and regulatory regions would follow.

The X-linked disease Duchenne muscular dystrophy (DMD) was among the first disease loci actually to be cloned using reverse genetic techniques. The early success in identifying this gene relied on evidence showing that a number of DMD patients have large deletions clustered in a region of the X chromosome known as Xp21. In addition, females having the disease were found to have a break in their active X chromosome at position Xp21. This suggested that the deletions were associated with pathology, causing loss of part of the normal gene at this locus. A combination of strategies, including RFLP analysis, was employed to isolate the disease gene.

The dystrophin gene, as it has become known, is one of the largest and most complex yet discovered. Encompassing more than 2,000,000 bp and possessing over 60 exons, it produces a 14,000-bp messenger RNA that codes for a protein containing 4000 amino acids. The dystrophin gene appears to be prone to damage; DMD patients show many different deletions in various regions of the gene. Interestingly, it now appears that a less severe form of the disease, Becker's muscular dystrophy, is also due to deletions in the dystrophin gene. These patients appear to produce low levels of a truncated and semifunctional dystrophin protein, suggesting that part of the coding region of the dystrophin gene has been lost.

Cloning the Cystic Fibrosis Gene

The 1989 isolation and analysis of the causative gene for cystic fibrosis (CF) on chromosome 7 was a case study of the practice and power of modern molecular genetics. First, it resulted from an interdisciplinary collaboration between more than 25 scientists at the Hospital for Sick Children and University of Toronto, the University of Michigan, and the University of Pittsburgh. Second, it was the first disease gene identified *entirely* by reverse genetics, including RFLP analysis and chromosome walking. Unlike muscular dystrophy, CF is not characterized by large-scale deletions or rearrangements that could be used to map the gene to its chromosomal location.

Researchers had to screen ten genomic libraries to isolate the CF gene, which spans approximately 250,000 nucleotides. The coding exons of the CF gene predict a protein of 1480 amino acids, called the cystic fibrosis transmembrane conductance regulator (CFTR). As in sickle cell anemia, the primary genetic lesion in CF appears to be a specific mutation affecting a single amino acid. Approximately 70% of CF patients show a 3-bp deletion that results in the loss of a single phenylalanine residue at amino acid position 508 of the CFTR polypeptide.

The CFTR protein includes 12 hydrophobic domains that are characteristic of proteins that weave through the plasma membrane and 2 folded structures that project into the cytoplasm and appear to bind ATP. CFTR may itself be a channel for the conductance of ions through the plasma membrane or a regulator of other ion channels. In either case, the hydrolysis of ATP appears to be directly required for this transport function. Interestingly, the position-508 deletion prevalent in CF patients occurs in one of the two suspected ATP-binding domains.

DNA Diagnosis

Accurate DNA diagnosis becomes feasible once a polymorphic (RFLP) marker has been located within 5 cM of the disease gene. At this distance, the recombination frequency (and the probability that the marker and the gene become unlinked) is 5%. Conversely, the co-inheritance of the

Some Diseases That Have Been Linked to RFLP Markers

Disease	Chromosome
Adenomatous polyposis (familial colon cancer)	5
Alpha-1 antitrypsin deficiency	14
Alzheimer's disease	21
Cleft palate	X
Cystic fibrosis	7
Fragile X syndrome (X-linked mental retardation)	X
Hemophilia A (Factor VIII deficiency)	X
Hemophilia B (Factor IX deficiency)	X
Huntington's disease	4
Lesch-Nyan syndrome	X
Malignant hyperthermia	19
Muscular dystrophy (Becker)	X
Muscular dystrophy (Duchenne)	X
Muscular dystrophy (myotonic)	19
Ornithine transcarbamylase deficiency	X
Phenylketonuria	12
Polycystic kidney disease	16
Retinoblastoma	13
Sickle cell anemia	11
Thalassemias	11
Wilms' tumor	13
Wiskott-Aldrich syndrome	2

marker and the disease gene, as well as the accuracy of diagnosis, is 95%. The addition of a second flanking marker, within 5 cM on the other side of the gene, theoretically increases the accuracy of predictions to 99.75%. (The chance of both markers becoming unlinked is $0.05 \times 0.05 = 0.0025$.) Using a battery of markers currently available, Huntington's disease and DMD can each be diagnosed with an accuracy of greater than 95%.

In general, the closer the marker to the disease locus, the more accurate the diagnosis. A marker and a gene in extremely close proximity may be in *linkage disequilibrium*, in which case they are *always* co-inherited. Such markers, including those actually located within the disease gene, can provide accurate diagnosis *without* a family history. (The sickle-cell polymorphism discussed earlier falls into this category.) However, markers located within a very large disease locus may not even be in linkage disequilibrium with the gene itself. For example, markers located at the 5' end of the dystrophin gene have a recombination frequency of 5%, thus an apparent recombination distance of 5 cM!

DNA diagnosis using RFLPs relies on linking one allele of a polymorphic marker to inheritance of a disease. Because the alleles present at the polymorphic locus may differ from family to family, it is necessary to follow a linked polymorphism through the pedigree of the family under study. This establishes which particular polymorphic allele is associated with the disease state in that particular family. It is also necessary to identify heterozygous carriers of the disease gene in whom one polymorphic allele segregates with the disease gene and a different polymorphism segregates with the normal gene. Diagnosis of an X-linked disease such as DMD ideally can be made from a carrier mother and a single afflicted son. In practice, though, diagnosis of DMD usually requires DNA analysis of six family members.

The use of polymorphic markers is already revolutionizing genetic counseling and prenatal diagnosis. In the past, members of families with a history of genetic illness often had no way of knowing if they were carriers of the disease gene—a carrier was only revealed upon the birth of an afflicted child. DNA diagnosis can now distinguish carriers from noncarriers, allowing couples to make informed decisions about childbearing. Noncarriers can know with certainty that their offspring will be free of disease, and carriers can avail themselves of early prenatal diagnosis.

DNA diagnosis should increase chances of detecting disease in its early stages, thus increasing therapy options and chances for a good prognosis. As mentioned in Chapter 5, DNA markers are already being used to monitor the progression of certain cancers and to predict the patient's response to alternative therapies. In the past, the diagnosis of chronic myelogenous leukemia relied on identification of the so-called Philadelphia (Ph^1) chromosome, which results from a translocation that fuses a chromosome-9 fragment with chromosome 22. This cytological workup is labor-intensive and time-consuming. Recently, several polymorphic markers have been identified that are specific for the Ph^1 translocation and can be rapidly detected by Southern blot analysis.

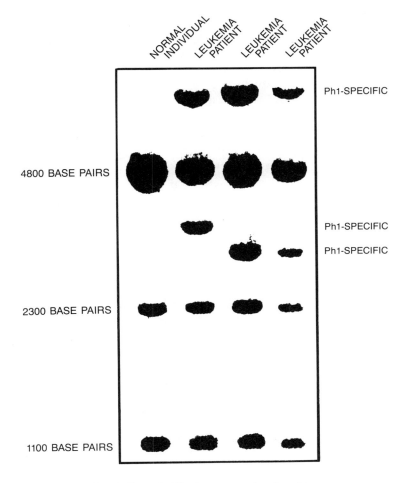

DNA Diagnosis of Chronic Myelogenous Leukemia
The probe used in this diagnosis spans the region of defects that result from the Philadelphia (Ph^1) chromosome translocation. (Courtesy of Oncogene Science, Inc.)

PCR provides a rapid means to diagnose activated oncogenes in human tumors—for example, the characteristic point mutations in codons 12, 59, and 61 of the activated *ras* oncogene. Oligonucleotides flanking these mutations are used as primers to amplify these sequences in DNA isolated from tumor cells. The same primers are used in dideoxy sequencing reactions of the amplified fragments, and the sequenced regions are screened for the activating mutations. This strategy can be employed for any oncogene whose activating mutations are known.

DNA diagnosis of cancer does not rely solely on RFLP markers. As mentioned previously, monitoring the number and types of oncogenes present in tumor cells should provide a powerful tool in cancer management. For example, the identification of cancer cells in underarm lymph nodes has been the major indicator of breast cancer recurrence. Obviously, the presence or absence of tumor cells in the lymph nodes is an extremely crude measure of tumor progression. However, there is mounting evidence that specific DNA lesions are associated with poor prognosis in breast cancer patients, including the amplification of two positive-acting oncogenes, *int*-2 and *erb*B-2 (also known as HER-2 or *neu*). Deletions of a specific region of chromosome 11 and diminished expression of the NM23 gene have also been linked to rapid recurrence, suggesting that loss or underexpression of tumor-suppressing genes is also an indicator of metastasis.

Approximately 70% of breast cancer patients who show no lymph node involvement will remain free of the disease for at least 5 years without preventive chemotherapy. However, in 1988, the National Cancer Institute recommended that *all* breast cancer patients receive chemotherapy as a means of lowering the recurrence rate. Oncogene markers may provide a more rational means of identifying those breast cancer patients who will benefit from aggressive chemotherapy and those for whom it is unnecessary.

Gene Therapy

Although the ability to diagnose genetic diseases improves almost daily, the means to treat these pathologies lags far behind. Gene enhancement or replacement therapy holds promise as the ultimate cure for genetic diseases, many of which are largely untreatable at this time. However, there is as yet no practical approach to augment or replace damaged genes in the cells of a living human being.

Provided that safe techniques are worked out, few people are likely to have serious objections to *somatic* gene therapy, which would treat genetic disorders within the somatic (body) cells of an individual. This would be only a further extension of current medical practice. The genetic changes induced by this type of therapy could not be inherited and would have no effect on future generations.

In contrast, *germ-line* gene therapy would change the genetic makeup of gamete-forming cells (sperm and ova) or of the developing embryo. Embryonic gene transfer, in fact, would likely pose fewer technical difficulties than somatic gene therapy. A gene that is successfully transfected or microinjected into a single-cell embryo subsequently would be cloned into all of the somatic cells and germ cells during embryogenesis.

In 1988, the Recombinant DNA Advisory Committee (RAC) of the National Institutes of Health gave the first government approval for an experiment to test the feasibility of gene transfer into humans. The experiment involves transfecting a neomycin resistance (*neo*) gene into tumor-infiltrating lymphocytes, which attack cancer cells. The transfected lymphocytes will then be injected into terminal cancer patients, where the *neo* gene will provide a marker to monitor their anti-cancer activity. In subsequent tests, the NIH research team, headed by W. French Anderson, hopes to transfect a gene for the growth hormone interleukin-2, which stimulates the replication of the killer cells.

In the last several chapters, we have discussed relatively routine techniques for introducing new genes into eukaryotic cells: calcium phosphate and viral transfection for cultured cells and microinjection for

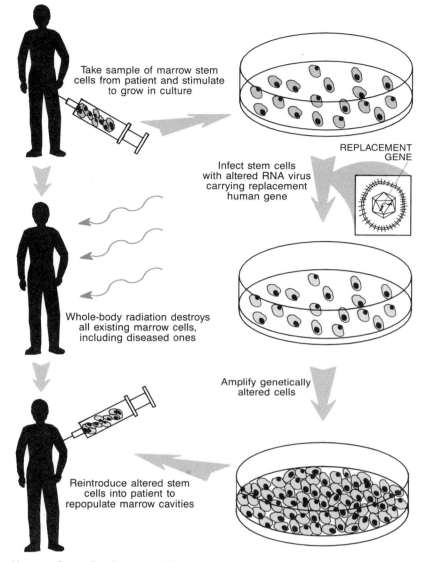

Human Gene Replacement Therapy

developing embryos. However, adaptation of these techniques for human gene therapy poses several serious technical problems:

1. How can the gene be introduced and incorporated into a large enough number of somatic cells?
2. How can the protein product be expressed in the proper tissues and in large enough quantities?
3. How can the chance of disruption or improper activation of resident genes be minimized?

Current research on gene therapy focuses on diseases of the bone marrow and blood, including sickle cell anemia, thalassemia, and adenosine deaminase deficiency (ADA, a severe immune deficiency). Each of these diseases is caused by a single-gene defect that affects production of a single protein. The genes involved are all expressed in marrow cells, which are relatively easy to manipulate and can be maintained in culture outside the patient's body. Also, there is hope that even partial expression of a single normal gene would have a therapeutic effect for these and other genetic illnesses. For example, β globin expression 20–50% of normal is thought to be sufficient to correct β-thalassemia and sickle cell anemia.

A gene therapy protocol for these diseases would begin with the withdrawal of a bone marrow sample from the patient. The marrow cells would then be cultured and transfected with a vector carrying the replacement gene. Following whole-body radiation, which destroys the patient's diseased marrow, the transfected marrow cells would be reinjected into the patient. These "repaired" cells would return through general circulation to repopulate the marrow cavity, giving rise, in the case of sickle cell anemia and thalassemia, to normal red blood cells. Since red blood cells have a life span of approximately 120 days, it would not take long to entirely replace defective cells with normal cells.

Much effort has so far focused on a vector system to deliver the repaired gene into the target marrow cells. The likely candidates are altered retroviruses whose oncogenes have been deleted. The altered viruses also lack genes coding for proteins necessary to package new viruses and therefore can only undergo one round of infection. This would prevent the virus from reinfecting when the manipulated cells are replaced in the body. Since red blood cells are thought to arise from a small population of stem cell clones, achieving nearly 100% infection of the stem cells in the marrow sample may prove critical to successful therapy. In theory, even a small number of unrepaired stem cells could be those that are activated to produce the new (but still defective) population of red blood cells.

The random integration of the provirus in the stem cell chromosome poses several problems. Since expression of the repair gene might well vary according to integration site, preferential integration at a down-regulated site could lead to inadequate production of the therapeutic protein. Integration might disrupt or alter the activity of resident genes or bring them under the control of regulatory sequences in the vector molecule. DNA alterations at the integration site might lead to deactivation of a recessive (growth-suppressing) oncogene or activation of a dominant (growth-inducing) oncogene, thus initiating or contributing to oncogenesis.

Although transfected DNA usually integrates randomly into the genome, it occasionally integrates at the site of homologous DNA sequences on the chromosome. Whereas homologous recombination occurs at very low frequency in somatic cells, it is the mechanism through which chromatids exchange relatively frequently during meiotic crossing over. Recombination is believed to occur through a reciprocal single-strand-DNA exchange that results in the formation of two heteroduplex DNA molecules, each containing one strand of its own and one strand from its homologous allele.

"Gene targeting" or "trapping" using homologous recombination would avoid the problems of random integration and make it possible to actually replace a defective gene with a normal one. However, the rarity of homologous recombination in cells transfected with plasmid DNA (1 in 1000 to 10,000) is especially serious in light of the need to ensure that nearly all reimplanted cells incorporate the replacement gene. Although there is no current means to improve the rate of homologous recombination, various selection procedures have been developed to identify cells that have homologously incorporated a transfected gene.

Mario Capecchi and his co-workers at the University of Utah have developed a selection system in which the replacement gene is ligated into a plasmid vector along with positive and negative markers. A neomycin resistance gene is inserted *within* the flanking sequences of the replacement gene, and the thymidine kinase gene (*tk*) from herpes simplex virus (HSV-*tk*) is inserted *outside* the replacement gene se-

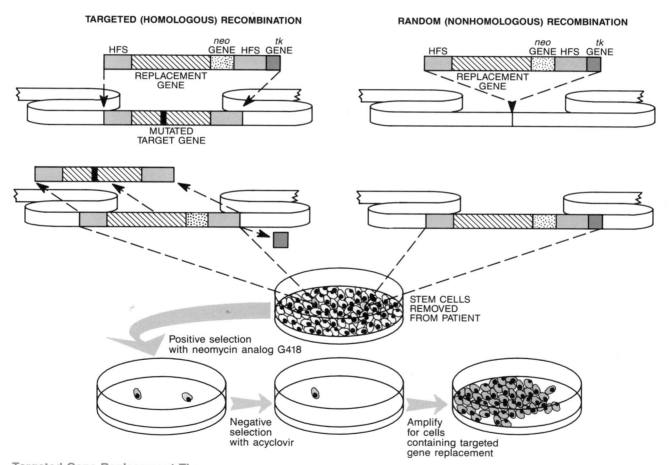

Targeted Gene Replacement Therapy
The replacement gene construct contains two marker genes, neomycin resistance (*neo*) and thymidine kinase (*tk*), and is bordered by homologous flanking sequences (HFS). Following transfection, the replacement gene almost always integrates randomly into the genome (at *left*). These cells integrate the *tk* gene and are sensitive to acyclovir. Rarely, the replacement gene integrates specifically at the site of its cellular homolog, eliminating the mutated gene (at *right*). These cells do not integrate the *tk* gene. Although every transfected cell is neomycin-resistant, only those where the gene recombines homologously are acyclovir-resistant.

quences. Following DNA uptake by electroporation, the cells undergo positive selection in medium containing the neomycin analog G418. Cells that have stably integrated the plasmid grow, and those that have not are killed.

Negative selection with the nucleoside analog acyclovir kills all cells that have integrated the HSV-*tk* gene. During random recombination, the HSV-*tk* gene remains attached to the replacement gene and is incorporated into the host chromosome. However, during homologous recombination, the HSV-*tk* gene is lost, because it lies outside the ends of the homologous target region. Thus, double selection with G418 and acyclovir enriches for cells that have properly integrated the replacement gene by homologous recombination. These selected cells can be further propagated in culture and then used to repopulate the bone marrow.

DNA Fingerprinting

The application of DNA polymorphisms is also revolutionizing forensic medicine, paternity testing, and identification of disaster victims. The term "DNA fingerprinting" was coined to allude to the traditional use of fingerprints as the most unique means of human identification. However, whereas classic fingerprinting analyzes a phenotypic trait, DNA typing directly analyzes genotypic information. When properly conducted, DNA-based testing can provide positive evidence of a person's identity. In contrast, the phenotypes detected by blood grouping and leukocyte antigen testing are shared by sufficiently large numbers of individuals that they are not, in the strictest sense, tests of identity. Rather, they are exclusionary tests that can only prove that forensic evidence does *not* match a suspect or that persons are *not* related.

All that is required for DNA fingerprinting is a small tissue sample from which DNA can be extracted. This can be blood samples in a paternity case, a sperm sample from a rape victim, dried blood or semen from fabric, skin fragments from under the fingernails of a victim after a struggle, or even several hairs (with the attached roots) combed from a crime scene. The time is approaching when a criminal will not be able to afford to leave even a single cell at the crime scene. PCR, in theory, makes it possible to obtain a DNA fingerprint by amplification of a single DNA molecule.

As a type of RFLP analysis, DNA fingerprinting is likewise based on Southern blotting. Samples of the DNAs from the relevant parties are digested with one or more restriction enzymes and electrophoresed on an agarose gel. The DNA is then transferred to a nylon membrane (or other solid support) and probed with a radioactive sequence that hybridizes to one or more polymorphic regions of the human genome. The best probes hybridize to "hypervariable" regions of DNA that are altered or rearranged frequently within the human population.

Two types of probes are now in use. A *single-locus* probe hybridizes to a unique hypervariable region of the genome and generates a pattern consisting of one or two bands from an individual's DNA—depending on

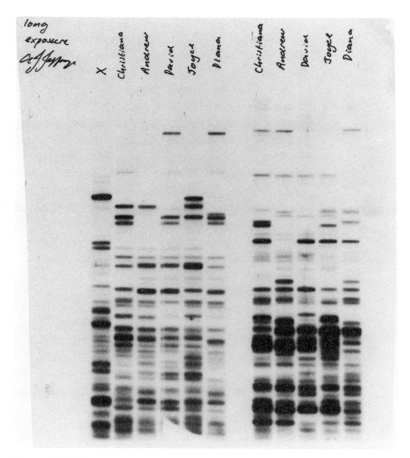

Use of Multilocus Probe to Show Family Relationship
This Southern blot shows the first use of DNA fingerprints as evidence in a court of law (1985). The match between many bands in their DNA fingerprints proved a family relationship between a Ghanian boy, Andrew, who wished to immigrate to England to join his mother (Christiana) and his siblings (David, Joyce, and Diana). Bands not shared by a child and the mother were inherited from the father. The DNA fingerprint of an unrelated person is shown in lane X. The evidence was prepared by Alec Jeffreys, who originated the use of multilocus probes and coined the term DNA fingerprint. (Courtesy of A. Jeffreys, University of Leicester.)

whether they are homogeneous or heterogeneous at that locus. Used alone, a single-locus probe only detects one or two differences, so a "cocktail" of several probes is typically used for forensic purposes. A *multilocus* probe, on the other hand, hybridizes to repetitive DNA sequences that occur at thousands of sites in the human genome—of which Southern blot analysis typically detects 20–30 interpretable bands.

Multilocus probes were discovered in 1984 by British researcher Alec Jeffreys, who coined the term DNA fingerprinting and was the first to use DNA polymorphisms in paternity, immigration, and murder cases. The discovery of the so-called Jeffreys' probes arose from the investigation of hypervariable regions composed of short repeated sequences of DNA, first described in 1980 by Arlene Wyman and Ray White, of the University of Utah. Working at the University of Leicester, Jeffreys found two "core" sequences that are common to many of the repeated sequences. He discovered that probes for these core sequences hybridize to digested human DNA, creating distinctive banding patterns that are inherited in a Mendelian fashion. Thus, one half of the bands in a child's DNA fingerprint is inherited from the mother and one half is inherited from the father.

Occasionally, a spontaneous mutation during embryogenesis creates a band in a child's fingerprint that has no match in the parental DNA. In the case of a multilocus probe, this would at most represent only 1 out of 20–30 interpretable bands. However, this occurrence could completely eliminate the usable data from a single-locus probe. In the absence of real data from large-scale population genetics studies, assumptions have been made about the discriminating power of DNA fingerprints. With the Jeffreys' multilocus probes or a well-constructed cocktail of single-locus probes, the chance of two people having the same DNA fingerprint is in the range of 1 in 3–30 billion.

Rapid protocols, not requiring the use of radioactivity, have been developed to amplify polymorphisms for DNA fingerprinting and disease diagnosis. A small sample of blood or other cells is lysed by boiling, the cell debris removed by centrifugation, and the PCR reagents are added directly to the crude cell extract. Following the appropriate number of synthesis cycles, the amplified DNA is run on an agarose gel, stained with ethidium bromide, and visualized directly under ultraviolet light.

PCR is the most sensitive biological assay for DNA yet devised and can, in theory, detect the presence of even a single DNA molecule present in the initial reaction mixture. Thus, there is always the risk of co-amplifying a contaminating DNA molecule in the mixture. Stringent precautions must be taken to ensure that stray DNA molecules are not introduced into the assay.

Alec Jeffreys, 1989
(Courtesy of A. Jeffreys, University of Leicester.)

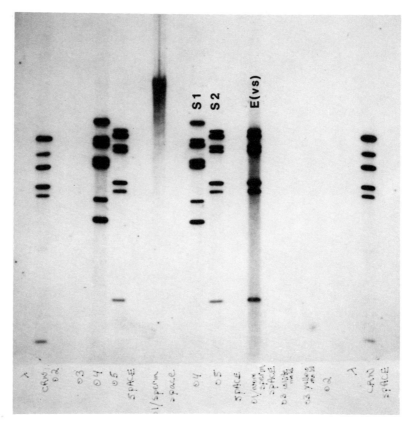

Use of Single-locus Probes in a Criminal Case, 1988
This Southern blot is from a case in which two friends, Randall Jones (S2) and Chris Reesh (S1), were accused in the rape-murder of a Florida woman. A cocktail of single-locus probes showed an exact match between DNA fingerprint of semen obtained from the victim, E(vs), and DNA fingerprint from Jones' blood sample. Jones received the death penalty—the first time this sentence was handed down on the strength of DNA fingerprint evidence. Reesh received an 8-year sentence as accessory. (Courtesy of Cellmark Diagnostics.)

For Further Reading

Erlich, Henry A., ed. 1989. *PCR technology*. Stockton Press, New York.

Fuhrmann, Walter and Friedrich Vogel. 1983. *Genetic counseling*. Springer-Verlag, New York.

Lawn, Richard M. and Gordon A. Vehar. 1986. The molecular genetics of hemophilia. *Sci. Am.* 254/3: 48–65.

Muench, Kary H. 1988. *Genetic medicine*. Elsevier, New York.

Mullis, Kary B. 1990. The unusual origin of the polymerase chain reaction. *Sci. Am.* 262/4: 56–65.

Murray, Andrew W. and Jack W. Szostak. 1987. Artificial chromosomes. *Sci. Am.* 257/5: 62–87.

National Institute of General Medical Sciences. 1984. *The new human genetics: How gene splicing helps researchers fight inherited disease*. National Institutes of Health, Washington, D.C.

National Research Council. 1988. *Mapping and sequencing the human genome*. National Academy Press, Washington, D.C.

Patterson, David. 1987. The causes of Down's Syndrome. *Sci. Am.* 257/2: 52–61.

White, Ray and Jean-Marc Lalouel. 1988. Chromosome mapping with DNA markers. *Sci. Am.* 258/2: 40–49.

Egyptian Hieroglyph of Wheat Harvest

CHAPTER 8
Applying DNA Science in Agriculture, Medicine, and Industry

Since earliest times, humans have nurtured plants and animals that are good to eat or useful for work and clothing. The domestication of plants and animals is perhaps the single greatest civilizing factor in human history. Increased production and performance of domesticated organisms made possible urbanization and task specialization in human society. Thus, the labor of fewer and fewer farmers produced enough food and clothing materials to satisfy the needs of growing numbers of nonfarmers—artisans, engineers, and merchants—freeing them to develop other elements of culture.

Biotechnology is a sophisticated extension of this age-old penchant to adapt living things to serve our own purposes. The advances in agriculture, medicine, and industry discussed in this chapter offer a glimpse of our latest attempts at organismic manipulation. Since these manipulations are targeted at specific DNA and protein interactions within living cells, the topic of this chapter can be thought of as "molecular domestication."

The Green Revolution

Improvements in domestic plants and animals accumulated slowly over centuries. Selective breeding—the enforced mating of parents with superior characteristics—played an important role in the development of agricultural animals. However, too little was known of plant reproduction for selective breeding to play a role in the development of the staple cereal crops: corn, wheat, barley, rice, and millet. These plants were "open-pollinated" in the field by wind and insects. The major genetic improvement in plants resulted from the farmer's practice of planting the largest seeds from the previous year's crop, thus producing larger fruits and seeds. Until the second quarter of this century, virtually all corn and wheat plants produced in the United States were open-pollinated.

In 1908, George Shull demonstrated the phenomenon of "hybrid vigor," which helped to explain the genetic basis of plant productivity. Working at the Carnegie Station for Experimental Evolution at Cold Spring Harbor, New York, Shull self-pollinated plants for several generations to produce pure-breeding lines that showed uniform kernel characteristics. However, these highly inbred plants were less productive and so weakened that they were difficult to propagate. Fortunately, Shull

Shull's Cornfield, 1905

George Shull, about 1935

Donald Jones, about 1910

Fathers of the Green Revolution
In this corn field at Cold Spring Harbor, George Shull crossed inbred corn plants to produce the first high-yielding hybrids. Donald Jones further increased productivity with a "double cross" between two hybrids. (Courtesy of Cold Spring Harbor Laboratory Archives; (*Bottom right*) Connecticut Agricultural Experiment Station.)

found that crossing two inbred lines produced vigorous offspring that displayed the uniform characteristics of the parents, but had yields 20% greater than open-pollinated varieties. Donald Jones, at the Connecticut Agricultural Experiment Station, subsequently showed that grain productivity and vigor is further increased by a "double cross" between two hybrids, each of which is the product of two inbred lines.

The work of Shull and Jones laid the scientific base of modern agricultural genetics. The development of new plant hybrids specifically adapted to regional growing conditions, coupled with increased use of machinery and petrochemical fertilizers, pesticides, and herbicides, led to the so-called "Green Revolution." The past several decades have seen a marked increase in crop productivity worldwide. For example, development of a high-yielding dwarf rice plant so dramatically helped the nutritional status of millions of people in Southeast Asia that it has been called the "miracle rice."

Why Plant Molecular Biology Lags Behind

Unfortunately, nothing is known about the genetic or molecular basis of the phenomenon of hybrid vigor. Moreover, molecular analysis of plant cells has lagged behind investigation of prokaryotic and animal cells. Although this may partly be due to research priorities and allocation of federal funds, several features of major crop plants complicate their genetic analysis:

1. Plants grow slowly and have relatively long generation times.
2. Plants have large genomes, often including large numbers of polyploid chromosomes. The presence of four or more copies of a particular chromosome confuses genetic analysis and makes recessive mutations difficult to observe. For example, at 15 billion base pairs, the haploid corn genome is more than five times larger than the human genome.
3. Plant cells are surrounded by a "wooden box." The cellulose outer wall makes plant cells highly resistant to the introduction of foreign DNA.

The last characteristic is most significant from the standpoint of molecular genetic analysis. As detailed throughout this book, the analysis of gene structure and function has benefited enormously from a variety of techniques used to insert cloned genes into cultured bacterial and mammalian cells. However, the plant cell wall has proved a real barrier to transformation. A reliable transformation system for certain dicot plants was only developed in the early 1980s. None yet exists for monocot plants, which constitute all of the world's cereal crops.

Ti Plasmid Transformation

As did animal cell biologists before them, plant biologists borrowed their transformation system from a tumor-causing microbe—in this case, a bacterium. *Agrobacterium tumefaciens* causes crown gall disease, a tumorous growth at the growing tip, or crown, of leguminous dicot plants (including beans, peas, and clover).

Agrobacterium gains entry through wounds or abrasions in the plant epidermis. A DNA sequence from the invading bacterium integrates into the host-cell DNA, where it induces proliferation of cells and diverts their metabolic machinery to produce a specific food molecule required for bacterial energy and growth. The food substance produced by the transformed cells is called an opine, a derivative of the amino acid arginine.

The tumor-causing functions are carried on a giant tumor-inducing (Ti) plasmid that is separate from the main bacterial chromosome. A 30,000-nucleotide region of the Ti plasmid, the T-DNA, carries specific genes needed for transformation and opine synthesis. During infection, the T-DNA excises from the Ti plasmid and moves into the host-cell nucleus, where it integrates into the chromosome. *vir* genes on the Ti plasmid encode enzymes that are needed for T-DNA excision and integration.

To make a Ti-plasmid vector, the transforming region of the T-DNA and the coding region of the opine synthetase gene are deleted. The foreign gene of interest is then linked directly to the synthetase promoter, which ensures high-level expression. The engineered Ti plasmid is typically used to transform cultured plant cells growing in an undifferentiated "callus" or simple leaf disks, whose cut edges provide entry sites for *Agrobacterium*. In either case, plantlets can be regenerated that have incorporated the foreign gene.

T-DNA transformation works only for plants capable of infection by *Agrobacterium*, but there is hope that the bacterium may be engineered to increase its host range to include important monocot plants. Other transformation systems that have been successful with one or more plant species include (1) microinjection; (2) electroporation, which utilizes surges of high voltage; and (3) particle bombardment, in which microscopic DNA-coated pellets are shot through the cell wall.

Cloning a Gene Using Ti Plasmid Transformation

(*1*) Insert cloned gene into T DNA region of Ti plasmid. (*2*) Infect leaf discs with *Agrobacterium* containing recombinant Ti plasmid and regenerate plantlets in tissue culture. (*3*) T DNA with insert gene is integrated into host chromosome.

Microprojectile Bombardment with a "Gene Shotgun"
A DNA preparation carrying a cloned gene is coated onto tungsten microprojectiles and loaded into a plastic case (foreground), which is then positioned in the gene gun (at rear) in the path of a .22-caliber blank cartridge. When detonated by an electronic firing pin, the gunpowder propels the microprojectiles down the gun barrel toward the target. (Courtesy of the Agricultural Research Service, USDA.)

Viral Resistance

Over the years, agriculturists have inoculated plants with a mild strain of virus or viroid in an attempt to induce resistance to more virulent strains. This sort of "immunization," called cross-protection, has reduced yield loss from viral infection in a number of crops, notably potatoes, tomatoes, and citrus. Although the mechanism of cross-protection is unknown, infection by the inoculated virus suppresses or delays onset of symptoms caused by the second (superinfecting) virus.

In 1986, a research team led by Roger Beachy, at Washington University, used *Agrobacterium* transformation to induce cross-protection against tobacco mosaic virus (TMV). The gene encoding the TMV coat protein was fused to the cauliflower mosaic virus promoter, and the fusion gene was inserted into a Ti plasmid. Transgenic tobacco plants created by infection with the recombinant Ti plasmid expressed TMV coat protein. The coat protein gene was stably integrated into the genome of the transformed plants and was inherited in a Mendelian fashion by their progeny. The transgenic offspring showed delayed onset of symptoms following inoculation with live TMV, and up to 60% of transgenic plants showed no symptoms at all.

Coat-protein-mediated protection reduced infection frequency, as well as local and systemic spread of infection. Subsequent experiments showed that the TMV coat protein gene also protects tomatoes and potatoes from TMV infection. Most surprising, the "vaccine" is multivalent—it also confers resistance to tomato mosaic virus.

Pesticide Resistance

Crop performance is based on such a large number of genes and interlocking metabolic pathways that it seems unlikely that a single molecular genetic basis of hybrid vigor will be uncovered. Therefore, scientists are concentrating on traits that may be governed by single genes. Initial success with gene-transfer experiments in plants has come in the areas of pesticide and herbicide resistance. Intensive work is also under way to identify genes for salt and drought tolerance.

The endotoxin gene from *Bacillus thuringiensis* (*Bt*) has been the transgene of choice in a number of pesticide resistance experiments. The *Bt* toxin, believed to be an environmentally safe insecticide, is active against a number of caterpillars, including the tobacco hornworm and gypsy moth. The strategy has been to link the toxin gene to a constitutive promoter that will express the toxin in all plant tissues. T-DNA transformation has been used to move the gene into tobacco and tomato plants, where it appears to express strongly enough to kill a large percentage of caterpillars. The *Bt* gene is inherited in a Mendelian fashion, indicating that it has been stably integrated into the plant genome.

An indirect approach to pest management bypasses the problem of plant transformation altogether. This involves inserting a toxin gene into the genome of a leaf- or root-colonizing bacterium, which synthesizes and secretes the pesticide in situ on the leaf surface. For example, *Bt* endotoxin and a biotoxin for root cutworm have been inserted into strains of *Pseudomonas fluorescens*.

Herbicide Resistance

Herbicides typically disable target enzymes in metabolic pathways unique to plants—such as those involved in photosynthesis or biosynthesis of essential amino acids. Glyphosate, the most widely used nonselective herbicide, inhibits 5-enolpyruvylshikimate-3-phosphate (EPSP) synthase, an enzyme necessary for the biosynthesis of aromatic amino acids in the chloroplast. Glyphosate and other broad-spectrum herbicides are often toxic to crop plants, as well as to the weeds they are intended to kill. Therefore, another major thrust is to identify and transfer herbicide resistance genes into major crop plants. The following three strategies are used for attaining herbicide resistance:

1. Stimulate overexpression of the target protein of the herbicide, so that enough of it will escape disablement by the herbicide.
2. Insert a genetically altered form of the target protein that is less sensitive to the herbicide.
3. Insert an enzyme that disables the herbicide.

The first and second approaches have been used to produce plants that are resistant to glyphosate. A resistant petunia, in which the EPSP gene is highly amplified, was isolated by growing cultured cells in increasing

Glyphosate-resistant Soybean Plant
These soybean plants have been treated with the herbicide glyphosate (Roundup). Plant at right was transformed by *Agrobacterium* with an inserted glyphosate resistance gene. (Courtesy of Monsanto Agricultural Company.)

levels of the herbicide. Petunias transformed with a fusion gene linking the viral promoter to a wild-type EPSP gene also show resistance. A mutant EPSP gene (*aroA*) isolated from *Escherichia coli* is 6000 times less sensitive to glyphosate than to the wild-type enzyme, but it lacks the plant "transit peptide" that is needed to transport EPSP into the chloroplast. To overcome this problem, a fusion gene was constructed, linking the *aroA* coding region with a dicot transit peptide sequence. Petunias and tobacco plants transformed with this fusion gene express the *aroA* protein in the chloroplast, in addition to their own EPSP.

The third approach to glyphosate resistance has yet to be achieved. However, metabolic pathways have been defined in species of *Pseudomonas* and other bacteria that allow the use of glyphosate as a sole phosphate source. Efforts are under way to clone the genes that encode these glyphosate-degrading enzymes.

Improving Plant Nutritional Value

Cereals are the staple food and major source of protein for a large percentage of the earth's population. Protein comprises on the order of 10% of the dry weight of grain. However, grains lack one or more essential amino acid and thus offer only incomplete nutrition. Therefore, the effort to engineer missing amino acids into cereal protein will potentially have the greatest impact on worldwide nutrition of any agricultural improvement.

Much attention has been focused on storage proteins, which are designed to provide nourishment to the embryo in the germinating seed. Storage proteins, such as zein in corn, constitute approximately half the total protein content of the grain seed. Because zein lacks one amino acid—lysine—we, in theory, need only to employ site-directed mutagenesis to insert nucleotides containing the lysine codon into the protein-coding region of the zein gene.

However, this is not an easy task. First, the additional codon must be inserted in such a way that the reading frame of the zein gene and construction of the protein are not disrupted. Second, the altered zein gene must be introduced into the corn genome and transgenic plants recovered. The lack of a reproducible means for transforming and regenerating corn plants makes this a serious obstacle. Finally, accurate tissue-specific expression of the transgene must be achieved.

Antisense RNA

An unusual approach to the control of gene expression has scored preliminary success in changing an important commercial trait of tomatoes. Fresh tomatoes must be shipped while still green, because ripe fruit is too soft and bruises easily. The enzyme polygalacturonase, which breaks down plant cell walls, is primarily responsible for fruit softening. Two research groups have found that transformation of an "antisense" copy of the galacturonase gene in tomato plants decreases expression of the softening protein by 90%.

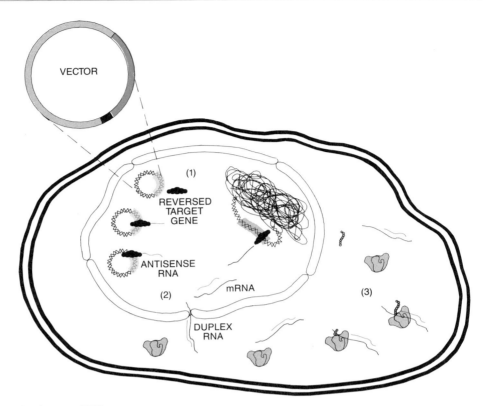

Antisense RNA
(1) Target gene on plant chromosome expresses normal mRNA chromosome, and plasmid vector expresses complementary antisense RNA. (2) Antisense RNA hybridizes to normal mRNA. (3) Translation is halted when ribosome encounters duplex RNA.

To make the antisense gene, a cDNA clone of the polygalacturonase gene was fused, *in reverse orientation*, to a constitutive promoter. This reverse, or antisense, gene was linked to T-DNA and introduced into tomato plants. Each transformed plant thus carried an antisense gene, as well as a normal copy of the polygalacturonase gene. During gene transcription, antisense messenger RNA (mRNA) molecules are produced that are complementary to normal mRNA molecules. Presumably, the antisense mRNAs bind to a proportion of normal RNAs, making them unavailable for translation into protein.

Manipulating Livestock

Much of the increase in livestock production, witnessed over the last quarter century, is attributed to the fact that improved nutrition allows a pure-bred animal to better fulfill its genetic potential. Examples of gene transfer into domestic animals underscore the fact that expression of a foreign gene is unlikely to provide a simple means to improve the productivity of livestock.

Microinjection of fusion genes, introduced in Chapters 5 and 6, provides a means to target expression of genes in particular tissues of the host mammal. In one of the earliest gene-transfer experiments, Richard Palmiter, at the Washington University, and Ralph Brinster, at the University of Pennsylvania, linked the coding sequences of rat growth hormone gene to the mouse metallothionein promoter. This chimeric plasmid was microinjected into one of the two pronuclei of fertilized mouse eggs. The resulting transgenic mice grew faster than normal mice. Examination of tissues from the experimental mice revealed that the metallothionein promoter had directed the expression of growth hormone in the liver, rather than in the pituitary gland, which is the normal site of synthesis. This apparently frees the transgene from normal regulatory control of the hypothalamus, allowing it to produce large amounts of growth hormone. Transgenic mice grow to twice normal size, but they encounter a number of health problems—females are often sterile and most transgenic animals succumb to kidney disease.

Conversion of feed to body weight, not absolute size, is a primary economic consideration in meat production. Thus, in terms of livestock management, big is not necessarily better. For example, mice of equal or larger stature than the transgenic mouse described above have been produced by only 15 generations of selective breeding. However, their rate of food conversion is only 2% improved over mice of normal size. On the other hand, transgenic pigs expressing human growth hormone do not grow any larger, but they produce leaner meat with less back fat and consume 20–30% less feed. Transgenic animals, however, appear to

"Supermouse"
Multiple copies of the rat growth hormone gene were injected into a fertilized mouse egg to produce a supermouse 50% larger than its normal sibling. The rat growth hormone gene is stably integrated into germ cells and is inherited in a Mendelian manner by subsequent generations of offspring. (Courtesy of R.L. Brinster, University of Pennsylvania.)

be susceptible to arthritis, stress, and reproductive failure and they lack motor coordination. Such findings reinforce the realization that biological systems are rarely as simple as they might first appear.

As explained in the next section, biologically active human growth hormone is efficiently manufactured by *E. coli*. Large quantities of a cow growth hormone, bovine somatotrophin (BST), became available in the late 1970s, when its gene was cloned into bacteria. In addition to stimulating growth and regulating carbohydrate and fat metabolism, BST also increases milk production. Although injection of the hormone increases milk production 10–25%, it may increase susceptibility to infection and decrease reproductive success.

Cysteine is necessary for wool growth. Since sheep lack the ability to synthesize this amino acid and since microorganisms in the sheep's gut remove most of the cysteine that is ingested, a lack of the amino acid is sometimes the limiting factor in wool growth. Researchers in Australia have cloned two bacterial genes coding for enzymes that convert serine to cysteine. Establishing these genes in transgenic sheep could allow the sheep to synthesize their own cysteine, thus ensuring a fuller coat.

Products from Cloned Genes: Using Organisms to Produce Therapeutic Proteins

Advances in our understanding of physiology and cell biochemistry have made it possible to identify the precise protein deficiencies responsible for a variety of metabolic disorders, including diabetes, hemophilia, and dwarfism. Such deficiencies can be corrected by supplying the missing or underproduced proteins—insulin for diabetes, clotting factors VIII and IX for hemophilia, and human growth hormone for dwarfism. However, ensuring adequate, contagion-free supplies of such therapeutic proteins proved difficult in the past. These proteins (often hormones) are produced in very limited quantities in the body and thus are laborious to isolate. For this reason, these therapeutic proteins have been expensive, and in the case of human growth hormone, the number of patients who could be treated was limited by availability.

Clotting factors are fractionated from whole human blood: approximately 8000 pints of blood must be processed to yield enough clotting factor to treat one hemophiliac for 1 year, and 7–10 pounds of pancreas from approximately 70 pigs or 14 cows are needed to purify enough insulin for 1 year's treatment of a single diabetic. The extraction of human growth factor was perhaps most onerous, requiring the pituitary glands from approximately 80 human cadavers to produce enough for 1 year's therapy. The magnitude of the supply problem becomes obvious when one considers that patients suffering from these diseases require long-term treatment lasting a minimum of 8–10 years (for dwarfism).

The risk of virus contamination is a most important consideration in any therapeutic product purified from mammalian cells. Simian virus 40 (SV40), which has proved so important in cancer research, was, in fact, first isolated as a contaminant in polio vaccine produced in monkey cells. Although there is no evidence of illness as a result of SV40-contaminated polio vaccines, supplies of both human growth hormone and clotting factors have at one time or another been infected with life-threatening microbes.

Prior to the identification of human immunodeficiency virus type 1 (HIV-1) and the development of virtually foolproof screening procedures, hemophiliacs had a significant risk of contracting AIDS (acquired immune deficiency syndrome) from transfusion of contaminated clotting factors. Pathogen contamination of human growth hormone was uncovered when a statistically significant number of patients developed symptoms of Creutzfeldt-Jakob disease, an enigmatic affliction that causes degeneration of tissues of the brain. Creutzfeldt-Jakob and two related diseases—kuru in humans and scrapie in sheep—are caused by what has been termed a *proteinaceous infectious particle* (prion) that appears to contain no nucleic acid.

Some therapeutic proteins isolated from animals, including porcine or bovine insulin, differ in amino acid makeup from the human protein they replace. The biological activity of an animal substitute may differ slightly from the native human protein, or it may elicit an immune response. Some diabetics have allergic reactions to porcine or bovine insulin, although this may be due to impurities in the preparations and not necessarily differences in the amino acid sequence.

Producing therapeutic proteins from cloned human genes inside bacterial hosts eliminates the risk of virus contamination and allergic sensitivity. Mammalian viruses cannot reproduce inside *E. coli* and hence could not be co-isolated with the protein from a bacterial culture. The protein harvested from the bacterial culture has been expressed from a human coding region and is identical (or very nearly so) to the native protein. Thus, diabetics sensitive to bovine or porcine insulin do not have an adverse reaction to human insulin of recombinant origin (brand name, Humulin).

HUMAN INSULIN

The production of Humulin illustrates that expression of a human protein in *E. coli* often requires detailed understanding of its biological synthesis and of the biochemical limitations of the bacterial cell. *E. coli* is incapable of processing eukaryotic pre-mRNAs or of performing the several posttranslational modifications needed to produce a biologically active insulin from its protein precursors.

Insulin is a small protein produced in patches of cells (the islets of Langerhans) in the endocrine pancreas. The mature insulin molecule consists of two polypeptide chains—an A chain of 30 amino acids and a B chain of 21 amino acids—held together by disulfide linkages. However, this active insulin results from sequential modifications in two precursor molecules: preproinsulin and proinsulin.

The insulin gene consists of two coding exons separated by a single intron. Following splicing of the pre-mRNA, a functional transcript is translated into a large polypeptide called preproinsulin. The molecule includes a 24-amino-acid signal peptide at its amino terminus, a feature

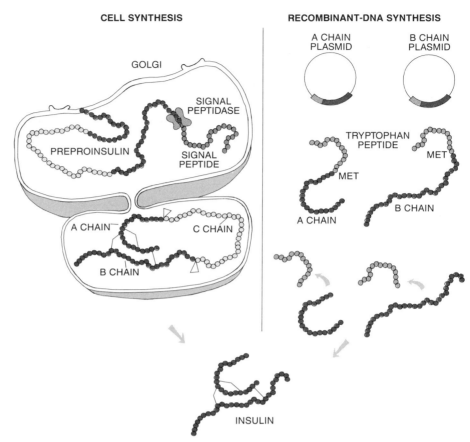

Cellular vs. Recombinant DNA Synthesis of Insulin
In the pancreas, insulin is synthesized as preproinsulin. Within the Golgi apparatus, the signal peptide is removed to yield inactive proinsulin. The C-chain peptide is subsequently removed to produce active insulin; A and B chains are linked by two disulfide bonds. Insulin of recombinant DNA origin (Humulin) is derived from plasmids in which the coding information for the A or B chain is fused to the promoter and first few codons of the *E. coli* tryptophan gene (*trp*). Separate cultures of *E. coli* are transformed with the A or B constructs and produce large amounts of the fusion peptides: either *trp*/A or *trp*/B. The tryptophan sequences are removed by treatment with cyanogen bromide (CNBr), which cleaves at a methionine (MET) residue at the junction of the insulin gene. The A and B chains are mixed together at the junction and disulfide bonds are formed by a chemical process.

of many secreted proteins that is necessary for their proper transport through the cytoplasm. The signal peptide, which is the first part of the preproinsulin molecule produced, anchors the free-floating ribosome to the endoplasmic reticulum (ER) and is subsequently clipped off as the molecule passes through the ER membrane.

The result is a molecule of 84 amino acids called proinsulin, whose looped shape is maintained by cross-linking disulfide bonds. Proinsulin makes its way to the Golgi apparatus, where a converting enzyme removes 33 amino acids from the middle of the connecting loop (the C chain)—leaving the remaining A and B chains held together by the disulfide linkages. This yields active insulin, which is stored in a secretory granule for eventual release into the bloodstream.

The human genomic sequence could not be used directly to produce active insulin in *E. coli*, because the bacterium lacks the enzyme systems needed (1) to splice out the intron sequence to produce a mature mRNA, (2) to remove the signal sequence from preproinsulin, and (3) to remove the C chain from proinsulin. Although others have been employed, the strategy first employed in 1979 by Eli Lilly and Company to produce Humulin neatly side-steps these constraints by simply omitting the above sequences! The nucleotide sequences coding for the A and B chains of active insulin were chemically synthesized and cloned into separate expression plasmids.

A bacterial strain containing each expression vector produces a fusion bacterial/human polypeptide, which is harvested and subsequently treated with cyanogen bromide to remove the bacterial amino acids. Cyanogen bromide cleaves the fusion polypeptide at the methionine residue that begins the human sequence. This treatment, which also alters tryptophan, is only useful because, by happenstance, neither the A nor B insulin chain includes tryptophan or additional methionine residues. Purified A and B chains are then mixed in equal portions and incubated under conditions that form the disulfide linkages.

HUMAN GROWTH HORMONE

Human growth hormone, a polypeptide of 191 amino acids, is also being produced in *E. coli* by use of recombinant DNA techniques. The coding sequence for the first 24 amino acids of the expressed gene was synthesized chemically, whereas amino acids 25 through 191 were derived from a cDNA copy of human growth hormone mRNA isolated from pituitary cells. The recombinant human growth hormone differs by one amino acid from normal human growth hormone due to the fact that *E. coli* is unable to remove the initiator methionine residue that is removed posttranslationally in human cells.

Both insulin and human growth hormone are relatively simple proteins that do not undergo glycosylation or other extensive posttranslational modifications. Extensively modified human proteins cannot as yet be produced in *E. coli*; they must be cloned into eukaryotic expression systems. However, expressing proteins in most eukaryotic cells is much more costly than expressing proteins in bacteria. Whereas a cloned gene can be engineered to produce up to 8% of total protein output in eukaryotic yeast cells, a cloned product may account for up to 40% of protein production of *E. coli*.

TISSUE PLASMINOGEN ACTIVATOR

Tissue plasminogen activator (t-PA) is an example of a therapeutic protein that must be expressed in mammalian cells. Marketed under the name Activase, t-PA is a protease that attacks fibrin, a major protein involved in forming blood clots. Patients that demonstrate early signs of heart attack or stroke are administered t-PA, which acts by destroying small blood clots that can potentially form blockages of arteries. Until the overexpression of cloned t-PA was attained in eukaryotic cells, the major clot destroyer in use was streptokinase, a protease isolated from *Streptococcus*. As a foreign protein, streptokinase may cause immune reactions, hemorrhaging, and other side effects. For these reasons, t-PA was initially hailed as a major improvement in the treatment of heart attack. However, retrospective studies showed little difference in the recovery of patients treated with t-PA versus those treated with streptokinase.

Another eukaryotic expression system is the brewer's yeast, *Saccharomyces cerevisiae*. Yeast fermentation combines the ease of microbial culture with the expression fidelity of mammalian systems. Protein yield is many times greater than that from mammalian cells, and in many cases, an expressed human protein is properly modified to its biologically active form. If the cloned gene is properly engineered, the expressed protein may be secreted into the medium where it can be harvested.

One means of obtaining extracellular secretion of a foreign protein relies on knowledge of the yeast mating-type system (discussed in Chapter 4). Prior to fusing to form a diploid cell, each mating type (**a** and α) releases a specific pheromone that signals readiness to mate. The gene is expressed as a large precursor molecule that is cut to produce four active pheromone molecules, which are then secreted from the cell. A signal peptide in the precursor is essential for secretion of the pheromone.

Secretion of a foreign protein can thus be ensured by linking the protein-coding sequences to the signal peptide sequence of the pheromone precursor. Several therapeutic proteins have been produced in this manner, including interferons (presumed antivirus agents), interleukin-2 (a white blood cell growth hormone), and β-endorphin (a pain-killing neuropeptide).

Recent experiments have shown that it is feasible to use the mammary glands of transgenic animals as factories to produce rare human proteins, such as clotting factor IX, which is used in the treatment of some types of hemophilia. Researchers in Scotland have made a fusion gene composed of the protein-coding region of the factor IX gene and the regulatory region of β-lactoglobulin, a milk protein in sheep. Transgenic animals secrete low levels of human factor IX into their milk and appear perfectly healthy.

Recombinant Vaccines and Antivirus Therapy

Modern vaccination dates to 1798, when Edward Jenner used the nonlethal cowpox virus to induce immunity against structurally similar but deadly variola virus that causes smallpox in humans. The success of the World Health Organization's global campaign to eradicate smallpox was based on a freeze-dried vaccine, developed in 1950, that used another member of the poxvirus family, vaccinia. With the elimination of smallpox by the late 1970s, vaccinia virus is now being engineered as a recombinant vehicle to elicit immunity to a number of viral diseases, including hepatitis B, herpes, influenza, rabies, dengue fever, and AIDS.

During an immune response, clones of B lymphocytes secrete different antibodies that recognize many antigenic features (epitopes) on the surface of an invading virus. In theory, stimulation with only one or several such antigenic subunits should result in immunity to the live virus. Thus, modern vaccine construction depends on cloning the genes for viral surface antigens that elicit a strong immune response. The cloned gene can then be used to prepare several different types of vaccines.

A subunit vaccine of the purified antigen can be prepared by expressing the cloned gene in *E. coli*. Immunization with this type of vaccine only stimulates B lymphocytes to produce antibodies, called a humoral response. Alternatively, one or more antigen genes can be cloned into the genome of a nonlethal carrier virus, such as vaccinia, which then

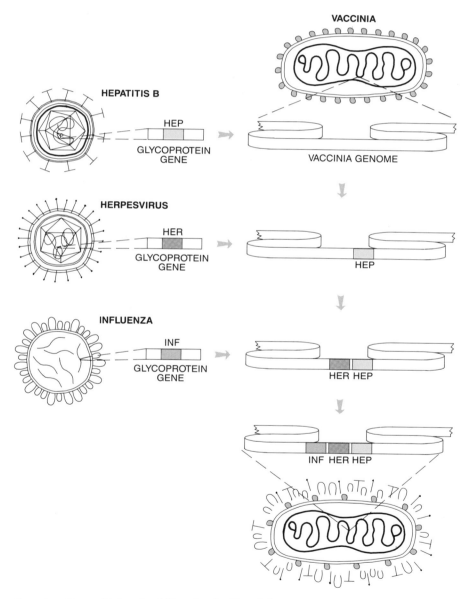

Constructing a Polyvalent Vaccine in Vaccinia
Genes for surface glycoproteins from viral pathogens are cloned into the vaccinia genome. The engineered vaccinia correctly expresses and glycosylates the foreign coat proteins, which are capable of inducing an immune response.

expresses the subunits along with its own genes. Such a live/subunit vaccine gives a stronger and longer-lasting effect because relatively large amounts of the antigen molecule are expressed *within* the cells of the inoculated patient. This type of vaccine elicits both humoral (B-cell-mediated) and cellular (T-cell-mediated) immunity.

Vaccinia is the logical choice for the construction of recombinant virus vaccines. Experience over many years has demonstrated that vaccinia rarely causes serious complications in humans and that it does not form an infectious "reservoir" in animals or humans. Vaccinia can easily accommodate an additional 25,000 nucleotides of sequence and can thus potentially carry antigen genes from several infectious organisms. For example, a *polyvalent* vaccinia vaccine containing genes from influenza, hepatitis B, and herpes simplex has already been used to induce formation of protective antibodies in laboratory animals.

A procedure using homologous recombination has been employed to engineer antigen genes into vaccinia, whose large genome (187,000 nucleotides) is rather difficult to manipulate by standard cloning techniques. First, a plasmid *insertion vector* is constructed in which one or more viral antigen genes are fused between vaccinia sequences and the promoter and initiation sequences of the vaccinia thymidine kinase gene (*tk*). The recombinant insertion vector is amplified in *E. coli* and then transfected into vaccinia-infected mammalian cells.

During virus replication, the insertion vector occasionally recombines with homologous flanking sequences in the vaccinia genome, inserting the antigen genes into the *tk* locus. Because of the reciprocal nature of recombination, insertion of the viral sequences leads to the loss of the vaccinia *tk* gene; recombinant viruses therefore lack thymidine kinase activity. Cells containing recombinant tk^- vaccinia are selected by their ability to grow in the presence of acyclovir, which kills cells infected by *tk*-expressing viruses. The recombination actually produces a safer vaccine; tk^- recombinants are less infectious and less pathogenic than normal vaccinia.

NUCLEOSIDE ANALOGS

While a host of antibiotics have been discovered since World War II, antivirus therapy lags behind. However, a detailed understanding of nucleic acid structure and virus replication has led to the development of a new class of effective antivirus compounds: the nucleoside analogs. These synthetic molecules mimic the structure of the nucleoside building blocks of DNA molecules, and their erroneous incorporation into growing nucleic acid chains selectively disrupts virus replication.

The use of *di*deoxynucleotides in DNA sequencing illustrates the general mechanism through which nucleoside analogs poison DNA replication. Recall that during the chain termination sequencing protocol described in Chapter 4, DNA polymerase binds a *di*deoxynucleotide and incorporates it into an elongating DNA chain. However, no further nucleotides can be added to that chain, because the *di*deoxynucleotide lacks a 3' hydroxyl necessary to form an ester linkage with an adjacent nucleotide.

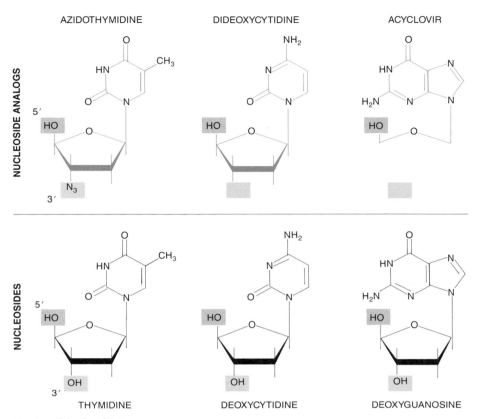

Nucleoside Analogs

Azidothymidine, dideoxycytidine, and acyclovir mimic the structures of corresponding natural nucleosides. Analogs are incorporated into a DNA molecule because each has the 5′ hydroxyl group needed to form a phosphodiester linkage to the previous nucleotide in an elongating DNA chain. However, chain elongation is halted, because the analog lacks a 3′ hydroxyl group needed to form a 3′-5′ linkage with an incoming nucleotide.

Ironically, nucleoside analogs were first synthesized in the 1960s as potential anti-cancer agents to slow the replication of tumor cells. When it was discovered that viruses often induce rapid DNA synthesis in infected cells, nucleoside analogs were tested for antivirus activity. Acyclovir, an analog of guanosine, showed marked activity against herpes simplex virus, and research over a 10-year period at Wellcome Research Laboratories revealed the subtleties of its action.

Three quirks of biochemistry markedly increase acyclovir's ability to disrupt viral DNA replication. First, before acyclovir can be incorporated into a growing DNA chain, three phosphate groups must be added to convert it into acyclovir triphosphate. However, the initial, and key, phosphorylation occurs preferentially in herpesvirus-infected cells. Although the viral thymidine kinase protein efficiently converts acyclovir into acyclovir monophosphate, cellular kinases are very limited in this activity. Second, acyclovir is preferentially incorporated into viral DNA, because the herpesvirus DNA polymerase has a much higher affinity for the drug than does the cellular DNA polymerase. Third, acyclovir binds irreversibly to the viral DNA polymerase, blocking it from further use in replication of viral DNA.

Nucleoside analogs have dramatically improved survival rates for patients with herpes simplex encephalitis, a viral infection of the brain. The survival rate doubled to 50% with the introduction of vidarabine, a deoxyadenosine analog (1977). The survival rate rose to 74% after the introduction of acyclovir in 1985. Acyclovir is also widely prescribed for the treatment of genital herpes and for infections of the varicella-zoster virus (shingles and chickenpox).

Azidothymidine (AZT), a nucleoside analog that also originated from the anti-cancer effort of the 1960s, has rapidly become the primary therapy for AIDS. The drug works by poisoning reverse transcription—HIV is a retrovirus and must make a DNA copy of its RNA genome prior to replication. Within infected cells, AZT is preferentially bound by reverse transcriptase, whose affinity for the drug is 100 times greater than the cellular DNA polymerases. Incorporation of an AZT unit into the growing DNA molecule halts DNA synthesis; the 3′ azido group cannot form an ester linkage with the 5′ carbon of the incoming nucleoside.

Unfortunately, AZT-resistant viruses are found in many patients after as few as 6 months of AZT therapy. Scientists at the Wellcome Research Laboratories found that resistant strains accumulate three or four specific mutations in the coding region of the reverse transcriptase (*pol*) gene that lead to amino acid substitutions in the enzyme. HIVs with these mutations show a 100-fold decrease in sensitivity to AZT. Although the molecular mechanism of resistance is still unknown, polymerase chain reaction (PCR) amplification of these mutations from lymphocytes should provide a rapid test for AZT resistance.

Protein Engineering and X-ray Crystallography

During a chemical reaction, the structures of one or more participating molecules (the reactants) are altered to produce new structures (the products). These molecular transformations are achieved by the making and breaking of bonds. Interaction between two or more molecules entails two basic requirements: proximity and energy. First, interacting molecules must be brought close together so that their reactive regions align. Second, reactants must gain sufficient energy to destabilize existing chemical bonds so that new ones can re-form. When the point of highest energy input is reached, the reactants exist briefly as a high-energy intermediate, in which bonds are partially broken or formed. The energy input needed to reach this "transition state" is termed the activation energy.

The activation energies of many reactions are so high that they will occur spontaneously only at vanishingly slow rates. (At the extreme, this could be on the order of years.) However, the addition of heat and pressure can overcome the proximity and energy barriers. In addition, catalysts often exist that speed up the rate of the reaction by reducing the activation energy.

Catalysts are critical to a number of industrial processes, including the manufacture of pharmaceuticals, fertilizers, pesticides, and textiles. However, most industrial catalysts are inefficient in comparison to biological catalysts (enzymes). For example, the fixation of nitrogen gas (N_2) and hydrogen gas (H_2) into ammonia (NH_3) is the critical step in the manufacture of nitrogen fertilizers. Even in the presence of the catalyst platinum, industrial nitrogen fixation still requires conditions of high temperature and pressure. However, the enzymes of *Rhizobium*, a symbiotic bacterium that colonizes root nodules of leguminous plants, fix nitrogen at ambient pressure and temperature.

The goal of protein engineering is to mimic the functional elegance of biological catalysts—to alter existing enzymes and create entirely new ones. Designer enzymes may incorporate a number of useful attributes, such as broadened substrate range, ability to catalyze several different reactions, increased number of active sites per molecule, and increased pH or temperature tolerance. Key to this endeavor is the means to distinguish the structural features that determine an enzyme's reactivity and specificity.

Unlike DNA, which has only several very similar structural forms, each protein molecule has its own unique three-dimensional pattern of folds, twists, and turns. Of special interest are the pockets or clefts that clasp the substrate molecule(s) and the active sites where catalysis occurs. Protein structure can be defined at four levels of complexity:

- Primary structure describes a linear sequence of amino acids.

- Secondary structure describes local three-dimensional folding, such as the α helix, β-pleated sheet, helix-turn-helix, and leucine zipper.

- Tertiary structure describes the complete three-dimensional folding of a polypeptide.

- Quaternary structure describes the assembly of two or more polypeptide subunits.

Unfortunately, no adequate rules have yet been proposed to allow us to predict tertiary structure from linear amino acid sequence. Furthermore, the general structural properties of active sites have not been determined. In the absence of predictors of overall folding, understanding protein structure-function still relies on data derived from X-ray photographs of protein crystals.

X-ray crystallography begins with the purification of a large enough mass of protein from which crystals can be grown. In the past, this arduous task was accomplished by biochemical means, but large amounts of protein are now obtained relatively easily from cloned genes expressed in cultured cells. Once the protein is obtained in at least 99% pure form, it must be coerced to form crystals large enough for crystallography. This is an empirical process in which saturated solutions of the protein are tested for crystal formation under differing conditions of solvent, salt, pH, and temperature.

The crystal is then mounted in front of an X-ray generator, and the emitted rays are diffracted (bent) as they pass through the crystal lattice.

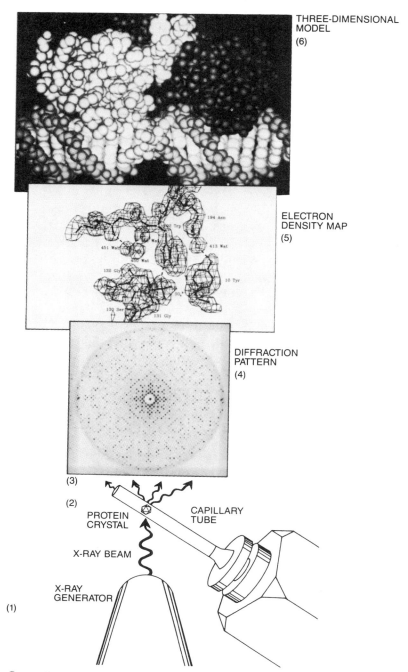

X-ray Crystallography

(*1*) Under appropriate conditions, highly purified protein can be induced to form crystals. (*2*) A suitably large crystal is mounted in a capillary tube, which is oriented in the path of an X-ray beam. (*3*) Some X-rays passing through the crystal are bent (diffracted) as they interact with atoms in the protein molecule. (*4*) Scattered X-rays strike a detector, creating a diffraction pattern that contains information about the location of individual atoms. (*5*) A computer performs the complex calculations that translate the diffraction pattern into an electron density map showing the relative positions of atoms that compose the protein molecule. Rotation of the crystal in three dimensions produces multiple electron density maps, representing different slices through the protein. (*6*) These are superimposed to give a three-dimensional model of the protein.

The resulting diffraction pattern is collected on a sheet of X-ray film or an electronic area detector located behind the crystal. Every atom within the protein molecule and every bond between atoms contributes to the overall diffraction pattern.

A computer is needed to solve the complex mathematical equations, called Fourier transformations, that translate the observed diffraction pattern into an electron density map. Rotation of the crystal in three dimensions produces multiple electron density maps, representing different slices through the molecule. Finally, computer programs are used to superimpose electron density maps to yield a three-dimensional model of the protein.

Site-directed Mutagenesis

Recombinant DNA technology has replaced many of the traditional, and tedious, methods for dealing with proteins. As we have discussed previously, protein sequence is now commonly derived from its encoding DNA sequence, and purified proteins are isolated from genes cloned into expression systems. Antibodies are made against proteins synthesized by cloned genes or by recombinant viruses.

Site-directed mutagenesis is one of the newest in this battery of protein-probing technologies and offers a means to alter protein structure by making precise changes in the sequence of a cloned gene. Earlier mutagenesis experiments relied on X-rays, ultraviolet light, chemicals, or transposons to make random changes in genes. Site-directed mutagenesis was made practical by the advent of computer-controlled DNA synthesizers; the synthetic oligonucleotides so produced are used to introduce specific nucleotide deletions, insertions, or substitutions in the protein-coding region of a gene. In each case, a new protein is produced by the altered gene. "Designer gene" is jargon for a gene that has been re-engineered in vitro.

Typically, each mutation is a single nucleotide substitution that will result in a different amino acid at that position of the encoded protein. To introduce such a point mutation, a single-stranded DNA molecule of 15–30 nucleotides is synthesized. This oligonucleotide is complementary to one of the DNA strands in the targeted region of the gene, except that a single nucleotide is substituted within the synthesized synthetic strand. The oligonucleotide is hybridized to the cloned gene contained in a plasmid, which has been denatured to single strands. Hydrogen bonding between complementary sequences positions the oligonucleotide on the cloned DNA.

DNA polymerase added to the reaction then uses the oligonucleotide as a primer to extend a new second strand. The entire complementary strand need not be synthesized in vitro; following transformation of the hybrid plasmid into *E. coli*, resident bacterial enzymes can complete the synthesis to produce an intact plasmid. The end result is a double-stranded plasmid containing a copy of the mutated gene. However, the site-specific change has only been introduced into one strand of the cloned gene, and the plasmid contains a single nucleotide mismatch.

Prior to or during replication of the plasmid, DNA repair enzymes within *E. coli* replace one of the mismatched nucleotides with the correct complementary one. Although either nucleotide can in theory be replaced, the repair mechanism greatly favors replacing the mutated nucleotide to complement the parental strand. However, in about 5% of transformants, the parental strand is changed to match the site mutation. This frequency is sufficient to recover the mutated gene, which is then used to synthesize the altered protein within *E. coli*.

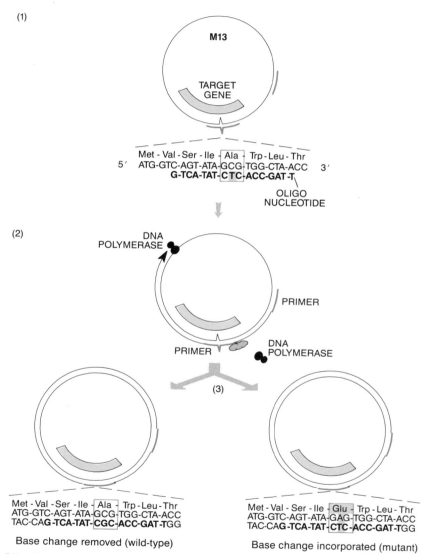

Site-directed Mutagenesis

Site-directed mutagenesis provides a means to introduce a specific nucleotide change in a gene of interest. The target gene to be mutated is cloned into the single-stranded DNA vector, M13. A 17-bp oligonucleotide primer is synthesized that is complementary to a region of the target gene, but contains a single nucleotide change. (*1*) The oligonucleotide is annealed to the M13 vector, creating a single base mismatch. A second primer complementary to a region of the M13 vector is also annealed. (*2*) DNA polymerase extends the second strand from the two primers. (*3*) DNA is transformed into *E. coli*, where second-strand synthesis is completed. The mismatch is repaired by resident *E. coli* enzymes to either eliminate or incorporate the nucleotide change.

Combined with X-ray crystallography, site-directed mutagenesis provides a powerful tool to determine the relationship between protein structure and function. By examining the three-dimensional structure obtained by X-ray crystallography, we can deduce regions of a protein that are potentially important to substrate binding and/or catalysis. Site-directed mutations are then systematically made in the coding sequences for these target regions of the protein. The enzymatic activity of the mutant proteins is compared to that of the native protein. By comparing the three-dimensional structures of the native versus mutant proteins, changes in protein folding that accompany specific amino acid changes can be correlated with altered catalytic function.

Such studies serve two purposes. First, they should reveal the rules that may ultimately lead to our ability to predict catalytic function from polypeptide structure or amino acid sequence. Second, they will eventually allow us to produce as yet unknown enzymes designed to carry out specific functions. For example, imagine making an enzyme that breaks down plastics into harmless byproducts. A super enzyme could theoretically be constructed by making site-directed mutations leading to amino acid substitutions at the active site that *increase* the rate of catalysis, or by cloning multiple active sites into the gene coding a single protein. A polyvalent enzyme might even be produced, similar to the polyvalent vaccines discussed earlier, by incorporating into a single enzyme the active sites that catalyze several different reactions or different steps in a reaction cascade.

Abzymes: Catalytic Antibodies

The term abzyme is a contraction of antibody and enzyme, and refers to an antibody that has been selected for its catalytic activity. In principle, antibodies and enzymes are quite similar. Both exert their effects by binding specific molecules. A specialized pocket or cleft in the antibody or enzyme molecule corresponds precisely with a protuberance on the surface of an antigen or substrate molecule. This "lock-and-key" fit allows for close molecular interaction between the two molecules.

In 1946, Linus Pauling, at the California Institute of Technology, proposed that an enzyme functions by binding most strongly to the transition state intermediate, rather than to the reactants of a chemical reaction. This stabilizes the transition state, thus lowering the energy of activation required to form it.

Richard Lerner and his associates at the Research Institute of Scripps Clinic reasoned that if an antibody could be developed against a transition-state intermediate, it should have the same stabilizing effect of an enzyme. Although naturally occurring enzymes have a limited range of specificities, the immune system can potentially be used to generate designer abzymes of virtually unlimited specificity. The validity of the hypothesis was demonstrated in 1986, when Lerner's group produced the first abzyme—one that catalyzes the hydrolysis of an ester bond.

The major difficulty to be overcome is the fact that the transition-state intermediate is so unstable that it effectively does not exist as a discrete molecule. Therefore, a stable molecule was identified (phosphonate

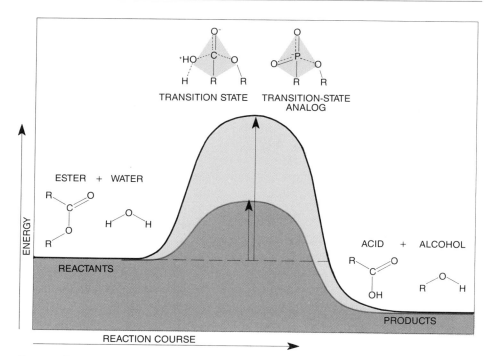

Energy Profile of the Hydrolysis of an Ester
Activation energy is the amount of energy required to drive a reaction to completion—in this case to convert an ester into its component acid and alcohol. The point of highest energy is the transition state, an ephemeral configuration the reactants assume on the way to making products. A transition state analog is a stable molecule that structurally mimics the shape and charge distribution of the transition state. An enzyme or a catalytic antibody lowers activation energy (red line) by stabilizing the transition state. (Adapted from R.A. Lerner and A. Tramontano. 1988. *Sci. Am.* 258/3: 66.)

ester) that mimics the transition state of ester hydrolysis. This transition-state analog was too small a molecule to function as a strong antigen on its own, so it was linked (conjugated) to a carrier protein. The conjugate was then injected into laboratory mice. Spleen cells from the immunized mice were fused to myeloma cells to produce clones of antibody-secreting cells (hybridomas).

Antibodies from isolated clones were tested for their ability to bind to the transition-state analog. Some of the antibodies that bound the transition-state analog accelerated the rate of ester hydrolysis 1000-fold over the uncatalyzed reaction. Furthermore, the addition of the transition-state analog to the reaction inhibited the catalytic action of the antibody, presumably by competing for binding to the active site. This competition by an analog also mimics the behavior of a natural enzyme.

Using this successful strategy, Lerner's group subsequently developed an abzyme that accelerates 250,000-fold the hydrolysis of an amide bond that joins a nitrogen atom to a carbon atom. The amide bond is extremely stable and requires approximately 10,000 times as much energy to break as an ester linkage. Thus, it is especially significant that the abzyme achieved a rate of catalysis within the realm of naturally occurring enzymes.

Amide linkages are of great biological importance in that they join amino acids to form polypeptide chains. This experiment was the first step toward developing abzymes that can cleave protein at specific amino acid positions. Since few naturally occurring proteases display such specificity, abzymes may become key tools for protein engineering.

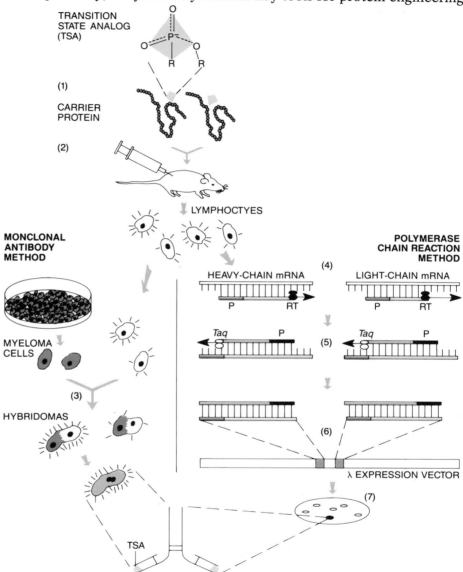

Monoclonal Antibody vs. PCR Methods for Producing Catalytic Antibodies to a Transition-state Analog

(*1*) Transition-state analog (TSA) is linked to a carrier protein. (*2*) Mouse is injected with the protein/TSA. Lymphocytes in mouse's spleen produce antibodies against the protein and TSA. In the monoclonal antibody method, (*3*) mouse lymphocytes are fused with cultured myeloma cells, and the resulting hybridomas are tested for catalytic activity. In the PCR method, (*4*) total mRNA is isolated from mouse lymphocytes. Primers (P) hybridize adjacent to variable regions of both heavy- and light-chain antibody mRNAs, and reverse transcriptase (RT) extends from primers to make cDNA copies of the variable region sequences. (*5*) *Taq* polymerase extends from a second primer to produce the second DNA strand. (*6*) PCR-amplified variable regions are ligated into a λ vector, which is designed to express the antibody-coding sequences in *E. coli*. (*7*) The λ library is screened for clones expressing antibodies that bind the transition-state analog.

The search for new abzymes has been slowed by the relative difficulty of creating and screening sufficiently large numbers of monoclonal antibodies for catalytic activity. In 1989, Lerner's group developed a method that makes it possible to screen rapidly a mouse's entire antibody "repertoire" and which may supersede hybridoma technology as a general means for the production of pure antibody preparations. This method relies on screening an *E. coli* expression library of "antigen-binding fragments" (Fab fragments), antibody fragments that retain the ability to recognize an antigen. Fab sequences were PCR-amplified from the mRNA of mice immunized with nitrophenol phosphonaminate (NPN), which is effective in generating catalytic antibodies. The amplified sequences were then ligated to λ vectors that express heavy and light chains of the cloned Fab fragments. The recombinant vectors were then used to infect *E. coli* cells, and the resulting plaques were screened for the ability to bind NPN. One hundred clones expressing putative catalytic antibodies were identified out of 1 million clones—some 1000 times more clones than can be effectively screened using hybridoma technology.

Another group at the Scripps Clinic demonstrated that transgenic plants can also express functional antibodies. mRNA from a mouse hybridoma expressing a catalytic antibody was used to construct complementary DNAs for κ and γ immunoglobulin chains. The cDNAs were ligated into a plant expression vector and transformed into tobacco plants using *Agrobacterium*. Plants expressing either κ or γ chains were crossed to produce F_1 offspring that expressed both chains simultaneously and accumulated antibody totaling 1.3% of leaf protein. Catalytic processing by plant abzymes may provide a means to introduce new catalytic properties into plant cells, such as detoxification of environmental pollutants.

Ribozymes: Catalytic RNA

The independent discovery of catalytic RNA (1983) by Thomas Cech, at the University of Colorado, and Sid Altman, at Yale University, destroyed the long-held dogma that only proteins are endowed with enzymatic activity. Altman's work on tRNA processing in *E. coli* led to the discovery of the enzyme RNase P, which removes the extra 5'-end sequences from pre-tRNA. He found that the enzyme has both a protein and an RNA component; the RNA component alone is sufficient to accurately and efficiently cleave pre-tRNA. Cech showed that ribosomal RNA precursors from the protozoan *Tetrahymena* catalyze elegant self-splicing reactions that cleave and rejoin RNA sequences at specific sites. The self-splicing reaction is an intramolecular event, since the cleavage site and region responsible for catalysis lie within the same molecule.

Since these initial discoveries, self-splicing RNA systems have been identified in several organisms, including two types of small infectious RNA particles: viroids and satellite RNAs. Among this enigmatic group of plant pathogens are the avocado sunblotch viroid and tobacco ringspot virus. However, it is not clear whether it is appropriate to call these pathogens "organisms," since they are naked RNA molecules that appear not to code for any proteins.

Autocatalytic cleavage has an essential role in viroid replication. The prevailing hypothesis is that new viroids are produced by a "rolling circle" mechanism in which a polymerase produces a concatemer of many linked copies of viroid RNA. Autocatalytic cleavage then splits the multiple copies into RNA monomers.

Recent experiments with the satellite RNA of tobacco ringspot virus defined the minimum sequences necessary for RNA enzymatic activity and showed that designer RNA enzymes can be made with new cutting specificities. It was demonstrated that the catalytic unit, referred to as the ribozyme, is confined to a 41-nucleotide region and that cleavage always occurs at a position 3' of a GUC target sequence. Flanking the catalytic site of the ribozyme are "positioning sequences" that base-pair to complementary sequences on either side of the target GUC site. This aligns the ribozyme catalytic site with the target cleavage site.

It was reasoned that by changing the positioning sequences, the specificity of the ribozyme could be altered. Using site-directed mutagenesis, the positioning sequences of the tobacco ringspot virus ribozyme were altered to make them complementary to the flanking RNA sequences of three specific GUC sites in the RNA that codes for chloramphenicol acetyltransferase (CAT). Incubation with any of the three new ribozymes leads to specific and efficient cleavage of the CAT RNA at the predicted GUC sites. Ribozymes with several cutting specificities first became commercially available in 1989.

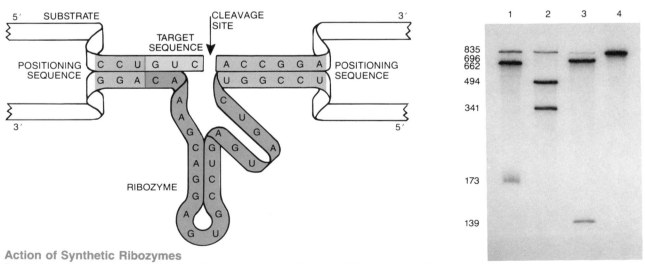

Action of Synthetic Ribozymes
Positioning sequences hybridize the ribozyme to the substrate RNA at a site adjacent to the GUC target sequence. The autoradiogram shows cleaveage of the 835-nucleotide transcript of the chloramphenicol acetyltransferase (CAT) gene by three ribozymes, each constructed with different positioning sequences. Lane *1* shows a single band of uncut mRNA (835 nucleotides). Lanes *2–4* show that each ribozyme cuts the mRNA at a single location, producing two fragments totaling 835 nucleotides. (Adapted from J. Haseloff and W.L. Gerlach. 1988. *Nature* 334: 587. Courtesy of J. Haseloff and W.L. Gerlach.)

For Further Reading

Antebi, Elizabeth and David Fishlock. 1986. *Biotechnology: Strategies for life*. The Cambridge: MIT Press, Cambridge, Massachusetts.

Cech, Thomas R. 1986. RNA as an enzyme. *Sci. Am. 255/5:* 64–75.

Chilton, Mary-Dell. 1983. A vector for introducing new genes into plants. *Sci. Am. 248/6:* 51–59.

Hall, Stephen S. 1987. *Invisible frontiers: The race to synthesize a human gene*. Tempus Books, Redmond, Washington.

Hirsch, Martin S. and Joan C. Kaplan. 1987. Antiviral therapy. *Sci. Am. 256/4:* 76–85.

Hogle, James M., Marie Chow, and David J. Filman. 1987. The structure of poliovirus. *Sci. Am. 256/3:* 42–49.

Olson, Steve. 1986. *Biotechnology: An industry comes of age*. National Academy Press, Washington, D.C.

Pestka, Sidney. 1983. The purification and manufacture of human interferons. *Sci. Am. 249/2:* 36–43.

Tramontano, Alfonzo and Richard Lerner. 1988. Catalytic antibodies. *Sci. Am. 258/3:* 58.

Wambaugh, Joseph. 1989. *The blooding*. William Morrow and Co., New York.

Weintraub, Harold M. 1990. Antisense RNA and DNA. *Sci. Am. 262/1:* 40–46.

What Science Knows About AIDS. (Special issue.) 1988. *Sci. Am. 259/4*.

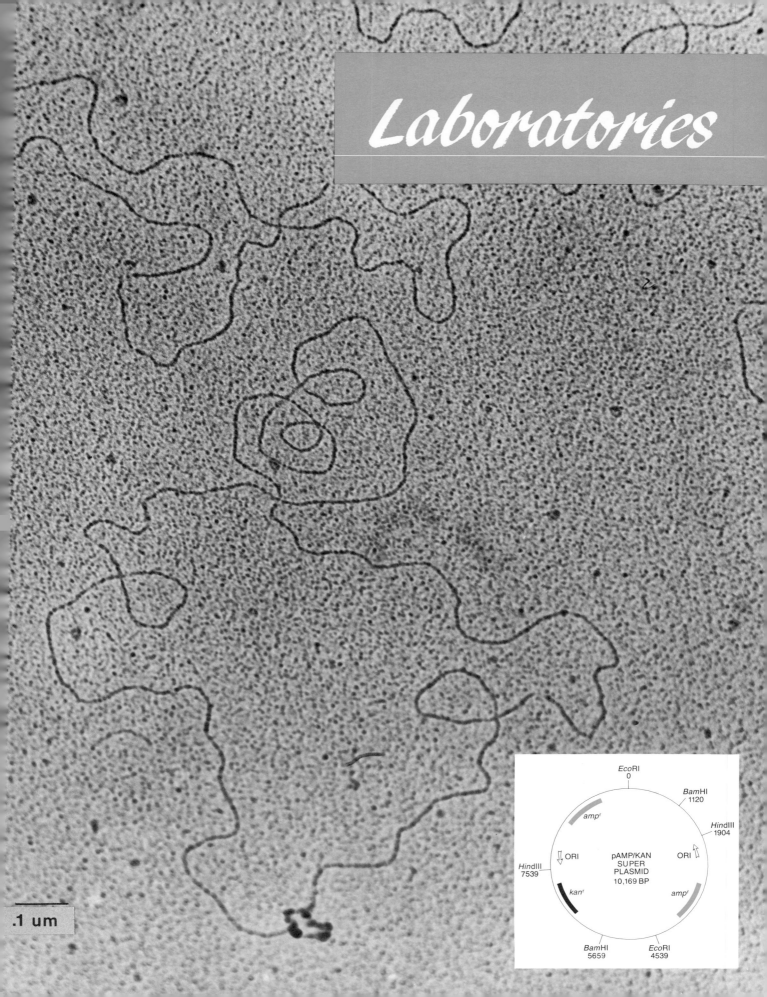

At top is a 10,169-bp pAMP/KAN "superplasmid" constructed from four *Bam*HI/*Hin*dIII fragments and having two ampicillin resistance genes, one kanamycin resistance gene, and two origins (magnification, 126,700×). At bottom is a 20,338-bp dimer composed of two units of the superplasmid linked by head-to-tail recombination. (*Inset*) Restriction map of the monomer. (Constructed, photographed, and restriction mapped by A. Bauman as part of an undergraduate research project in the laboratory of J. Perreault, Lawrence University.)

Lab Safety and Adherence to National Institutes of Health (NIH) Guidelines

Safety Precautions

> **WARNING**
>
> Individuals should use this manual only in accordance with prudent laboratory safety precautions and under the supervision of a person familiar with such precautions. Use of this manual by unsupervised or improperly supervised individuals could result in serious injury.

Instructors and students are urged to pay particular attention to specific cautions placed throughout this manual and to follow carefully the instructions contained in these caution boxes.

LABORATORY WASTE DISPOSAL

Alternative procedures for the handling and disposal of laboratory wastes can and should be used instead of those suggested in this manual when local requirements, conditions, or practices dictate. *Instructors should be familiar with and follow all national, state, local, or institutional regulations or practices pertaining to the use and disposal of materials utilized in this manual.*

WORKING WITH *ESCHERICHIA COLI* AND ANTIBIOTIC RESISTANCE GENES

E. coli is part of the normal bacterial flora of the large intestines, and it is rarely associated with severe illness in otherwise healthy individuals. However, one might be concerned about the consequences of accidentally ingesting some *E. coli*, especially those that have been transformed with plasmids. Research has shown that the *E. coli* strain MM294 used in this laboratory sequence is unlikely to grow inside the human intestines and is incapable of transferring plasmid DNA to *E. coli* living there.

MM294 and other *E. coli* strains commonly used in research today are derivatives of a wild-type *E. coli* strain K-12, which was originally isolated from the feces of a diphtheria patient at Stanford University in 1922. The work of Edward Tatum in the 1940s popularized its use in biochemical and genetic studies. Research in the 1970s showed that strain K-12 cannot effectively colonize the human gut, probably because of genetic changes accumulated during decades of in vitro culture (Bachman 1987). *The Prelab Notes of Laboratory 2 contain detailed instructions for the responsible handling and disposal of* E. coli *which should be followed throughout this laboratory course.*

Plasmid DNA can be transferred from one *E. coli* cell to another during conjugation. Transport requires a mobility protein (encoded by a plasmid-borne *mob* gene) and a specific site (*nic*) on the plasmid. The *mob* protein nicks the plasmid at *nic* and then attaches to the nicked strand to conduct the plasmid through a mating channel into a recipient cell. Most newer plasmids, including pAMP and pKAN, lack both the *mob* gene and the *nic* site and cannot be mobilized for transport (Lauria and Suit 1987; Sambrook et al. 1989).

USE OF ETHIDIUM BROMIDE

Ethidium bromide is the most rapid and sensitive means to stain DNA. However, it is a mutagen by the Ames microsome assay and a suspected carcinogen. For this reason, instructors may elect (or be required) to use methylene blue—a safe but less-sensitive stain that can be used successfully for all staining procedures in this manual. We have provided directions for both staining methods, and the pros and cons of each are discussed in the Prelab Notes of Laboratory 3. *These Notes also contain instructions for the responsible handling and decontamination of ethidium bromide solutions, which should be reviewed prior to performing any staining procedure using ethidium bromide.*

The greatest risk is to inhale ethidium bromide powder when mixing a 5 mg/ml stock solution, and therefore we strongly recommend purchasing a *pre-mixed* 5mg/ml solution from a supplier. The stock solution should be handled according to the instructions contained in the Material Safety Data Sheet provided by the supplier. The 5 mg/ml stock solution is diluted to make a staining solution with a final concentration of 1 μ/ml according to the procedure described in Appendix 2. With responsible handling, the dilute solution (1 μg/ml) used for gel staining poses minimal risk.

NIH Guidelines

Instructors and students should be aware that the National Institutes of Health oversees all research involving interspecific DNA transfer. Experiments that entail moving a gene or DNA sequence from one species to another are regulated by the NIH *Guidelines for Research Involving Recombinant DNA Molecules.*

Laboratory 6, Recombination of Antibiotic Resistance Genes, has been designed so that it falls below the threshold for NIH regulation. Provision III-D-3 of the Guidelines states: "The following molecules are exempt from these guidelines ... those that consist entirely of DNA from a prokaryotic host, including its indigenious plasmids or viruses when propagated only in the host (or a closely related strain of the same species) or when transferred to another host by well-established physiological means." (National Institutes of Health 1982.) As documented in Appendix 3, III and IV, the vectors pAMP and pKAN conform to this exemption because they are constructed entirely of plasmic and bacteriophage sequences.

All experiments suggested in the For Further Research section at the end of each laboratory also conform to the Guidelines. We encourage all instructors to design student experiments so that they conform to the letter of the NIH Guidelines. Students should not be allowed to attempt experiments involving interspecific mixing of DNA unless they are to be performed within the context of an NIH-approved research program.

REFERENCES

Bachman, Barbara. 1987. Derivations and genotypes of some mutant derivatives of *Escherichia coli* K-12. In Escherichia coli *and* Salmonella typhimurium: *Cellular and molecular biology* (ed. F.C. Neidhardt), vol. 2, p. 1190. American Society for Microbiology, Washington, D.C.

Luria, Salvador E., and Joan L. Suit. 1987. Colicins and Col plasmids. In Escherichia coli *and* Salmonella typhimurium: *Cellular and molecular biology* (ed. F.C. Neidhardt), vol. 2, pp. 1620–1621. American Society for Microbiology, Washington, D.C.

Sambrook, Joseph, Edward Fritsch, and Thomas Maniatis. 1989. *Molecular cloning: A laboratory manual,* volume I, p. 1.5. Cold Spring Harbor Laboratory Press, New York.

National Institutes of Health. 1982. Guidelines for research involving recombinant DNA molecules. *Federal Register 47/167:* 38051.

Laboratory 1

Measurements, Micropipetting, and Sterile Techniques

This laboratory introduces micropipetting and sterile pipetting techniques used throughout this course. Mastery of these techniques is important for good results in all of the experiments that follow. Most of the laboratories are based on *microchemical* protocols that use very small volumes of DNA and reagents. These require use of an adjustable micropipettor (or microcapillary pipet) that measures as little as one microliter (μl)—a millionth of a liter.

Many experiments require growing *Escherichia coli* in culture medium that provides an ideal environment for other microorganisms as well. Therefore, it is important to maintain sterile conditions to minimize the chance of contaminating an experiment with foreign bacteria or fungi. *Sterile conditions* must be maintained whenever living bacterial cells are to be used in further cultures. Use sterilized materials for everything that comes in contact with a bacterial culture: nutrient media, solutions, pipets, micropipet tips, inoculating and spreading loops, flasks, culture tubes, and plates.

Remember this rule of thumb: Use sterile technique if live bacteria are needed at the end of a manipulation (general culturing and transformations). Sterile technique is not necessary when the bacteria are destroyed by the manipulations in the experiment or when working with solutions for DNA analysis (plasmid isolation, DNA restriction, and DNA ligation).

LABORATORY 1
Measurements, Micropipetting, and Sterile Techniques

I. **Small-volume Micropipettor Exercise**

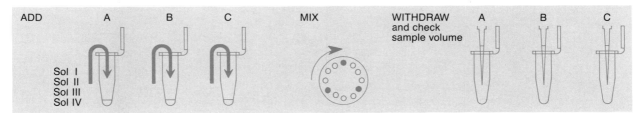

II. **Large-volume Micropipettor Exercise**

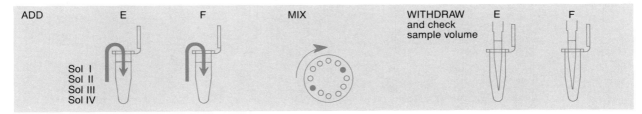

III. **Sterile Use of 10-ml Standard Pipet**

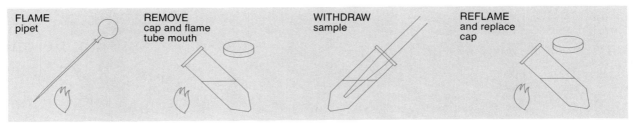

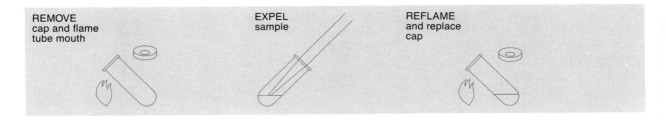

PRELAB NOTES

Digital Micropipettors

The volume range of digital micropipettors varies from manufacturer to manufacturer. Select both a small-volume micropipettor with a range of 0.5–10 µl or 1–20 µl and a large-volume micropipettor with a range of 100–1000 µl.

Microcapillary Pipets

Microcapillary pipets are an inexpensive alternative to adjustable micropipettors. These disposable glass capillary tubes come in sizes that cover the range of volumes used in this course. Several types of inexpensive micropipet aids are available. A thumbscrew micrometer may be easier to use than a pipettor bulb. A no-cost and easily controllable pipet bulb can be made by tying a knot in a length of latex tubing, which is usually provided with the capillary pipets.

Under conditions of high static electricity, capillary pipetting can be very difficult and, at times, impossible. The reagent droplet adheres stubbornly to the side of the pipet and cannot be transferred to the side of a polypropylene reaction tube. Even under the best of circumstances, microcapillary pipets are more difficult to master. Allow sufficient time to become competent with them before attempting any experiments.

Transfer Pipets

Small polypropylene transfer pipets are handy because they have an integrated bulb. The smallest size, which holds a *total* volume of approximately 1 ml, has a thin tip that can be used to measure microliter amounts. Before use, calibrate the transfer pipet using a digital micropipet or microcapillary pipet. Pressing on the pipet barrel, rather than the bulb, creates less air displacement and makes measuring small volumes easier.

10-ml Pipets

Presterilized, disposable 10-ml plastic pipets are most convenient and are supplied in bulk pack or individually wrapped. Bulk-packed pipets should be opened immediately before use. To dispense, cut one corner of the plastic wrapper at the end opposite the pipet tips. Avoid touching and contaminating the wrapper opening; tap bag to push the pipet end through the cut opening. Reclose with tape to keep sterile for future use. To use individually wrapped pipets properly, peel back only enough of the wrapper to expose the wide end of the pipet and affix end into the pipet aid or bulb. Completely peel back the wrapper immediately before use.

To Flame or Not to Flame?

There is general disagreement about whether or not it is necessary to flame pipets and mouths of tubes as part of the sterile technique. Flaming warms the air at the mouth of the container, creating an outward convection current that prevents microorganisms from falling in. Even so, the effect of flaming may be primarily psychological when fresh sterile supplies are used and manipulations are done quickly. Especially when using individually wrapped supplies, flaming can be omitted without compromise to sterility. When flaming plasticware, do so briefly to avoid melting the plastic.

Microfuge

Although not essential, a microfuge is very useful for pooling and mixing droplets of pipetted solutions in the bottom of a 1.5-ml reaction tube.

PRELAB PREPARATION

1. To simplify initial practice with a micropipettor, use colored solutions that are easily visible. Prepare five colored solutions using food coloring or other dyes mixed with water.

2. Prepare for each experiment:

    four 1.5-ml tubes, each containing 1 ml of a different colored solution, marked I, II, III, and IV.

    one 50-ml conical tube containing 25 ml of the colored solution, marked V.

Measurements, Micropipetting, and Sterile Techniques

REAGENTS

1 ml of Solution I, colored
1 ml of Solution II, colored
1 ml of Solution III, colored
1 ml of Solution IV, colored
25 ml of Solution V, colored

SUPPLIES AND EQUIPMENT

0.5–10-µl micropipettor + tips
100–1000-µl micropipettor + tips
10-ml pipet
pipet aid or bulb
50-ml conical tube
15-ml culture tube
1.5-ml tubes
beaker for waste/used tips
Bunsen burner (optional)
microfuge (optional)
permanent marker
test tube rack

Metric Conversions

Become familiar with metric units of measurement and their conversions. We will concentrate on liquid measurements based on the liter, but the same prefixes also apply to dry measurements based on the gram. The two most useful units of liquid measurement in molecular biology are the milliliter (ml) and microliter (µl).

1 ml = 0.001 liter   1,000 ml = 1 liter
1 µl = 0.000001 liter   1,000,000 µl = 1 liter

Complete the following conversions:

1 µl = _____ ml   _____ µl = 1 ml
10 µl = _____ ml
100 µl = _____ ml

Use of Digital Micropipettors   (10 minutes)

"Nevers"

- Never rotate volume adjustor beyond the upper or lower range of the pipet, as stated by manufacturer.

- Never use micropipettor without tip in place; this could ruin the precision piston that measures the volume of fluid.

- Never lay down pipettor with filled tip; fluid could run back into piston.
- Never let plunger snap back after withdrawing or ejecting fluid; this could damage piston.
- Never immerse barrel of pipettor in fluid.
- Never flame micropipettor tip.

Pipetting Directions

1. Rotate volume adjustor to desired setting. Note change in plunger length as volume is changed. Be sure to locate decimal point properly when reading volume setting.

2. *Firmly* seat proper-sized tip on end of micropipettor.

3. When withdrawing or expelling fluid, always hold tube firmly between thumb and forefinger. Hold tube at nearly eye level to observe the change in the fluid level in pipet tip. Do not pipet with tube in test tube rack or have another person hold tube while pipetting.

4. Each tube must be held in the hand during each manipulation. Grasping the tube body, rather than the lid, provides more control and avoids contamination from the hands.

5. Hold pipettor almost vertical when filling.

6. Most digital micropipettors have a two-position plunger with friction "stops." Depressing to the first stop measures the desired volume. Depressing to the second stop introduces an additional volume of air to blow out any solution remaining in the tip. Pay attention to these friction stops, which can be felt with the thumb.

7. To withdraw sample from reagent tube:
    a. Depress plunger *to first stop* and hold in this position. Dip tip into solution to be pipetted, and draw fluid into tip by *gradually* releasing plunger.
    b. Slide pipet tip out along inside wall of reagent tube to dislodge excess droplets adhering to the outside of tip.
    c. Check that there is no air space at the very end of the tip. To avoid future pipetting errors, learn to recognize the approximate level that particular volumes fill the pipet tip.

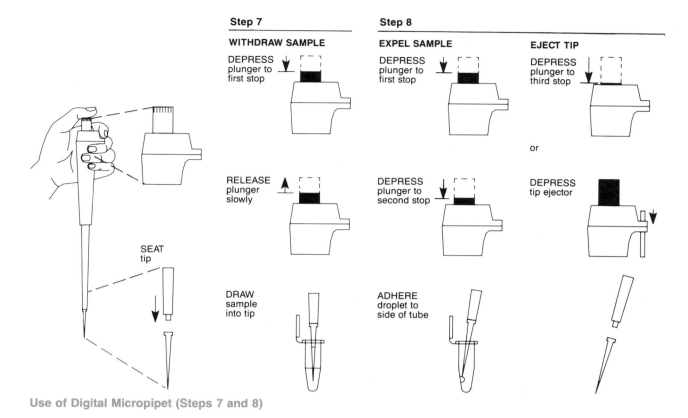

Use of Digital Micropipet (Steps 7 and 8)

8. To expel sample into reaction tube:

    a. Touch pipet tip to inside wall of reaction tube into which the sample will be emptied. This creates a capillary effect that helps draw fluid out of tip.

    b. *Slowly* depress plunger to the first stop to expel sample. Depress to second stop to blow out last bit of fluid. Hold plunger in depressed position.

    c. Slide pipet out of reaction tube with plunger depressed to avoid sucking any liquid back into tip.

    d. Manually remove or eject tip into a beaker kept on the lab bench for this purpose. The tip is ejected by depressing the measurement plunger beyond the second stop or by depressing a separate tip-ejection button.

9. To prevent cross-contamination of reagents:

    a. Always add appropriate amounts of single reagent sequentially to all reaction tubes.

    b. Release each reagent drop onto new location on inside wall, near bottom of reaction tube. In this way, the same tip can be used to pipet the reagent into each reaction tube.

c. Use *fresh tip* for each new reagent to be pipetted.

d. If tip becomes contaminated, switch to a new one.

10. Eject used tips into a beaker kept on the lab bench for this purpose.

I. Small-volume Micropipettor Exercise (15 minutes)

This exercise simulates setting up a reaction, using a micropipettor with a range of 0.5–10 µl or 1–20 µl.

1. Use permanent marker to label three 1.5-ml tubes A, B, and C.

2. Use matrix below as a checklist while adding solutions to each reaction tube.

Tube	Sol. I	Sol. II	Sol. III	Sol. IV
A	4 µl	5 µl	1 µl	—
B	4 µl	5 µl	—	1 µl
C	4 µl	4 µl	1 µl	1 µl

3. Set micropipettor to 4 µl and add Solution I to each reaction tube.

4. Use *fresh tip* to add appropriate volume of Solution II to a clean spot on reaction tubes A, B, and C.

5. Use *fresh tip* to add 1 µl of Solution III to tubes A and C.

6. Use *fresh tip* to add 1 µl of Solution IV to tubes B and C.

7. Close tops. Pool and mix reagents by one of the following methods:

a. Sharply tap tube bottom on bench top. Make sure that the drops have pooled into one drop at the bottom of the tube.

or

b. Place in microfuge and apply a short, several-second pulse. Make sure reaction tubes are placed in a *balanced* configuration in the microfuge rotor. Spinning tubes in an unbalanced position will damage microfuge motor.

An empty 1.5-ml tube can be used to balance a sample with a volume of 20 µl or less.

8. A total of 10 µl of reagents was added to each reaction tube. To check that your measurements were accurate, set pipet to 10 µl and very carefully withdraw solution from each tube.

a. Is the tip just filled?

or

b. Is a small volume of fluid left in tube?

   or

   c. After extracting all fluid, is an air space left in tip end? (The air can be displaced and actual volume determined simply by rotating volume adjustment to push fluid to very end of tip. Then, read volume directly.)

9. If several measurements were inaccurate, repeat exercise to obtain a near-perfect result.

II. Large-volume Micropipettor Exercise (10 minutes)

This exercise simulates a bacterial transformation or plasmid preparation, for which a 100–1000-µl micropipettor is used. It is far easier to mismeasure when using a large-volume micropipettor. If the plunger is not released slowly, an air bubble may form or solution may be drawn into piston.

1. Use a permanent marker to label two 1.5-ml reaction tubes E and F.

2. Use matrix below as a checklist while adding solutions to each reaction tube.

Tube	Sol. I	Sol. II	Sol. III	Sol. IV
E	100 µl	200 µl	150 µl	550 µl
F	150 µl	250 µl	350 µl	250 µl

3. Set micropipettor to add appropriate volumes of Solutions I–IV to tubes E and F. Follow same procedure as for small-volume pipettor.

4. A total of 1000 µl of reactants was added to each tube. To check that your measurements were accurate, set micropipettor to 1000 µl and carefully withdraw solution from each tube.

   a. Is the tip just filled?

   or

   b. Is a small volume of fluid left in tube?

   or

   c. After extracting all fluid, is an air space left in tip end? (The air can be displaced and actual volume determined simply by rotating volume adjustment to push fluid to very end of tip. Then, read volume directly.)

5. If measurements were inaccurate, repeat exercise to obtain a near-perfect result.

III. Sterile Use of 10-ml Standard Pipet (10 minutes)

The following directions include flaming the pipet and tube mouth. It is probably best to learn to flame and then omit flaming when safety or situation dictates. The directions also assume one-person pipetting, which is rather difficult. The process is much easier when working as a team: One person handles the pipet, while the other removes and replaces caps of tubes.

The key to successful sterile technique is to work quickly and efficiently. Before beginning, clear off lab bench and arrange tubes, pipets, and culture medium within easy reach. Locate Bunsen burner in a central position on lab bench to avoid reaching over flame.

Loosen caps so that they are ready for easy removal. Remember, the longer the top is off the tube, the greater the chance of microbe contamination. Do not place sterile cap on nonsterile lab bench.

> **CAUTION**
>
> Always use a pipet aid or bulb to draw solutions up the pipet. Never pipet solutions using mouth suction: This method is not sterile and can be dangerous.

Nonsterile pipets may be used for this practice exercise.

1. Light Bunsen burner.

This expels contaminated air and prepares vacuum to withdraw fluid.

2. Set pipet aid to 5 ml or depress pipet bulb.

When using an individually wrapped pipet, be careful to open wrapper end opposite the pipet tip. Unwrap only enough of pipet to attach end into pipet aid or bulb.

3. Select sterile 10-ml pipet and insert into pipet aid or bulb. *Remember to handle only large end of pipet; avoid touching lower two thirds.*

4. Remove the remaining wrapper, and quickly pass lower two thirds of pipet cylinder through Bunsen flame several times. *Be sure to flame any portion of pipet that will enter sterile container.* Pipet should become warm, but not hot enough to melt plastic pipet or to cause glass pipet to crack when immersed in solution to be pipetted.

5. Hold 50-ml conical tube containing Solution V in free hand and remove cap using little finger of hand holding pipet aid or bulb. *Do not place cap on lab bench.*

6. Quickly pass mouth of conical tube through Bunsen flame several times, being careful not to melt plastic.

7. Withdraw 10 ml of Solution V from conical tube.

8. Reflame mouth of tube and replace top.

9. Remove top of sterile 15-ml culture tube with little finger of hand holding pipet. Quickly flame mouth of tube.

10. Expel fluid into culture tube. Reflame mouth of tube and replace top.

RESULTS AND DISCUSSION

Inaccurate pipetting and improper sterile technique are the chief contributors to poor laboratory results. If you are still uncomfortable with micropipettors and/or sterile technique, take time now for additional practice. These techniques will soon become second nature to you.

1. Why must tubes be balanced in a microfuge rotor?

2. Use rotor diagrams below to show how you would balance 3–11 tubes (for the 12-place rotor) or 3–15 tubes (for the 16-place rotor). In balancing odd numbers of tubes, begin with a balanced triangle of three tubes, then add balanced pairs. Which number of tubes cannot be balanced in the 12- or 16-place rotors?

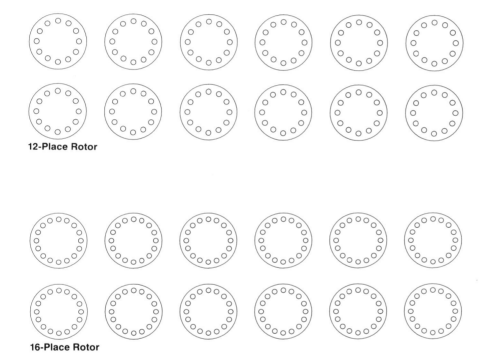

12-Place Rotor

16-Place Rotor

3. What common error in handling a micropipettor can account for pipetting too much reagent into a tube? What errors account for underpipetting?

4. When is it necessary to use sterile technique?

5. What does flaming accomplish?

FOR FURTHER RESEARCH

Devise a method to determine the percentage of error in micropipetting. Then, determine the percentage of error when you purposely make pipetting mistakes with small- and large-volume micropipettors.

Laboratory 2

Bacterial Culture Techniques

This laboratory contains most of the culture techniques that are used throughout the course. We suggest that Part A, Isolation of Individual Colonies, be done in sequence between Laboratory 1 and Laboratory 3. Part B, Overnight Suspension Culture, need be done only in conjunction with plasmid purification in Laboratories 6 and 9. Part C, Mid-log Suspension Culture, is done in preparation for making competent cells by the standard calcium chloride procedure in Laboratory 8.

In Part A, *E. coli* cells are streaked onto LB agar plates such that single cells are isolated from one another. Each cell then reproduces to form a visible colony composed of genetically identical clones. Streaking cells to obtain individual colonies is usually the first step in genetic manipulations of microorganisms. Using cells derived from a single colony minimizes the chance of using a cell mass contaminated with a foreign microorganism. To demonstrate antibiotic resistance, the growth of wild-type *E. coli* and of an *E. coli* containing an ampicillin resistance gene are compared, using LB medium containing ampicillin. The resistant strain contains the plasmid pAMP, which produces an enzyme that destroys the ampicillin in the medium, thus allowing these cells to grow.

Part B provides a protocol for growing small-scale suspension cultures of *E. coli* that reach stationary phase with overnight incubation. Overnight cultures are used for purification of plasmid DNA and for inoculating mid-log cultures. When growing *E. coli* strains that contain a plasmid, it is best to maintain selection for antibiotic resistance by growing in LB broth containing the appropriate antibiotic. Strains containing an ampicillin resistance gene (such as pAMP) should be grown in LB broth plus ampicillin.

Part C provides a protocol for preparing a mid-log culture of *E. coli*. Cells in mid-log growth can generally be rendered more competent to uptake plasmid DNA than can cells at stationary phase. Mid-log cells are used in the classic transformation protocol described in Laboratory 8. The protocol begins with an overnight suspension culture of *E. coli*. Incubation with agitation has brought the culture to stationary phase and ensures a large number of healthy cells capable of further reproduction. The object is to subculture (reculture) a small volume of the overnight culture in a large volume of fresh nutrient broth. This "re-sets" the culture to zero growth, where after a short lag phase, the cells enter the log-growth phase. As a general rule, 1 volume of overnight culture (the *inoculum*) is added to 100 volumes of fresh LB broth in an Erlenmeyer flask. To provide good aeration for bacterial growth, the flask volume should be at least four times the total culture volume.

LABORATORY 2/PART A
Isolation of Individual Colonies

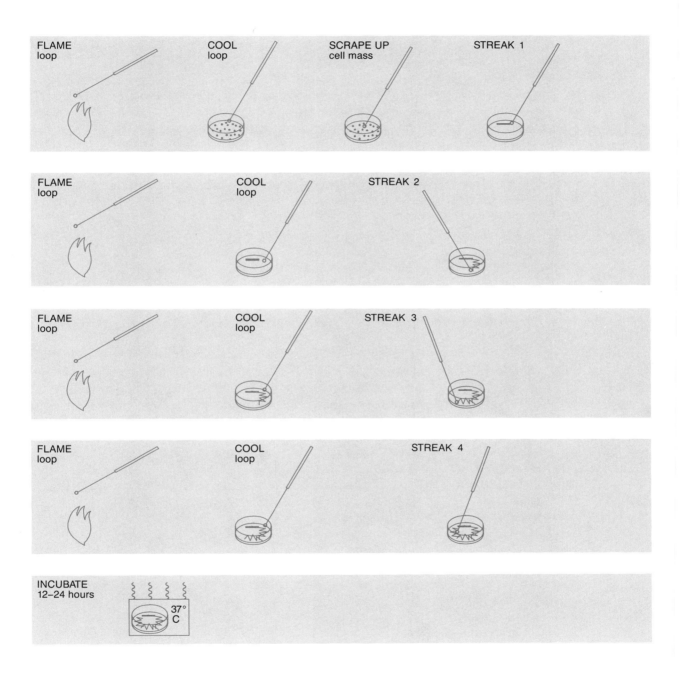

PRELAB NOTES

E. coli Strains

All protocols involving bacterial growth, transformation, and plasmid isolation have been tested and optimized with E. coli strain MM294, derived in the laboratory of Matthew Meselson at Harvard University. MM294/pAMP has been transformed with pAMP, an ampicillin resistance plasmid constructed at Cold Spring Harbor Laboratory. Other strains commonly used for molecular biological studies should give comparable results. However, growth properties of other E. coli strains in suspension culture may differ significantly. For example, the time needed to reach mid-log phase and the cell number represented by specific optical densities differ from strain to strain.

Nutrient Agar

We prefer LB (Luria-Bertani) agar, but almost any rich nutrient agar can be used for plating cells. Presterilized, ready-to-pour agar is a great convenience. It needs only to be melted in a microwave oven or boiling water bath, cooled to approximately 60°C, and poured onto sterile culture plates.

> **CAUTION**
>
> To prevent boiling over, the container should be no more than half full. Loosen cap to prevent bottle from exploding.

Ampicillin

Plasmids containing ampicillin resistance are most commonly used for cloning DNA sequences in E. coli. Ampicillin is very stable in agar plates, thresholds for selection are relatively broad, and contaminants are infrequent. Despite its stability, ampicillin, like most antibiotics, is inactivated by prolonged heating. Therefore, it is important to allow agar solution to cool until container can be held comfortably in the hand (about 60°C) before adding antibiotic. Use the sodium salt, which is very soluble in water, instead of the free acid form, which is difficult to dissolve.

Responsible Handling and Disposal of *E. coli*

A commensal organism of *Homo sapiens*, *E. coli* is a normal part of the bacterial flora of the human gut. It is not considered pathogenic and is rarely associated with any illness in healthy individuals. Furthermore, K-12 *E. coli* strains, including MM294 and all commonly used lab strains, are ineffective in colonizing the human gut. Adherence to simple guidelines for handling and disposal makes work with *E. coli* a nonthreatening experience.

1. To avoid contamination, always reflame inoculating loop or cell spreader one final time before placing it on lab bench.

2. Keep nose and mouth away from tip end when pipetting suspension culture to avoid inhaling any aerosol that might be created.

3. Do not overincubate plates. Because a large number of cells are inoculated, *E. coli* is generally the only organism that will appear on plates incubated for 12–24 hours. However, with longer incubation, contaminating bacteria and slower-growing fungi can arise. If plates cannot be observed following initial incubation, refrigerate them to retard growth of contaminants.

4. Collect for treatment—bacterial cultures *and* tubes, pipets, and micropipettor tips that have come into contact with the cultures. Disinfect these materials as soon as possible after use. Contaminants, often odorous and sometimes potentially pathogenic, are readily cultured over a period of several days at room temperature. Disinfect bacteria-contaminated materials in one of two ways:

    a. Autoclave materials at 121°C for 15 minutes. Tape three to four culture plates together and loosen tube caps before autoclaving. Collect contaminated materials in a "bio bag" or heavy-gauge trash bag; seal bag before autoclaving. Dispose of autoclaved materials in regular garbage.

    or

    b. Treat with solution containing 5000 parts per million (ppm) available chlorine (10% bleach solution). Immerse contaminated pipets, tips, and tubes (open) directly into sink or tub containing bleach solution. Plates should be placed, with lids open, in sink or tub, and flooded with bleach solution. Allow materials to stand in bleach solution for 15 minutes or more. Then drain excess bleach solution, seal materials in plastic bag, and dispose in regular garbage.

5. Wipe down lab bench with soapy water, 10% bleach solution, or disinfectant (such as Lysol) at the end of lab.

6. Wash hands before leaving lab.

PRELAB PREPARATION

1. One day to 3 weeks before class, MM294 must be streaked onto several LB agar plates and MM294/pAMP must be streaked onto several LB + ampicillin (LB/amp) plates. Following overnight incubation at 37°C, wrap the plates in Parafilm or plastic wrap to prevent drying and store at 4°C (in a refrigerator) until they are needed. Alternately, streak directly from stab or slant cultures.

2. Prepare for each experiment:

    two LB agar plates.
    two LB + ampicillin (LB/amp) plates.

3. Make sure plate type is clearly marked on the bottom of the plate (not on the top).

4. Prewarm incubator to 37°C.

Isolation of Individual Colonies

CULTURES AND PLATES

MM294 culture
MM294/pAMP culture
two LB agar plates
two LB + ampicillin (LB/amp) plates

SUPPLIES AND EQUIPMENT

inoculating loop
"bio bag" or heavy-duty trash bag
10% bleach or disinfectant
Bunsen burner
37°C incubator
permanent marker

Plate-streaking Technique (15 minutes)

As with sterile pipetting, plan out manipulations before beginning to streak plates. Organize lab bench to allow plenty of room and work quickly. If working from a stab or slant culture, loosen cap before starting.

1. Use permanent marker to label *bottom* of each agar plate with your name and the date. Each plate will have been previously marked to indicate whether it is plain LB agar (LB) or LB agar + ampicillin (LB/amp).

2. Select the two LB plates. Mark one −pAMP for cells without plasmid and the other +pAMP for cells with plasmid.

3. Select the two LB/amp plates. Mark one −pAMP for cells without plasmid and the other +pAMP for cells with plasmid.

4. Hold inoculating loop like a pencil, and sterilize loop in burner flame until it glows red hot. Then continue to pass lower half of shaft through flame.

5. Cool for 5 seconds. *To avoid contamination, do not place inoculating loop on lab bench.*

6. *If working from culture plate:*
    a. Remove lid from *E. coli* culture plate with free hand. *Do not place lid on lab bench*. Hold lid face down just above culture plate to help prevent contaminants from falling on plate or lid.
    b. Stab inoculating loop into a clear area of the agar several times to cool.

c. Use loop tip to scrape up a visible cell mass from a colony. Do not gouge agar. Replace culture plate lid, and proceed to Step 7.

If working from stab culture:

a. Grasp bottom of *E. coli* culture vial between thumb and two fingers of free hand. Remove vial cap using little finger of same hand that holds inoculating loop. *Avoid touching rim of cap.*

b. Quickly pass mouth of vial several times through burner flame.

c. Stab inoculating loop into side of agar several times to cool.

d. Stab loop several times into area of culture where bacterial growth is apparent. Remove loop, flame vial mouth, and replace cap. Proceed to Step 7.

7. Select LB − pAMP plate and lift top only enough to perform streaking as shown below. *Do not place top on lab bench.*

    a. *Streak 1:* Glide inoculating loop tip back and forth across the agar surface to make a streak across top of the plate. Avoid gouging agar. Replace lid of plate between streaks.

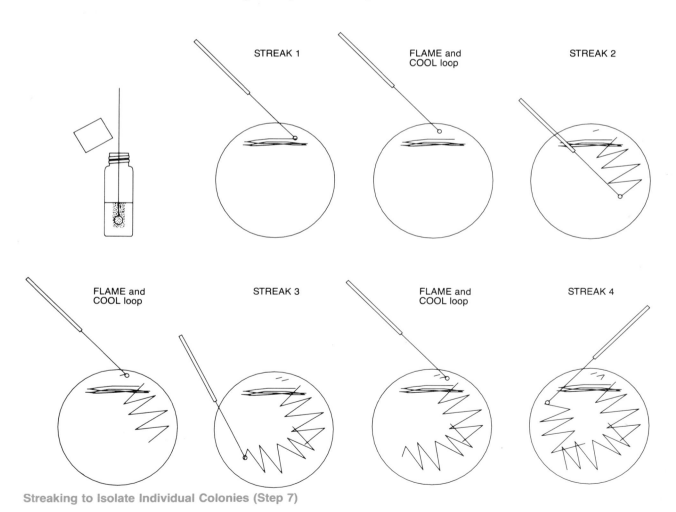

Streaking to Isolate Individual Colonies (Step 7)

b. *Streak 2:* Reflame inoculating loop and *cool* by stabbing it into the agar away from the first (primary) streak. Draw loop tip *once* through primary streak and, without lifting loop, make a zigzag streak across one quarter of the agar surface. *Replace plate lid.*

c. *Streak 3:* Reflame loop and *cool* in the agar as before. Draw loop tip *once* through the last line of the secondary streak, and make another zigzag streak in the adjacent quarter of the plate *without touching the previous streak.*

d. *Streak 4:* Reflame loop and *cool*. Draw tip *once* through tertiary streak, and make a final zigzag streak in remaining quarter of plate.

8. Repeat Steps 4–7 to streak *E. coli* onto LB/amp +pAMP plate.

9. Repeat Steps 4–7 to streak *E. coli*/pAMP onto LB −pAMP plate.

10. Repeat Steps 4–7 to streak *E. coli*/pAMP onto LB/amp +pAMP plate.

11. Reflame loop, and allow it to cool, before placing it on lab bench. Make it a habit to always flame loop one last time.

12. Place plates upside down in 37°C incubator and incubate 12–24 hours. (Plates are inverted to prevent condensation that might collect on the lids from falling back on agar and causing colonies to run together.)

If cells are to be used for colony transformation, continue incubating plates at room temperature for 2–3 days. During this time, colonies grow large and become sticky, making them easier to pick up with an inoculating loop.

13. Optimal growth of well-formed, individual colonies is achieved in 12–24 hours. At this point, colonies should range in diameter from 0.5 mm to 3 mm.

14. Take time for responsible cleanup.

   a. Segregate bacterial cultures for proper disposal.

   b. Wipe down lab bench with soapy water, 10% bleach solution, or disinfectant (such as Lysol) at end of lab.

   c. Wash hands before leaving lab.

RESULTS AND DISCUSSION

This laboratory demonstrates a method to streak bacteria to single colonies. It also introduces antibiotics and plasmid-borne resistance to antibiotics—topics that will be important in several laboratories that follow.

There are two classes of antibiotics: *bacteriostats,* which prevent cell growth, and *bacteriocides,* which kill cells outright. The antibiotics used in this course, ampicillin and kanamycin, are both bacteriocides, although their modes of action are quite different. Ampicillin, a derivative of penicillin, blocks synthesis of the peptidoglycan layer that lies between the *E. coli* inner and outer cell membrane. Thus, it does not affect existing cells with intact cell envelopes, but kills dividing cells as they synthesize new peptidoglycan. Kanamycin (introduced in later laboratories) is a member of the aminoglycoside family of antibiotics, which block protein synthesis by covalently modifying the bacterial ribosome. Thus, kanamycin quickly kills both dividing and quiescent cells.

The ampicillin resistance gene carried by the plasmid pAMP produces a protein, β-lactamase, that disables the ampicillin molecule. β-Lactamase cleaves a specific bond in the β-lactam ring, a four-membered ring in the ampicillin molecule that is essential to its antibiotic action. β-Lactamase not only disables ampicillin within the bacterial cell, but because it leaks through the cell envelope, it also disables ampicillin in the surrounding medium. The enzyme kanamycin phosphotransferase prevents kanamycin from interacting with the ribosome.

Antibiotic-resistant Growth

If you are not able to observe plates on the day after streaking, store plates at 4°C to arrest *E. coli* growth and to slow growth of any contaminating microbes. Wrap in Parafilm or plastic wrap to retard drying.

Observe plates and use the matrix below to record which plates have bacterial *growth* and which have *no growth*. On plates with growth, distinct, individual colonies should be observed within one of the streaks.

On the LB/amp plate, growth must be observed in the secondary streak to count as antibiotic-resistant growth. In a heavy inoculum, nonresistant cells in the primary streak may be isolated from the antibiotic on a bed of other nonresistant cells.

	Transformed cells + pAMP	Wild-type cells − pAMP
LB/amp	experiment	negative control
LB	positive control	positive control

On the LB/amp + pAMP plate, tiny "satellite" colonies may be observed radiating from the edges of large, well-established colonies. These satellite colonies are not ampicillin-resistant, but grow in an "antibiotic shadow," where ampicillin in the media has been broken down by the large resistant colony. Satellite colonies are generally a sign of antibiotic weakened by not cooling medium enough before adding antibiotic, long-term storage of more than 30 days, or overincubation.

1. Were results as expected? Explain possible causes for variations from expected results.

2. In Step 7:

   a. What is the reason for the zigzag streaking pattern?

   b. Why is the inoculating loop resterilized between each new streak?

   c. Why should a new streak intersect the previous one only at a single point?

3. Describe the appearance of a single *E. coli* colony. Why can it be considered genetically homogeneous?

4. Upcoming laboratories use cultures of *E. coli* cells derived from a single colony or from several discrete parental colonies isolated as described in this experiment. Why is it important to use this type of culture in genetic experiments?

5. *E. coli* strains containing the plasmid pAMP are resistant to ampicillin. Describe how this plasmid functions to bring about resistance.

LABORATORY 2/PART B
Overnight Suspension Culture

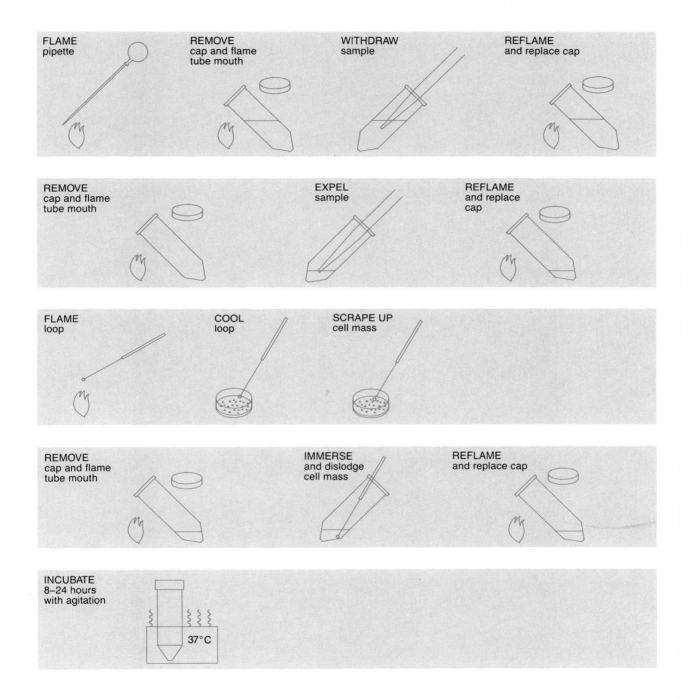

PRELAB NOTES

Review Prelab Notes in Laboratory 2A, Isolation of Individual Colonies.

Although it is always best to grow any overnight culture with shaking, it is not absolutely essential when growing cells for purifying plasmid or for inoculating a larger culture. For these purposes, suspensions can be incubated, without shaking, in a rack within a 37°C incubator. However, the cultures will need to incubate for 2–3 days to obtain an adequate number of cells.

A 50-ml conical tube is preferable for growing overnight cultures. It provides greater surface area for aeration than does a 15-ml culture tube, although a 15-ml tube can also be used. In overnight cultures for plasmid preparations, it is best, but not essential, to maintain antibiotic selection of the transformed strain by growing in medium containing the appropriate antibiotic. It is prudent to inoculate a "back up" overnight culture, in case the first was not inoculated properly.

PRELAB PREPARATION

1. For plasmid purification: Make sure a freshly streaked plate (less than 1 week old) of a transformed or plasmid-bearing *E. coli* strain is available.

2. For mid-log culture: Make sure a freshly streaked plate of wild-type MM294 or other *E. coli* strain is available. It is convenient to make the overnight culture exactly the size of the inoculum needed to start mid-log culture: 1-ml overnight per 100-ml mid-log culture. The entire overnight culture is then simply poured into a flask of fresh, sterile LB broth.

3. Prewarm shaking water bath or incubator to 37°C.

Overnight Suspension Culture

CULTURE AND MEDIA

E. coli plate
LB broth with appropriate antibiotic

SUPPLIES AND EQUIPMENT

inoculating loop
10-ml standard pipet
50-ml conical tube, sterile
pipet aid or bulb
10% bleach or disinfectant
Bunsen burner
permanent marker
37°C shaking water bath (or 37°C dry shaker or dry shaker +37°C incubator)

Prepare Overnight Culture (10 minutes)

Review Sterile Use of 10-ml Standard Pipet, Section III in Laboratory 1, Measurements, Micropipetting, and Sterile Techniques. Think sterile! A pipet should be considered contaminated whenever the tip end comes into contact with anything in the environment—lab bench, hands, or clothing. When contamination is suspected, discard pipet and start with a fresh one. Plan out steps to perform, organize lab bench, and work quickly.

1. Label sterile 50-ml tube with your name and the date. The large tube provides a large surface area for good aeration of culture.

2. Use 10-ml pipet to sterilely transfer 5 ml of LB broth into tube.

    a. Attach pipet aid or bulb to pipet. Briefly flame pipet cylinder.

    b. Remove cap of LB bottle using little finger of hand holding pipet bulb. Flame mouth of LB bottle.

    c. Withdraw 5 ml of LB. Reflame mouth of bottle, and replace cap.

    d. Remove cap of sterile 50-ml culture tube. Briefly flame mouth of culture tube.

    e. Expel sample into tube. Briefly reflame mouth of tube, and replace cap.

3. Locate well-defined colony 1–4 mm in diameter on a freshly streaked plate.

If working as a team, one partner should handle pipet and the other should handle tubes and caps.

Pipet flaming can be eliminated if individually wrapped pipets are used.

Loop flaming is eliminated if individually wrapped, sterile plastic loop is used.

4. Sterilize inoculating loop in burner flame until it glows red hot. Then, continue to pass lower half of its handle through flame.

5. Cool loop tip by stabbing it several times into agar near edge of plate.

6. Use loop to scrape up a visible cell mass from selected colony.

7. Sterilely transfer colony into culture tube:
   a. Remove cap of culture tube using little finger of hand holding loop.
   b. Briefly flame mouth of culture tube.
   c. Immerse loop tip in broth, and agitate to dislodge cell mass.
   d. Briefly reflame mouth of culture tube, and replace cap.

8. Reflame loop before placing it on lab bench.

9. Loosely replace cap to allow air to flow into culture. Affix a loop of tape over cap to prevent it from becoming dislodged during shaking.

10. Incubate 8–24 hours at 37°C, preferably with continuous agitation. Shaking is not essential for a culture to be used for plasmid purification. The culture can be incubated at 37°C, without shaking, for 1 or more days. Following initial incubation, culture can be stored at room temperature for several days, until ready for use.

11. Take time for responsible cleanup.
    a. Segregate for proper disposal—bacterial cultures and tubes and pipets that have come into contact with the cultures.
    b. Disinfect overnight culture and pipets and tips with 10% bleach or disinfectant.
    c. Wipe down lab bench with soapy water, 10% bleach solution, or disinfectant (such as Lysol).
    d. Wash hands before leaving lab.

RESULTS AND DISCUSSION

E. coli has simple nutritional requirements and grows slowly on "minimal" medium containing (1) an energy source such as glucose, (2) salts such as NaCl and $MgCl_2$, (3) the vitamin biotin, and (4) the nucleoside thymidine. *E. coli* synthesizes all necessary vitamins and amino acids from these precursors. *E. coli* grows rapidly in a complete medium, such as LB. Yeast extract and hydrolyzed milk protein (casein) provide a ready supply of vitamins and amino acids.

A liquid bacterial culture goes through a series of growth phases. For approximately 30 minutes following inoculation, there is a *lag phase* during which there is no cell growth. The bacteria begin dividing rapidly during *log phase*, when the cell number doubles every 20–25 minutes. As nutrients in the media are depleted, the cells stop dividing and the culture enters *stationary phase*. During *death phase*, waste products accumulate and cells begin to die.

Optimum growth in liquid culture is achieved with continuous agitation, which aerates the cells, facilitates the exchange of nutrients, and flushes away waste products of metabolism. It can be safely assumed that a culture in complete medium has reached stationary phase following overnight incubation with continuous shaking.

A stationary-phase culture will look very cloudy and turbid. Discard any overnight culture where vigorous growth is not evident. Expect less growth in cultures incubated for several days *without continuous shaking*. To gauge growth, shake tube to suspend cells that have settled at the bottom of the tube.

1. Why is 37°C the optimum temperature for *E. coli* growth?

2. Give two reasons why it is ideal to provide continuous shaking for a suspension culture.

3. What growth phase is reached by a suspension of *E. coli* following overnight shaking at 37°C?

4. Approximately how many *E. coli* cells are in a 5-ml suspension culture at stationary phase?

LABORATORY 2/PART C

Mid-log Suspension Culture

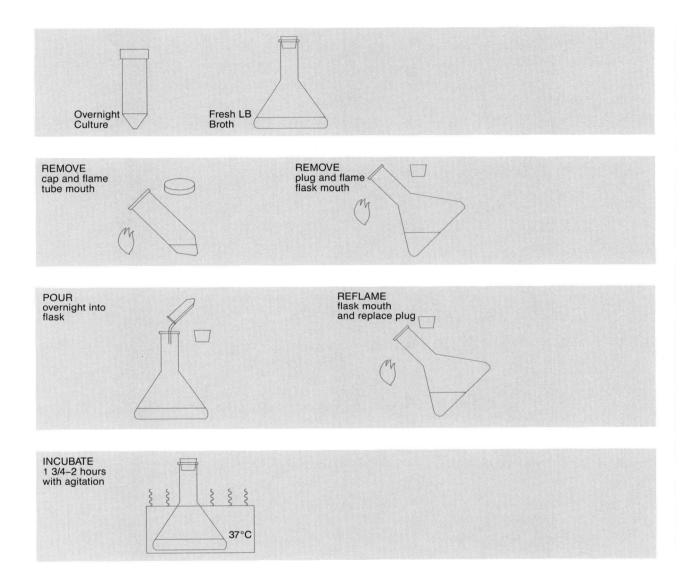

PRELAB NOTES

Competent Cell Yield

In Laboratory 8A, Classic Procedure for Preparing Competent Cells, each experiment will require 20 ml of mid-log cells, which yields 2 ml of competent cells. If competent cells are being prepared in large quantity for group use, remember that the ratio of mid-log cells to competent cells is 10 to 1. A 100-ml mid-log culture will yield 10 ml of competent cells, sufficient for 50–200-µl transformations. If preparing cells for freezer storage, aliquot 50–200-µl samples in sterile tubes.

Sterile Technique

Scrupulous sterile technique must be used when preparing overnight and mid-log cultures. No antibiotic is used, and any contaminant will multiply as cells are repeatedly manipulated and/or stored for future use.

Aeration of Culture

A shaking incubator is necessary for growing *E. coli* for competent cells. Proper aeration and nutrient exchange are essential to achieve vigorous growth; only cells collected during the middle part of log (mid-log) phase will produce competent cells with a high transformation frequency. An economical alternative to a shaking water bath or temperature-controlled dry shaker is to place a small platform shaker inside a 37°C incubator. For adequate surface-to-volume ratio for the exchange of air, cells should be grown in an Erlenmeyer flask with a volume of LB broth not exceeding one third of the total volume of the flask.

Timing of Culture

Inoculate a mid-log culture 2–4 hours before period during which you will perform Laboratory 8. Using the protocol below, *E. coli* strain MM294 reaches mid-log phase after 1.75–2.0 hours of incubation. Cells can be used immediately or held on ice for up to 2 hours before beginning Laboratory 8.

Timing of culture to reach mid-log phase is likely to be affected by any change in protocol. For example, a culture inoculated with an overnight culture that was grown without shaking will take longer to reach mid-log phase. A culture begun by inoculating into LB broth prewarmed to 37°C will reach mid-log phase more quickly than one begun at room temperature. Different strains of *E. coli* display different growth properties. Strain MM294 may exhibit different growth properties in a nutrient broth other than LB broth.

PRELAB PREPARATION

1. The day before performing this protocol, start a culture of MM294 or other *E. coli* strain to be transformed with plasmid DNA, according to instructions in Laboratory 2B, Overnight Suspension Culture.

2. It is convenient to make the overnight culture exactly the size of the inoculum needed to start a mid-log culture, for example, 1 ml of overnight culture for 100 ml of mid-log culture. Then, simply pour the entire overnight culture into a flask of fresh LB broth. Pipetting is not necessary.

3. Sterile flasks of LB broth can be prepared weeks ahead of time. Add 100 ml of LB broth to a 500-ml Erlenmeyer flask, plug neck with gauze-covered cotton or synthetic foam as a microbe barrier, and cap with aluminum foil. Autoclave at 121°C (250°F) for 15 minutes. Cool and store at room temperature, until ready for use.

Mid-log Suspension Culture

CULTURES AND MEDIA

MM294 overnight culture
LB broth, sterile

SUPPLIES AND EQUIPMENT

500-ml Erlenmeyer flask, sterile
10-ml pipets, sterile (optional)
10% bleach or disinfectant
Bunsen burner
37°C shaking water bath (or 37°C dry shaker or dry shaker +37°C incubator)
spectrophotometer (optional)

Prepare Culture   (2 hours, including incubation)

1. Sterilely transfer 1 ml of overnight culture into 100 ml of LB broth at *room temperature*.

2. *If using a 1-ml overnight culture:*

   a. Remove cap from overnight culture tube, and flame mouth. *Do not place cap down on lab bench.*

   b. Remove plug from flask, and flame mouth.

   c. Pour entire overnight culture into flask. Reflame mouth of flask, and replace plug.

   If transferring only a portion of larger overnight culture:

   a. Flame pipet cylinder.

   b. Remove cap from overnight culture tube, and flame mouth of tube.

   c. Withdraw 1 ml of overnight suspension. Reflame mouth of overnight culture tube, and replace cap.

   d. Remove plug from flask, and flame mouth.

   e. Expel overnight sample into flask. Reflame mouth of flask, and replace plug.

3. Incubate at 37°C with continuous shaking.

4. *If a spectrophotometer is available:* Approximately 1 hour after inoculating, sterilely withdraw a 1-ml sample of the culture and

Cells will reach mid-log phase more quickly if overnight culture is inoculated into LB prewarmed to 37°C. Time estimate in Step 4 is based on inoculation of room-temperature LB.

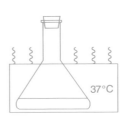

measure absorbance (optical density at 550 nm). Repeat procedure at approximately 20-minute intervals. An MM294 culture should be grown to an OD_{550} of 0.3–0.5.

If a spectrophotometer is not available: It can be safely assumed that an MM294 culture has reached an OD_{550} of 0.3–0.5 after 1 hour 45 minutes of incubation with continuous shaking.

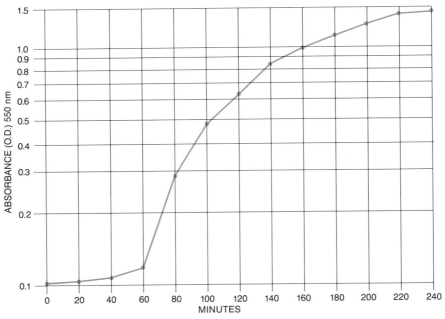

Growth Curve for *E. coli* Strain MM294

5. Store mid-log culture on ice until ready to begin Laboratory 8. This arrests cell growth. Cells can be stored on ice for up to 2 hours prior to use.

6. Take time for responsible cleanup.

   a. Segregate for proper disposal—bacterial cultures and tubes and pipets that have come into contact with the cultures.

   b. Disinfect overnight culture, mid-log culture, tubes, and pipets with 10% bleach or disinfectant.

   c. Wipe down lab bench with soapy water, 10% bleach solution, or disinfectant (such as Lysol).

   d. Wash hands before leaving lab.

RESULTS AND DISCUSSION

1. What variables influence the length of time for an *E. coli* culture to reach mid-log phase?

2. What are the disadvantages of beginning a mid-log culture from a colony scraped off a plate, as opposed to an inoculum of overnight culture?

This experiment can be started in the morning by one experimenter and continued by others throughout the day, until late afternoon.

FOR FURTHER RESEARCH

1. Start a 500-ml *E. coli* culture as described in protocol above. Determine the optical density of samples withdrawn at 20-minute intervals, from time zero for as many hours as possible. Plot a graph of time *versus* OD_{550}.

   a. What is the slope of the curve at a point that corresponds to an OD_{550} of 0.3?

   b. Describe the growth of the culture at this point.

2. Perform the following experiment to correlate the optical density of culture to actual number of viable *E. coli* cells. Observe sterile technique.

   a. Inoculate 500 ml of LB broth with 5 ml of *E. coli* overnight culture. Swirl to mix.

   b. Immediately remove a 10-ml aliquot (time = 0) of the culture, and place on ice to arrest growth. Then incubate remaining culture at 37°C with vigorous shaking.

   c. Remove additional aliquots from shaking culture every 20 minutes for a total of 4 hours. Hold each aliquot on ice until you are ready to perform Steps d–f.

   d. Determine the OD_{550} of each aliquot.

   e. Make a 10^2 dilution by mixing 10 µl of the aliquot with 990 µl of fresh LB broth. Prepare three serial dilutions of each aliquot for plating in Step f:

   10^4 = 10 µl of 10^2 culture + 990 µl of LB
   10^5 = 100 µl of 10^4 culture + 900 µl of LB
   10^6 = 100 µl of 10^5 culture + 900 µl of LB

   f. Spread 100 µl of each dilution onto an LB agar plate, for a total of three plates for each time point (aliquot). *Label each plate bottom with time point and dilution.* Invert plates, and incubate for 12–24 hours at 37°C.

   g. For each time point, select a dilution plate that has between 30 and 300 colonies. Multiply the number of colonies by the appropriate dilution factor to give cell number per milliliter in the original aliquot.

   h. Plot two graphs:

   time (*x* axis) *versus* OD_{550} and cell number (*y* axis)
   cell number (*x* axis) *versus* OD_{550} (*y* axis)

   i. An OD_{550} of 0.3–0.4 corresponds to what number of cells?

j. What is the average cell number at each of the following points:

　　lag phase
　　first third of log phase (early log)
　　second third of log phase (mid log)
　　final third of log phase (late log)
　　stationary phase

k. Do OD_{550} measurements distinguish between living and dead cells?

Laboratory 3

DNA Restriction Analysis

This laboratory introduces the genotypic analysis of DNA using restriction enzymes and gel electrophoresis. Three samples of purified DNA from bacteriophage λ (48,502 bp in length) are incubated at 37°C, each with one of three restriction endonucleases: *Eco*RI, *Bam*HI, and *Hin*dIII. Each enzyme has five or more restriction sites in λ DNA and therefore produces six or more restriction fragments of varying lengths. A fourth sample, the negative control, is incubated without an endonuclease and remains intact.

The digested DNA samples are then loaded into wells of an 0.8% agarose gel. An electric field applied across the gel causes the DNA fragments to move from their origin (the sample well) through the gel matrix toward the positive electrode. The gel matrix acts as a sieve through which smaller DNA molecules migrate faster than larger ones; restriction fragments of differing sizes separate into distinct bands during electrophoresis. The characteristic pattern of bands produced by each restriction enzyme is made visible by staining with a dye that binds to the DNA molecule.

LABORATORY 3
DNA Restriction Analysis

I. **Set Up Restriction Digest**

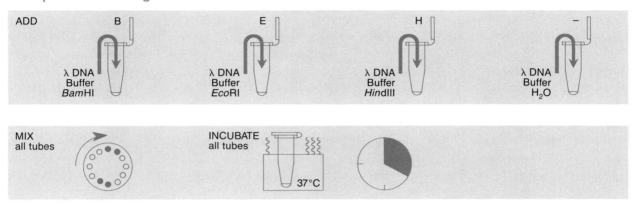

II. **Cast 0.8% Agarose Gel**

III. **Load Gel and Electrophorese**

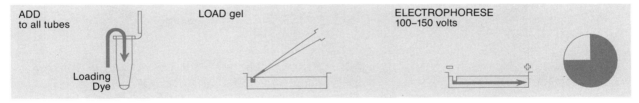

IV. **Stain Gel and View**

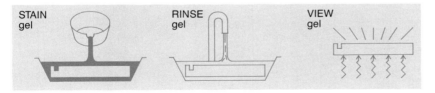

V. **Photograph Gel**

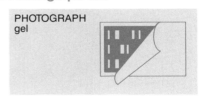

PRELAB NOTES

Storing and Handling Restriction Enzymes

Restriction enzymes, like many enzymes, are most stable at cold temperatures and lose activity if warmed for any length of time. Since maintaining these enzymes in good condition is critical to the success of experiments in this course, follow the guidelines listed below for handling.

1. Always store enzymes in a NON-frost-free freezer that maintains a constant temperature between $-10°C$ and $-20°C$. NON-frost-free freezers typically develop a layer of frost around the chamber that acts as an efficient insulator and helps maintain constant temperature. Frost-free freezers go through freeze-thaw cycles that subject enzymes to repeated warming and subsequent loss of enzymatic activity. If a frost-free freezer must be used, store enzymes in their Styrofoam shipping container within the freezer. The container will help to maintain constant temperature during the thaw cycle.

2. When a large shipment of an enzyme is received, split it into several smaller aliquots of 50–100 µl in 1.5-ml tubes. Use a permanent marker on tape to clearly identify aliquots with enzyme type, concentration in units/µl, and date received. Use up one aliquot before starting another.

3. Remove restriction enzymes from freezer directly onto crushed or cracked ice in an insulated ice bucket or cooler. Make sure that the tubes are pushed down into the ice and not just sitting on top. Keep enzymes on ice at all times during preparation, and return to freezer immediately after use.

4. Keep aliquots of enzymes, buffer, and DNA in a cooler filled with ice while in use; in this way, unused aliquots will remain fresh.

5. Although it is good technique to set up restriction digests on ice, it is much simpler to set up reactions in a test tube rack at room temperature. Little loss of enzyme activity occurs during the brief time it takes to set up the reaction.

But take heart—the enzymes used in this course are all remarkably stable. Those used in training workshops presented by us have survived multiple day-long shipments on ice, freezer power failures, and various abuses by student interns. In five years and more than 50 training workshops, we have yet to experience enzyme failure. Most restriction enzymes can survive several hours left out on the lab bench, but don't try your luck!

Storing DNA and Restriction Buffer

Purified DNA is generally stored in the refrigerator (approximately 4°C). DNA can be kept at $-10°C$ to $-20°C$ for long-term storage of several months or more. However, repeated freezing and thawing damages DNA. Restriction buffer is best kept frozen and is not affected by freeze-thawing.

Buffers

Several types of buffers are used in this course: restriction buffer, electrophoresis buffer, and ligation buffer. Each has a different chemical composition and use. Always double-check to ensure that the proper buffer is being used.

Tris/Borate/EDTA (TBE) electrophoresis buffer can be reused several times. Collect used buffer and store in a large carboy. If different gels will be run over a period of several days, store buffer in an electrophoresis box with the cover in place to retard evaporation. Prior to reusing buffer stored in an electrophoresis box, rock the chamber back and forth to mix buffer at either end. This reequilibrates ions that differentially accumulate at either end during electrophoresis.

Groups of restriction enzymes operate under different conditions of salt and pH. For optimal activity, several different buffers would be needed for the enzymes used in this course. To simplify procedures, we use a "compromise" restriction buffer—a universal buffer that is a compromise between the conditions preferred by various enzymes.

All buffers are used at a final concentration of $1\times$. Rely on the standard $C_1V_1 = C_2V_2$ formula to determine how much buffer to add to obtain a $1\times$ solution:

(vol. buffer)	(conc. of buffer)	=	(total vol. of reaction)	($1\times$ buffer)
(5 μl)	($2\times$)	=	10 μl	($1\times$)
(1 μl)	($10\times$)	=	10 μl	($1\times$)

For convenience, use $2\times$ restriction buffer whenever possible—it can save a pipetting step to add water to bring a reaction up to 10 μl total volume. It is also easier and more accurate to pipet 5 μl than to pipet 1 μl. Compare a typical restriction reaction using $2\times$ versus $10\times$ restriction buffer:

	2× Buffer	10× Buffer
DNA	4 µl	4 µl
Enzyme	1 µl	1 µl
Buffer	5 µl	1 µl
Water	—	4 µl
Total solution	10 µl	10 µl

Bacteriophage λ DNA

Because it is inexpensive and readily available, purified DNA from bacteriophage λ is most suitable for demonstrating the concept of DNA restriction. λ DNA costs approximately $0.10 per microgram, compared to plasmid DNA ranging typically from $2.00–3.00 per microgram. Most commercially available λ is derived from a temperature-sensitive lysogen of *E. coli* called *cI857* and is 48,502 bp in length. Restriction of chromosomal DNA, even from a simple organism such as *E. coli*, will generate thousands of DNA fragments that appear as a smear in an agarose gel.

Diluting DNA

DNA for near-term use can be diluted with distilled or deionized water. However, dilute with Tris-EDTA (TE) buffer for long-term storage. EDTA in the buffer binds divalent cations, such as Mg^{++}, that are necessary cofactors for DNA-degrading nucleases. Always dilute DNA to the concentration specified by the protocol.

1. Determine the total volume of DNA required by multiplying the number of experiments times the total volume of DNA per experiment, including overage.

    (10 experiments) (20 µl DNA) = 200 µl DNA

2. Use this number in the $C_1V_1 = C_2V_2$ formula, along with the desired final DNA concentration and the concentration of the stock DNA. Solve for V_1, the volume of stock DNA needed in the dilution.

    (C_1 stock DNA) (V_1) = (C_2 final DNA) (V_2 total volume)
    (0.5 µg/µl) (V_1) = (0.1 µg/µl) (200 µl)
    (V_1) = (0.1 µg/µl) (200 µl)/(0.5 µg/µl) = 40 µl stock DNA

3. Add water or TE to make total volume of final solution.

    40 µl stock DNA + 160 µl H_2O or TE = 200 µl final solution

Aliquoting Reagents

1. We find it safest to prepare separate aliquots of enzyme, DNA, and buffer in 1.5-ml tubes for each experiment. Each aliquot should

contain slightly more than will be required for the lab. Following the experiment, the tubes are discarded, and new aliquots are made for the next experiment. Although this procedure appears wasteful, it avoids cross-contamination that invariably occurs if aliquots are reused or shared between groups. It is a small price to pay for consistent results.

4. For aliquots of restriction enzymes, add 1 μl extra when 1–3 μl are called for and 2 μl extra when 4–6 μl are actually needed. The overage aids in visualizing the reagent in the tube and allows for small pipetting errors.

5. It is probably unwise to make small aliquots of enzymes more than 1 or 2 days in advance. A several-microliter droplet, clinging to the side of a 1.5-ml tube, has a large surface-to-volume ratio. For this reason, it may be affected by temperature fluctuations. Also, make sure that the tubes are fully submerged in ice while awaiting use.

6. Aliquots of DNA and buffer should be approximately 20% larger than volume actually needed in the experiment. This allows for overpipetting and other mishaps. Considering that DNA is generally the most expensive component of an experiment, you may prefer to aliquot the exact amount, and add other reagents directly to DNA tube.

7. *Reagent volumes listed in the Prelab Preparation section of each laboratory have been scaled to include the overage suggested here.*

8. Large aliquots of distilled water and loading dye can be used for several experiments.

9. Colored 1.5-ml tubes are very handy for color-coding each reagent aliquot.

Pooling Reagents

During aliquoting and movement to and from freezer or refrigerator and ice bucket, reagent aliquots often become spread in a film around the sides or caps of the 1.5-ml tubes. Use one of the following methods to pool reagent droplets just prior to use to make them easier to find in tube.

- Spin tubes briefly in a microfuge.

   or

- Spin tubes briefly in a preparatory centrifuge, using adapter collars for 1.5-ml tubes. Alternately, spin tubes within 15-ml tube, and remove carefully.

   or

- Tap tubes sharply on bench top.

Restriction Enzyme Activity

The "unit" is the standard measure of restriction enzyme activity and is defined as the amount of enzyme needed to digest to completion 1 μg of λ DNA in a 50-μl reaction in 1 hour. The unit strength of various restriction enzymes varies from batch to batch and from manufacturer to manufacturer. Typical batches of commercially available enzymes have activities in the range of 10 units/μl.

We suggest using enzymes at full strength as supplied—a working concentration of approximately 1 unit per microliter of reaction mix. Although this is technically far more enzyme than required, such "overkill" assures complete digestion by compensating for the following experimental conditions:

1. As a time saver, reaction times for restriction digests have been shortened to 20 minutes. Complete digestion of the DNA would not occur during an abbreviated incubation if the restriction enzyme was used at the standard condition of 1 unit/μg of DNA.

2. Many of the enzymes do not exhibit 100% activity in a compromise buffer.

3. Enzymes lose activity over time, due to imperfect handling.

4. It is easy enough to underpipet when measuring 1 μl of enzyme, especially considering that the pipettor's mechanical error is greatest at the low end of its volume range.

Incubating Restriction Reactions

A constant-temperature water bath for incubating reactions can be made by maintaining a trickle flow of tap water into a Styrofoam box. Monitor temperature with a thermometer. An aquarium heater can be used to maintain constant temperature.

Twenty minutes is the bare minimum incubation time for the restriction reaction to go to completion. If electrophoresis is to be done the following day, the reactions can be incubated for 1 to 24 hours. After several hours, enzymes lose their activity, and the reaction simply stops. Stop incubation whenever it is convenient; reactions may be stored in freezer ($-20°C$) until ready to continue. Thaw reactions before adding loading dye.

Storing Cast Agarose Gels

Gels can be cast a day or two before use. Keep gels covered with TBEelectrophoresis buffer to prevent drying.

Electrophoresing

Shortly after current is applied to the electrophoresis system, loading dye should be seen moving through gel toward positive side of gel apparatus.

It will appear as a blue band, eventually resolving into two bands of color. The faster-moving, purplish band is bromophenol blue. The slower-moving, aqua band is xylene cyanol. In a 0.8% gel, bromophenol blue migrates through gel at same rate as a DNA fragment of approximately 300 bp. Xylene cyanol migrates at a rate equivalent to approximately 9000 bp. Best separation for analysis of plasmid DNA is achieved when the bromophenol blue band migrates 4–7 cm or more from the origin. However, be careful not to let the bromophenol blue band run off the end of the gel.

The migration of DNA through the agarose gel is dependent on voltage—the higher the voltage the faster the rate of migration. Refer to chart below for approximate running times at various voltages. Times below are for "mini-gel" system with 84 mm × 96 mm gel; times will vary according to apparatus.

Voltage	150	125	100	75	50	25	12.5
Time	0:40	0:50	1:20	1:40	3:20	6:40	13:00

Responsible Handling of Ethidium Bromide

The protocols in this manual limit the use of ethidium bromide to a single procedure that can be performed by the instructor in a controlled area. With responsible handling, the dilute solution (1 µg/ml) used for gel staining poses minimal risk. Ethidium bromide, like many natural and man-made substances, is a mutagen by the Ames microsome assay and a suspected carcinogen.

> **CAUTION/HANDLING AND DECONTAMINATION OF ETHIDIUM BROMIDE**
> 1. Always wear gloves when working with ethidium bromide solutions or stained gels.
> 2. Limit ethidium bromide use to a restricted sink area.
> 3. Following gel staining, use funnel to decant as much as possible of the ethidium bromide solution into storage container for reuse or decontamination and disposal.
> 4. Disable stained gels and used staining solution according to accepted laboratory procedure. The method given below is from Quillardet and Hofnung (1988).
>   a. If necessary, add sufficient water to reduce the concentration of ethidium bromide to less than 0.5 mg/ml.
>   b. Add 1 volume of 0.5 M $KMnO_4$, and *mix carefully*.
>   c. Add 1 volume of 2.5 N HCl, and *mix carefully*.
>   d. Let stand at room temperature for several hours.
>   e. Add 1 volume of 2.5 N NaOH, and *mix carefully*.
>   f. Discard disabled solution down sink drain. Drain disabled gels, and discard in regular trash.
>
> Caution: $KMnO_4$ is an irritant and is explosive. Solutions containing $KMnO_4$ should be handled in a chemical hood.

The greatest risk is to inhale ethidium bromide powder when mixing a 5 mg/ml stock solution. Therefore, we suggest purchasing a ready-mixed stock solution from a supplier. The stock solution is diluted to make a staining solution with a final concentration of 1 µg/ml.

DNA Staining with Methylene Blue

The volumes and concentrations of DNA used in these experiments have been optimized for staining with ethidium bromide, which is the most rapid and sensitive method. If methylene blue staining is preferred, increase stated concentrations 4–5 times for λ DNA and 2 times for plasmid DNA. If DNA *concentration* is increased, volumes used in laboratories remain as stated.

Viewing Stained Gels

Transillumination, where light passes up through gel, gives superior viewing of gels stained with either ethidium bromide or methylene blue.

A mid-wavelength (260–360 nm) ultraviolet (UV) lamp emits in the optimum range for illuminating ethidium-bromide-stained gels. Avoid short-wavelength lamps, whose radiation is most dangerous. Long-wavelength ("black light") lamps, although safe, give less-intense illumination.

> **CAUTION**
>
> Ultraviolet light can damage the retina of the eye. Never look at unshielded UV light source with naked eyes. View only through filter or safety glasses that absorb harmful wavelengths.

A fluorescent light box for viewing slides and negatives provides ideal illumination for methylene-blue-stained gels. An overhead projector may also be used. Cover surface of light box or projector with plastic wrap to keep liquid off the apparatus.

Photographing Gels

Photographs of DNA gels provide a permanent record of the experiment, allowing time to analyze results critically, to discover subtleties of gel interpretation, and to correct mistakes. Furthermore, time exposure can record bands that are faint or invisible to the unaided eye.

A Polaroid "gun" camera, equipped with a close-up diopter lens, is used to photograph gels on either an ultraviolet or white-light transilluminator. A plastic hood extending from the front of the camera forms a mini-darkroom and provides correct lens-to-subject distance. Alternately, a close-focusing 35mm camera can be used. For UV photography, two filters are placed in front of the lens: a 23A orange is closest to the camera and a 2B UV-blocking filter (clear) is closest to the subject. Any yellow or orange filter will intensify contrast in gels stained with methylene blue. The UV filter set described above works well and can be left in place for both ethidium bromide and methylene blue photography.

For UV photography, use Polaroid Type 667 high-speed film (black and white). Regular color Polaroid film is less expensive for use with methylene-blue stained gels.

For Further Information

The protocol presented here is based on the following published methods:

Helling, R.B., H.M. Goodman, and H.W. Boyer. 1974. Analysis of *Eco*RI fragments of DNA from lambdoid bacteriophages and other viruses by agarose-gel electrophoresis. *J. Virol. 14:* 1235.

Sharp, P.A., B. Sugden, and J. Sambrook. 1973. Detection of two restriction endonuclease activities in *Haemophilus parainfluenzae* using analytical agarose-ethidium bromide electrophoresis. *Biochemistry 12:* 3055.

Quillardet, Phillipe and Maurice Hofnung. 1988. Ethidium bromide and safety—Readers suggest alternative solutions. Letter to editor. *Trends Genet. 4:* 89.

PRELAB PREPARATION

The volumes of agarose solution and Tris/Borate/EDTA (TBE) buffer needed vary according to electrophoresis apparatus used. The volumes quoted here are based on typical "mini-gel" systems.

1. Aliquot for each experiment:

    20 μl of 0.1 μg/μl λ DNA (store on ice).
    25 μl of 2× restriction buffer (store on ice).
    2 μl each of *Bam*HI, *Eco*RI, and *Hin*dIII (store on ice).
    500 μl of distilled water.
    500 μl of loading dye.

2. Prepare 0.8% agarose solution (40–50 ml per experiment). Keep agarose liquid in a hot-water bath (at about 60°C) throughout lab. Cover solution with aluminum foil to retard evaporation.

3. Prepare 1× TBE buffer for electrophoresis (400–500 ml per experiment).

4. Prepare ethidium bromide or methylene blue staining solution (100 ml per experiment).

5. Adjust water bath to 37°C.

DNA Restriction Analysis

REAGENTS

0.1 µg/µl λ DNA
restriction enzymes
   EcoRI
   BamHI
   HindIII
2× restriction buffer
distilled water
loading dye
0.8% agarose
1× Tris/Borate/EDTA
   (TBE) buffer
1 µg/ml ethidium bromide
   (or 0.025% methylene blue)

SUPPLIES AND EQUIPMENT

0.5–10-µl micropipettor + tips
1.5-ml tubes
aluminum foil
beaker for agarose
beaker for waste/used tips
10% bleach
camera and film (optional)
electrophoresis box
masking tape
microfuge (optional)
Parafilm or wax paper
   (optional)
permanent marker
plastic wrap (optional)
power supply
rubber gloves
test tube rack
transilluminator (optional)
37°C water bath

I. Set Up Restriction Digest   (30 minutes, including incubation)

1. Use permanent marker to label four 1.5-ml tubes, in which restriction reactions will be performed:

   B = BamHI
   E = EcoRI
   H = HindIII
   – = no enzyme

2. Use matrix below as a checklist while adding reagents to each reaction. Read down each column, adding the same reagent to all appropriate tubes. *Use a fresh tip for each reagent.* Refer to detailed directions that follow.

Tube	λ DNA	Buffer	BamHI	EcoRI	HindIII	H₂O
B	4 µl	5 µl	1 µl	—	—	—
E	4 µl	5 µl	—	1 µl	—	—
H	4 µl	5 µl	—	—	1 µl	—
–	4 µl	5 µl	—	—	—	1 µl

It is not necessary to change tips when adding the same reagent. Same tip may be used for all tubes, provided tip has not touched solution already in tubes.

3. Collect reagents, and place in test tube rack on lab bench.

4. Add 4 µl of DNA to each reaction tube. Touch pipet tip to side of reaction tube, as near to the bottom as possible, to create capillary action to pull solution out of tip.

5. Always add buffer to reaction tubes *before* adding enzymes. Use *fresh tip* to add 5 µl of restriction buffer to a clean spot on each reaction tube.

6. Use *fresh tips* to add 1 µl of *Eco*RI, *Bam*HI, and *Hin*dIII to appropriate tubes.

7. Use *fresh tip* to add 1 µl of deionized water to tube labeled "−."

8. Close tube tops. Pool and mix reagents by pulsing in a microfuge or by sharply tapping the tube bottom on lab bench.

9. Place reaction tubes in 37°C water bath, and incubate for a minimum of 20 minutes. Reactions can be incubated for a longer period of time.

After several hours, enzymes lose activity and reaction stops.

STOP POINT Following incubation, freeze reactions at −20°C until ready to continue. Thaw reactions before continuing to Section III, Step 1.

Gel is cast directly in box in some electrophoresis apparatuses.

II. Cast 0.8% Agarose Gel (15 minutes)

1. Seal ends of gel-casting tray with tape, and insert well-forming comb. Place gel-casting tray out of the way on lab bench so that agarose poured in next step can set undisturbed.

2. Carefully pour enough agarose solution into casting tray to fill to depth of about 5 mm. Gel should cover only about one-third the height of comb teeth. Use a pipet tip to move large bubbles or solid debris to sides or end of tray, while gel is still liquid.

3. Gel will become cloudy as it solidifies (about 10 minutes). *Be careful not to move or jar casting tray while agarose is solidifying.* Touch corner of agarose *away* from comb to test whether gel has solidified.

Too much buffer will chan.. current over the top rather than through the gel, increasing the time required to separate DNA. TBE buffer can be used several times, do not discard. If using buffer remaining in electrophoresis box from a previous experiment, rock chamber back and forth to remix ions that have accumulated at either end.

4. When agarose has set, unseal ends of casting tray. Place tray on platform of gel box, so that comb is at negative (black) electrode.

5. Fill box with TBE buffer, to a level that just covers entire surface of gel.

6. Gently remove comb, taking care not to rip wells.

7. Make certain that sample wells left by comb are completely submerged. If "dimples" are noticed around wells, slowly add buffer until they disappear.

Buffer solution helps to lubricate the comb. Some gel boxes are designed such that the comb must be removed prior to inserting the casting tray into the box. In this case, flood casting tray and gel surface with running buffer *before* removing the comb. Combs removed from a dry gel can cause tearing of the wells.

STOP POINT Cover electrophoresis tank and save gel until ready to continue. Gel will remain in good condition for at least several days if it is completely submerged in buffer.

III. Load Gel and Electrophorese (50–70 minutes)

1. Add loading dye to each reaction. Either

   a. Add 1 µl of loading dye to each reaction tube. Close tube tops, and mix by tapping tube bottom on lab bench, pipetting in and out, or pulsing in a microfuge. Make sure tubes are placed in a *balanced* configuration in rotor.

   or

   b. Place four individual droplets of loading dye (1 µl each) on a small square of Parafilm or wax paper. Withdraw contents from reaction tube, and mix with a loading dye droplet by pipetting in and out. Immediately load dye mixture according to Step 2. Repeat successively, with clean tip, for each reaction.

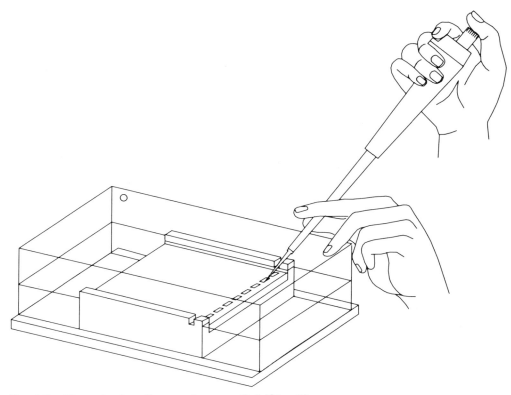

Hand Positions for Loading an Agarose Gel (Step 2)

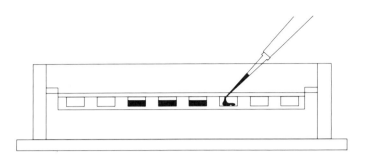

A piece of dark construction paper beneath the gel box will make the wells more visible.

2. Use micropipettor to load entire contents of each reaction tube into separate well in gel, as shown in diagrams. Use *fresh tip* for each reaction.

   a. Steady pipet over well using two hands.

   b. If there is air in the end of the tip, carefully depress plunger to push the sample to the end of the tip. (If air bubble forms "cap" over well, DNA/loading dye will flow into buffer around edges of well.)

   c. Dip pipet tip through surface of buffer, center it over the well, and gently depress pipet plunger to slowly expel sample. Sucrose in the loading dye weighs down the sample, causing it to sink to the bottom of the well. *Be careful not to punch tip of pipet through bottom of the gel.*

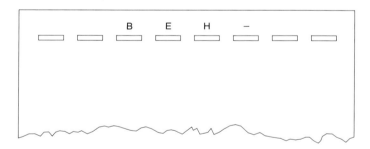

Alternately, set power supply on lower voltage, and run gel for several hours. When running two gels off the same power supply, current is double that for a single gel at the same voltage.

3. Close top of electrophoresis box, and connect electrical leads to a power supply, anode to anode (red-red) and cathode to cathode (black-black). Make sure both electrodes are connected to same channel of power supply.

4. Turn power supply on, and set to 100–150 volts. The ammeter should register approximately 50–100 milliamperes. If current is not detected, check connections and try again.

5. Electrophorese for 40–60 minutes. Good separation will have occurred when the bromophenol blue band has moved 4–8 cm from wells. If time allows, electrophorese until the bromophenol blue band nears the end of the gel. Stop electrophoresis before bromophenol blue band runs off end of gel.

6. Turn off power supply, disconnect leads from the inputs, and remove top of electrophoresis box.

7. Carefully remove casting tray from electrophoresis box, and slide gel into disposable weigh boat or other shallow tray. Label staining tray with your name.

STOP POINT   Cover electrophoresis tank and save gel until ready to continue. Gel can be stored in a zip-lock plastic bag and refrigerated overnight for viewing/photographing the next day. However, over longer periods of time, the DNA will diffuse through the gel, and the bands will become indistinct or disappear entirely.

---

8. Stain and view gel using one of the methods described in Sections IVA and IVB.

Staining may be performed by an instructor in a controlled area when students are not present.

IVA. Stain Gel with Ethidium Bromide and View   (10–15 minutes)

> **CAUTION**
>
> Review Responsible Handling of Ethidium Bromide (page 256). Wear rubber gloves when staining, viewing, and photographing gel and during clean up. Confine all staining to a restricted sink area.

1. Flood gel with ethidium bromide solution (1 µg/ml), and allow to -stain for 5–10 minutes.

2. Following staining, use funnel to decant as much ethidium bromide solution as possible from staining tray back into storage container.

3. Rinse gel and tray under running tap water.

4. If desired, gel can be destained in tap water or distilled water for 5 minutes or more to help remove background ethidium bromide from the gel.

Staining time increases markedly for thicker gels. Do not be tempted to use a higher concentration of ethidium bromide in the staining solution. This *will not* enhance the DNA bands, but only increases the background staining of the agarose gel itself.

Ethidium bromide solution may be reused to stain 15 or more gels. When staining time increases markedly, disable ethidium bromide solution as explained on page 256.

STOP POINT   Staining intensifies dramatically if rinsed gels set overnight at room temperature. Stack staining trays, and cover top gel with plastic wrap to prevent desiccation.

---

5. View under ultraviolet transilluminator or other UV source.

> **CAUTION**
>
> Ultraviolet light can damage eyes. Never look at unshielded UV light source with naked eyes. View only through a filter or safety glasses that absorb harmful wavelengths.

6. Take time for responsible cleanup.

   a. Wipe down camera, transilluminator, and staining area.

   b. Wash hands before leaving lab.

IVB. Stain Gel with Methylene Blue and View (30+ minutes)

1. Wear rubber gloves during staining and cleanup.

2. Flood gel with 0.025% methylene blue, and allow to stain for 20–30 minutes.

3. Following staining, use funnel to decant as much methylene blue solution as possible from staining tray back into storage container.

Destaining time is decreased by rinsing the gel in warm water, with agitation.

4. Rinse gel in running tap water. Let gel soak for several minutes in several changes of fresh water. DNA bands will become increasingly distinct as gel destains.

STOP POINT For best results, continue to destain overnight in a *small volume* of water. (Gel may destain too much if left overnight in large volume of water.) Cover staining tray to retard evaporation.

---

5. View gel over light box; cover surface with plastic wrap to prevent staining.

V. Photograph Gel (5 minutes)

Exposure times vary according to mass of DNA in lanes, level of staining, degree of background staining, thickness of gel, and density of filter. Experiment to determine best exposure. When possible, stop lens down (to higher f/number) to increase the depth of field and sharpness of bands.

1. *For ultraviolet (UV) photography of ethidium-bromide-stained gels:* Use Polaroid high-speed film Type 667 (ASA 3000). Set camera aperture to f/8 and shutter speed to B. Depress shutter for a 2–3-second time exposure.

For white-light photography of methylene-blue-stained gels: Use Polaroid film Type 667, with an aperture of f/8 and shutter speed of 1/125 second.

> **CAUTION**
>
> Avoid getting caustic developing jelly on skin or clothes. If jelly gets on skin, wash immediately with plenty of soap and water.

2. Place left hand firmly on top of camera to steady. Firmly grasp small white tab, and pull straight out from camera. This causes a large yellow tab to appear.

3. Grip yellow tab in center, and in one steady motion, pull it straight out from the camera. This starts development.

4. Allow to develop for recommended time (45 seconds at room temperature). Do not disturb print while developing.

5. After full development time has elapsed, separate print from negative by peeling back at end nearest yellow tab.

6. Wait to see the result of the first photo before making other exposures.

RESULTS AND DISCUSSION

Agarose gel electrophoresis combined with ethidium bromide staining allows for the rapid analysis of DNA fragments. However, prior to the introduction of this method in 1973, analysis of DNA molecules was a laborious task. The original separation method, involving ultracentrifugation of DNA in a sucrose gradient, gave only crude size approximations and took more than 24 hours to complete.

Electrophoresis using polyacrylamide gel in a glass tube was an improvement, but it could only be used to separate small DNA molecules of up to 2000 bp. Another drawback was that the DNA had to be radioactively labeled prior to electrophoresis. Following electrophoresis, the polyacrylamide gel was cut into thin slices, and the radioactivity in each slice was determined. The amount of radioactivity detected in each slice was plotted versus distance migrated, producing a series of radioactive peaks representing each DNA fragment.

DNA restriction analysis is at the heart of recombinant DNA technology and of the laboratories in this course. The ability to cut DNA predictably and precisely enables us to manipulate and recombine DNA molecules at will. The fact that discrete bands of like-sized DNA fragments are seen in one lane of an agarose gel shows that each of the more than 1 billion λ DNA molecules present in each restriction reaction were all cut in precisely the same place.

By convention, DNA gels are "read" from left-to-right, with the sample wells oriented at the top. The area extending from the well down the gel is termed a "lane." Thus, reading down a lane identifies fragments generated by a particular restriction reaction. Scanning across lanes identifies fragments that have comigrated the same distance down the gel and are thus of like size.

1. Why is water added to tube labeled " – " in Part I, Step 7?

2. What is the function of compromise restriction buffer?

3. What are the two functions of loading dye?

4. How does ethidium bromide stain DNA? How does this relate to the need to minimize exposure to humans?

5. Troubleshooting electrophoresis. What will occur

    a. if gel box is filled with water instead of TBE buffer?

    b. if water is used to prepare gel instead of TBE buffer?

    c. if electrodes are reversed?

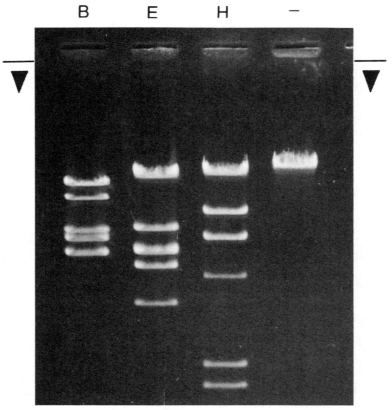

Ideal Gel

6. Examine the photograph of your stained gel (or view on a light box or overhead projector). Compare your gel with the ideal gel shown on facing page, and try to account for the fragments of λ DNA in each lane. How can you account for differences in separation and band intensity between your gel and the ideal gel?

7. Troubleshooting gels. What effect will be observed in the stained bands of DNA in an agarose gel

   a. if casting tray is moved or jarred while agarose is solidifying in Part II, Step 3?

   b. if gel is run at very high voltage?

   c. if a large air bubble or clump is allowed to set in agarose?

   d. if too much DNA is loaded in a lane?

8. Linear DNA fragments migrate at rates inversely proportional to the $\log_{10}$ of their molecular weights. For simplicity's sake, base-pair length is substituted for molecular weight.

   a. The matrix below gives the base-pair size of λ DNA fragments generated by a HindIII digest:

HindIII		EcoRI			BamHI		
Dis.	Act. bp	Dis.	Cal. bp	Act. bp	Dis.	Cal. bp	Act. bp
	27,491[a]						
	23,130[a]						
	9,416						
	6,557						
	4,361						
	2,322						
	2,027						
	564[b]						
	125[c]						

   [a] Pair appears as single band on gel.
   [b] Band may not be visible in methylene-blue-stained gel.
   [c] Band runs off end of gel when bromophenol blue is approximately 2 cm from end of gel. When present on gel, band is not detected by methylene blue and is usually difficult to detect with ethidium bromide staining.

   b. Using the ideal gel shown on facing page, carefully measure distance (in millimeters) each HindIII, EcoRI, and BamHI fragment migrated from the origin. Measure from front edge of well to front edge of each band. Enter distances into matrix. Alternately, measure distances on overhead-projected image of methylene-blue-stained gel.

c. Match base-pair sizes of *Hin*dIII fragments with bands that appear in the ideal digest. Label each band with kilobase pair (kbp) size. For example, 27,491 bp equals 27.5 kbp.

d. Set up semilog graph paper with distance migrated as the x (arithmetic) axis and log of base-pair length as the y (logarithmic) axis. Then, plot distance migrated versus base-pair length for each *Hin*dIII fragment.

e. Connect data points with a line.

f. Locate on x axis the distance migrated by the first *Eco*RI fragment. Using a ruler, draw a vertical line from this point to its intersection with the best-fit data line.

g. Now extend a horizontal line from this point to the y axis. This gives the base-pair size of this *Eco*RI fragment.

h. Repeat Steps f and g for each *Eco*RI and *Bam*HI fragment. Enter results in the calculated base-pair (Cal. bp) columns for each digest.

i. Enter the actual base-pair size of *Eco*RI and *Bam*HI fragments (as provided by your instructor) into Act. bp column.

j. For which fragment sizes was your graph most accurate? For which fragment sizes was it least accurate? What does this tell you about the resolving ability of agarose gel electrophoresis?

9. DNA fragments of similar size will not always resolve on a gel. This is seen in lane E, where *Eco*RI fragments of 5804 bp and 5643 bp migrate as a single heavy band. These are referred to as a doublet and can be recognized because they are brighter than similarly sized singlets. What could be done to resolve the doublet fragments?

10. Determine a range of sensitivity of DNA detection by ethidium bromide by comparing the mass of DNA in the bands of the largest and smallest detectable fragments on the gel. To determine the mass of DNA in a given band:

$$\frac{\text{fragment bp (conc. DNA) (vol. DNA)}}{\lambda \text{ bp}}$$

For example:

$$\frac{24{,}251 \text{ bp } (0.1 \text{ μg/μl}) (4 \text{ μl})}{(0.2 \text{ μg}) \, 48{,}502 \text{ bp}} = 1/2 \ (0.4 \text{ μg})$$

Now, compute the mass of DNA in the largest and smallest *singlet* fragments on the gel.

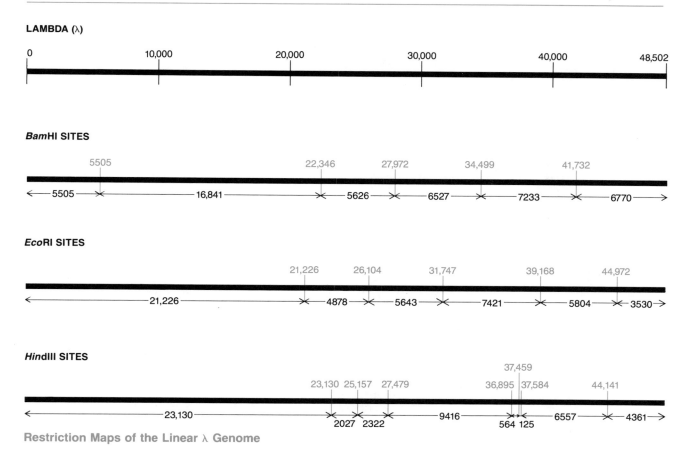

Restriction Maps of the Linear λ Genome

11. λ DNA can exist both as a circular molecule and as a linear molecule. At each end of the linear molecule is a single-stranded sequence of 12 nucleotides, called a COS site. The COS sites at each end are complementary to each other and thus can base pair to form a circular molecule. These complementary ends are analogous to the "sticky ends" created by some restriction enzymes. Commercially available λ DNA is likely to be a mixture of linear and circular molecules. This leads to the appearance of more bands on the gel than would be predicted from a homogeneous population of linear DNA molecules. This also causes the partial loss of other fragments. For example, the left-most HindIII site is 23,130 bp from the left end of the linear λ genome, and the right-most site is 4361 bp from the right end. The 4361-bp band is faint in comparison to other bands on the gel of similar size. This indicates that a percentage of the DNA molecules are circular—combining the 4361-bp terminal fragment with the 23,130-bp terminal fragment to produce a 27,491-bp fragment. However, the combined 27,491-bp fragment usually runs as a doublet along with the 23,130-bp fragment from the linear molecule.

   a. Use a protractor to draw three circles about 3 inches in diameter. These represent λ DNA molecules with base-paired COS sites.

b. Label a point at 12:00 on each circlet 48/0. This marks the point where the COS sites are joined.

c. Using data from the restriction maps of the linear λ genome on preceding page, make a rough map of restriction sites for *Hin*dIII on one of the circles. Note the situation described above.

d. Next make rough restriction maps of *Bam*HI and *Eco*RI sites on the remaining two circles.

e. What *Bam*HI and *Eco*RI fragments are created in the circular molecules? Why (or why not) can you locate each of these fragments on your gel or the ideal gel above?

FOR FURTHER RESEARCH

1. Some of the circular λ molecules are covalently linked at the COS sites. Other circlets are only hydrogen-bonded and can dissociate to form linear molecules. Heating λ DNA to 65°C for 10 minutes linearizes any noncovalent COS circlets in the preparation by breaking hydrogen bonds that hold the complementary COS sites together.

   a. Set up duplicate restriction digests of λ DNA with several enzymes. Then heat one reaction from each set at 65°C for 10 minutes, while holding the duplicates on ice. Relate changes in restriction patterns of heated versus unheated DNA to a restriction map of the circular λ genome as in Question 11 in Results and Discussion.

   b. How can the data generated by this experiment be used to quantify the approximate percentage of circular DNA in your preparation?

2. Design and carry out a series of experiments to study the kinetics of a restriction reaction.

   a. Determine approximate percentage of digested DNA at various time points.

   b. Repeat experiments with several enzyme dilutions and several DNA dilutions.

   c. In each case, at what time point does the reaction appear to be complete?

3. Design and test an assay to determine the relative stability of *Bam*HI, *Eco*RI, and *Hin*dIII at room temperature.

4. Determine the identity of an unknown restriction enzyme.

   a. Perform single digests of λ DNA with the unknown enzyme, as well as with several known restriction enzymes. Run the restriction fragments in an agarose gel at 50 volts to produce well-spread and well-focused bands.

b. For each fragment, plot distance migrated *versus* base-pair size, as in Question 8 in Results and Discussion. Use graph to determine the base-pair lengths of the unknown fragments and compare with restriction maps of commercially available enzymes.

5. Research the steps needed to purify a restriction enzyme from *E. coli* and characterize its recognition sequence.

Field Guide to Electrophoresis Effects

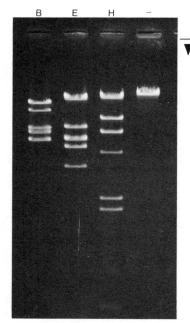

Ideal Gel

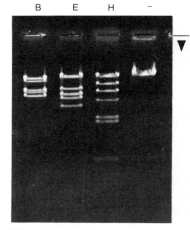

Short Run
Bands compressed. Short time electrophoresing.

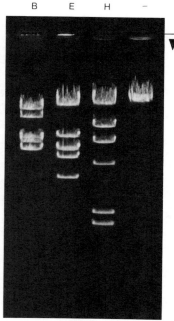

Overloaded
Bands smeared in all lanes. Too much DNA in digests.

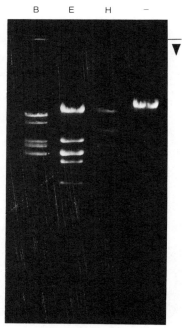

Punctured Wells
Bands faint in Lanes B and H. DNA lost through hole punched in bottom of well with pipet tip.

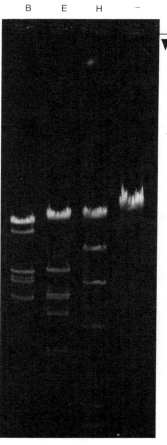

Long Run
Bands spread. Long time electrophoresing.

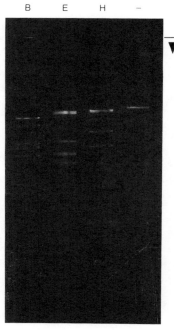

Underloaded
Bands faint in all lanes. Too little DNA in digests.

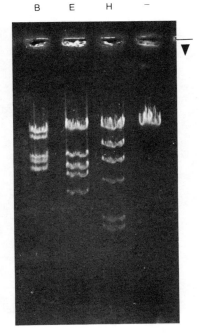

Poorly Formed Wells
Wavy bands in all lanes. Comb removed before gel was completely set.

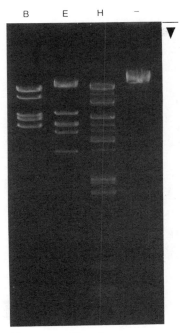

Enzymes Mixed
Extra bands in Lane H. *Bam*HI and *Hin*dIII mixed in digest.

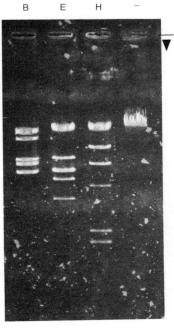

Precipitate
Precipitate in TBE buffer used to make gel.

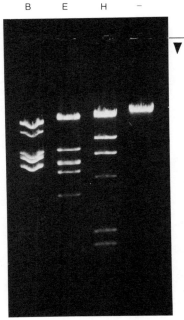

Bubble in Lane
Bump in band in Lane B. Bubble in lane.

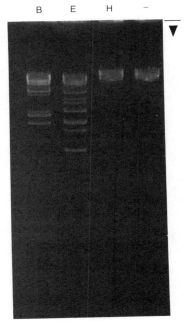

Incomplete Digest
Bands faint in Lane H. Very little *Hin*dIII in digest. Also, extra bands are present in lanes B and E.

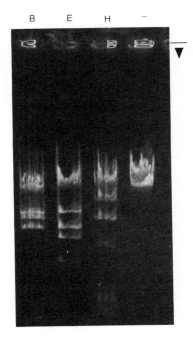

Gel Made with Water
Bands smeared in all lanes. Gel made with water or wrong concentration of TBE buffer.

275

Laboratory 4

Effects of DNA Methylation on Restriction

In this laboratory, the *Eco*RI methylation system is used to illustrate the sequence specificity of a modifying enzyme that protects DNA from restriction enzyme digestion. M.*Eco*RI methylase adds a methyl group to the second adenine residue in the *Eco*RI recognition site, thus preventing the endonuclease from binding and cutting the DNA. *S*-Adenosyl methionine (SAM), included in the methylation reaction, donates the methyl group that is attached to the DNA molecule by the methylase.

Three samples of bacteriophage λ DNA are incubated at 37°C with M.*Eco*RI methylase, one of which is subsequently incubated with *Eco*RI and another of which is incubated with *Hin*dIII. The third sample, a control, is incubated without a restriction enzyme. Three control samples of nonmethylated DNA are also incubated with *Eco*RI, *Hin*dIII, and no enzyme. All of the samples are electrophoresed in an agarose gel and stained. Comparison of the band patterns reveals that the methylated DNA is protected from digestion by *Eco*RI. However, methylation at the *Eco*RI site has no effect on the activity of the restriction enzyme *Hin*dIII.

LABORATORY 4
Effects of DNA Methylation on Restriction

I. Set Up Methylase Reaction

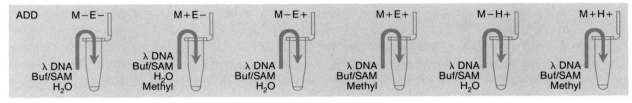

II. Cast 0.8% Agarose Gel

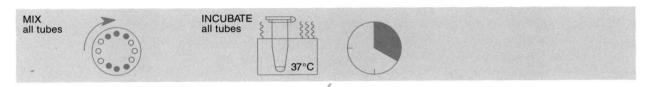

III. Set Up Restriction Reaction

IV. Load Gel and Electrophorese

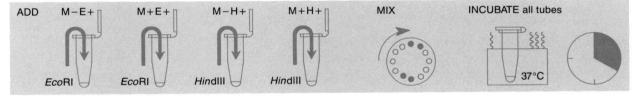

V. Stain Gel and View

VI. Photograph Gel

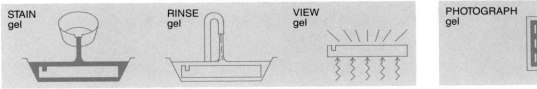
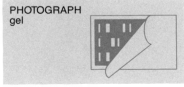

279

PRELAB NOTES

Review Prelab Notes in Laboratory 3, DNA Restriction Analysis.

S-Adenosyl Methionine

S-Adenosyl methionine (SAM) is incorporated into a 2× restriction buffer, so that the same buffer is used for *both* methylation and restriction reactions. Because SAM is not very stable, the buffer/SAM solution should be mixed just prior to the lab and discarded after use. Also, be sure to use a fresh stock of SAM not more than several months old.

To Avoid Confusion

This laboratory has two distinct steps involving two similar-sounding reagents. In the first step, DNA is preincubated with M.*Eco*RI methylase. In the second step, the methylated DNA is incubated with *Eco*RI restriction enzyme. To avoid mishaps, do not set out the *Eco*RI restriction enzyme until *after* the methylation reactions are set up.

For Further Information

The protocol presented here is based on the following published methods:

Greene, P.J., M.S. Poonian, A.L. Nussbaum, L. Tobias, D.E. Garfin, H.W. Boyer, and H.M. Goodman. 1975. Restriction and modification of a self-complementary octanucleotide containing the *Eco*RI substrate. *J. Mol. Biol.* 99: 237.

Helling, R.B., H.M. Goodman, and H.W. Boyer. 1974. Analysis of *Eco*RI fragments of DNA from lambdoid bacteriophages and other viruses by agarose-gel electrophoresis. *J. Virol.* 14: 1235.

Sharp, P.A., B. Sugden, and J. Sambrook. 1973. Detection of two restriction endonuclease activities in *Haemophilus parainfluenzae* using analytical agarose-ethidium bromide electrophoresis. *Biochemistry* 12: 3055.

PRELAB PREPARATION

1. Prepare 2× restriction buffer plus SAM within a day or two before the experiment.

2. Aliquot for each experiment:

    30 µl of 0.1 µg/µl λ DNA (store on ice).
    40 µl of 2× restriction buffer/SAM (store on ice).
    5 µl of M.*Eco*RI methylase (store on ice).
    3 µl each of *Eco*RI and *Hin*dIII (store on ice).
    500 µl of distilled water.
    500 µl of loading dye.

3. Prepare 0.8% agarose solution (40–50 ml per experiment). Keep agarose liquid in a hot-water bath (about 60°C) throughout lab. Cover with aluminum foil to retard evaporation.

4. Prepare 1× Tris/Borate/EDTA (TBE) buffer for electrophoresis (400–500 ml per experiment).

5. Prepare ethidium bromide or methylene blue staining solution (100 ml per experiment).

6. Adjust one water bath to 37°C and one to 70°C.

Effects of DNA Methylation on Restriction

REAGENTS

0.1 µg/µl λ DNA
restriction enzymes:
  EcoRI
  HindIII
  M.EcoRI methylase
2× restriction buffer/SAM
distilled water
loading dye
0.8% agarose
1× Tris/Borate/EDTA
  (TBE) buffer
1 µg/ml ethidium bromide
  (or 0.025% methylene blue)

SUPPLIES AND EQUIPMENT

0.5–10-µl micropipettor + tips
1.5-ml tubes
aluminum foil
beaker for agarose
beaker for waste/used tips
10% bleach
camera and film (optional)
electrophoresis box
masking tape
microfuge (optional)
Parafilm or wax paper
  (optional)
permanent marker
plastic wrap (optional)
power supply
rubber gloves
test tube rack
transilluminator (optional)
37°C water bath
70°C water bath

I. Set Up Methylase Reaction (30 minutes, including incubation)

1. Use permanent marker to label six 1.5-ml tubes, in which methylation and restriction reactions will be performed:

   M−E− = no methylase, no EcoRI
   M+E− = methylase, no EcoRI
   M−E+ = no methylase, EcoRI
   M+E+ = methylase, EcoRI
   M−H+ = no methylase, HindIII
   M+H+ = methylase, HindIII

2. Use matrix on following page as a checklist while adding reagents to each reaction. Read down each column, adding the same reagent to all appropriate tubes. *Use a fresh tip for each reagent.* Refer to detailed instructions that follow.

Tube	λ DNA	Buffer/SAM	H₂O	M.EcoRI methylase
M−E−	4 μl	5 μl	2 μl	—
M+E−	4 μl	5 μl	1 μl	1 μl
M−E+	4 μl	5 μl	1 μl	—
M+E+	4 μl	5 μl	—	1 μl
M−H+	4 μl	5 μl	1 μl	—
M+H+	4 μl	5 μl	—	1 μl

3. Collect reagents, and place in test tube rack on lab bench.

It is not necessary to change tips when adding the same reagent. Same tip may be used for all tubes, provided tip has not touched solution already in tubes.

4. Add 4 μl of DNA to each reaction tube. Touch pipet tip to side of reaction tube, as near to the bottom as possible, to create capillary action to pull solution out of tip.

5. Use *fresh tip* to add 5 μl of restriction buffer/SAM to a clean spot on each reaction tube.

6. Use *fresh tip* to add specified volume of distilled H₂O to appropriate tubes.

To avoid confusing methylase with reagents in Part III, discard after completing Step 7.

7. Use *fresh tip* to add 1 μl of EcoRI methylase to tubes labeled M+E−, M+E+, and M+H+.

8. Close tube tops. Pool and mix reagents by pulsing in a microfuge or by sharply tapping tube bottom on lab bench.

9. Place reaction tubes in 37°C water bath, and incubate for a minimum of 20 minutes. Reactions can be incubated for longer periods of time.

After several hours, methylase loses activity and reaction stops.

STOP POINT Following incubation, freeze reactions at −20°C until ready to continue. Thaw reactions before continuing to Section III, Step 1.

Gel is cast directly in box in some electrophoresis apparatuses.

II. Cast 0.8% Agarose Gel (15 minutes)

1. Seal ends of gel-casting tray with tape, and insert well-forming comb. Place gel-casting tray out of the way on lab bench, so that agarose poured in next step can set undisturbed.

2. Carefully pour enough agarose solution into casting tray to fill to a depth of about 5 mm. Gel should cover only about one-third the height of comb teeth. Use a pipet tip to move large bubbles or solid debris to sides or end of tray, while gel is still liquid.

3. Gel will become cloudy as it solidifies (about 10 minutes). *Be careful not to move or jar casting tray while agarose is solidifying.* Touch corner of agarose *away* from comb to test whether gel has solidified.

4. When agarose has set, unseal ends of casting tray. Place tray on platform of gel box, so that comb is at negative (black) electrode.

5. Fill box with Tris/Borate/EDTA (TBE) buffer, to a level that just covers entire surface of gel.

6. Gently remove comb, taking care not to rip wells.

7. Make certain that sample wells left by comb are completely submerged. If "dimples" are noticed around wells, slowly add buffer until they disappear.

Too much buffer will channel current over the top rather than through the gel, increasing the time required to separate DNA. TBE buffer can be used several times; do not discard. If using buffer in electrophoresis box from a previous experiment, rock chamber back and forth to remix ions that have accumulated at either end.

STOP POINT Cover electrophoresis tank and save gel until ready to continue. Gel will remain in good condition for at least several days if it is completely submerged in buffer.

Buffer solution helps to lubricate the comb. Some gel boxes are designed such that you must remove the comb prior to inserting the casting tray into the box. In this case, flood casting tray and gel surface with running buffer *before* removing the comb. Combs removed from a dry gel can tear the wells.

III. Set Up Restriction Reaction (30 minutes, including incubation)

1. Remove methylation reactions from water bath.

2. Use matrix below as a checklist while adding reagents to each reaction. Read down each column, adding the same reagent to all appropriate tubes. *Use a fresh tip for each reagent.* Refer to detailed instructions that follow.

Tube	*Eco*RI	*Hin*dIII
M−E−	—	—
M+E−	—	—
M−E+	1 μl	—
M+E+	1 μl	—
M−H+	—	1 μl
M+H+	—	1 μl

3. Collect *Eco*RI and *Hin*dIII, and place in test tube rack on lab bench.

4. Add 1 μl of *Eco*RI to tubes labeled M−E+ and M+E+.

5. Use *fresh tip* to add 1 μl of *Hin*dIII to tubes labeled M−H+ and M+H+.

6. Close tube tops. Pool and mix reagents by pulsing in a microfuge or by sharply tapping tube bottom on lab bench.

After several hours, enzymes lose activity and reaction stops.

7. Place reaction tubes in 37°C water bath, and incubate restriction reactions for minimum of 20 minutes. Reactions can be incubated for longer periods of time.

STOP POINT Following incubation, freeze reactions at −20°C until ready to continue. Thaw reactions before continuing to Section IV, Step 1.

IV. Load Gel and Electrophorese (50–70 minutes)

1. Add loading dye to each reaction. Either

   a. Add 1–2 µl of loading dye to each reaction tube. Close tube tops, and mix by tapping tube bottom on lab bench, pipetting in and out, or pulsing in a microfuge. Make sure tubes are placed in a *balanced* configuration in rotor.

   or

   b. Place six individual droplets of loading dye (1 µl each) on a small square of Parafilm or wax paper. Withdraw contents from reaction tube, and mix with a loading dye droplet by pipetting in and out. Immediately load dye mixture according to Step 2. Repeat successively, *with clean tip*, for each reaction.

2. Use micropipettor to load entire contents of each reaction tube into separate well in gel, as shown in diagram below. *Use fresh tip for each reaction.*

   A piece of dark construction paper beneath the gel box will make the wells more visible.

   a. Steady pipet over well using two hands.

   b. If there is air in the end of the tip, carefully depress plunger to push the sample to the end of the tip. (If air bubble forms "cap" over well, DNA/loading dye will flow into buffer around edges of well.)

   c. Dip pipet tip through surface of buffer, center it over the well, and gently depress pipet plunger to slowly expel sample. Sucrose in the loading dye weighs down the sample, causing it to sink to the bottom of the well. *Be careful not to punch tip of pipet through bottom of the gel.*

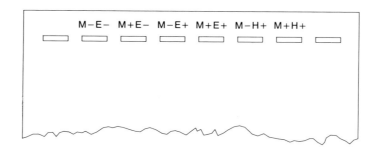

3. Close top of electrophoresis box, and connect electrical leads to a power supply, anode to anode (red-red) and cathode to cathode (black-black). Make sure both electrodes are connected to same channel of power supply.

4. Turn power supply on, and set to 100–150 volts. The ammeter should register approximately 50–100 milliamperes. If current is not detected, check connections and try again.

Alternately, set power supply on lower voltage, and run gel for several hours. When running two gels off the same power supply, current is double that for a single gel at the same voltage.

5. Electrophorese for 40–60 minutes. Good separation will have occurred when the bromophenol blue band has moved 4–8 cm from wells. If time allows, electrophorese until the bromophenol blue band nears the end of the gel. *Stop* electrophoresis before bromophenol blue band runs off end of gel.

6. Turn off power supply, disconnect leads from the inputs, and remove top of electrophoresis box.

7. Carefully remove casting tray from electrophoresis box, and slide gel into disposable weigh boat or other shallow tray. Label staining tray with your name.

STOP POINT Cover electrophoresis tank, and save gel until ready to continue. Gel can be stored in a zip-lock plastic bag and refrigerated overnight for viewing/photographing the next day. However, over longer periods of time, the DNA will diffuse through the gel, and the bands will become indistinct or disappear entirely.

8. Stain and view gel using one of the methods described in Sections VA and VB.

Staining may be performed by an instructor in a controlled area when students are not present.

VA. Stain Gel with Ethidium Bromide and View (10–15 minutes)

> CAUTION
>
> Review Responsible Handling of Ethidium Bromide (page 256). Wear rubber gloves when staining, viewing, and photographing gels and during cleanup. Confine all staining to a restricted sink area.

Staining time increases markedly for thicker gels.

Ethidium bromide solution may be reused to stain 15 or more gels. When staining time increases markedly, disable ethidium bromide solution as explained on page 256.

1. Flood gel with ethidium bromide solution (1 μg/ml), and allow to stain for 5–10 minutes.

2. Following staining, use funnel to decant as much ethidium bromide solution as possible from staining tray back into storage container.

3. Rinse gel and tray under running tap water.

4. If desired, gel can be destained in tap water or distilled water for 5 minutes or more to remove background ethidium bromide.

STOP POINT Staining intensifies dramatically if rinsed gels set overnight at room temperature. Stack staining trays, and cover top gel with plastic wrap to prevent desiccation.

5. View under ultraviolet transilluminator or other UV source.

> CAUTION
>
> Ultraviolet light can damage eyes. Never look at unshielded UV light source with naked eyes. View only through a filter or safety glasses that absorb harmful wavelengths.

6. Take time for responsible cleanup.

   a. Wipe down camera, transilluminator, and staining area.

   b. Wash hands before leaving lab.

VB. Stain Gel with Methylene Blue and View (30+ minutes)

1. Wear rubber gloves during staining and cleanup.

2. Flood gel with 0.025% methylene blue, and allow to stain for 20–30 minutes.

3. Following staining, use funnel to decant as much methylene blue solution as possible from staining tray back into storage container.

Destaining time is decreased by rinsing the gel in warm water, with agitation.

4. Rinse gel in running tap water. Let gel soak for several minutes in several changes of fresh water. DNA bands will become increasingly distinct as gel destains.

STOP POINT For best results, continue to destain overnight in a *small volume* of water. (Gel may destain too much if left overnight in large volume of water.) Cover staining tray to retard evaporation.

---

5. View gel over light box; cover surface with plastic wrap to prevent staining.

VI. Photograph Gel (5 minutes)

Exposure times vary according to mass of DNA in lanes, level of staining, degree of background staining, thickness of gel, and density of filter. Experiment to determine best exposure. When possible, stop lens down (to higher f/number) to increase depth of field and sharpness of bands.

1. *For ultraviolet (UV) photography of ethidium-bromide-stained gels:* Use Polaroid high-speed film Type 667 (ASA 3000). Set camera aperture to f/8 and shutter speed to B. Depress shutter for a 2–3-second time exposure.

   For white-light photography of methylene-blue-stained gels: Use Polaroid film Type 667, with an aperture of f/8 and shutter speed of 1/125 second.

   > CAUTION
   >
   > Avoid getting caustic developing jelly on skin or clothes. If jelly gets on skin, wash immediately with plenty of soap and water.

2. Place left hand firmly on top of camera to steady. Firmly grasp small white tab, and pull straight out from camera. This causes a large yellow tab to appear.

3. Grip yellow tab in center, and in one steady motion, pull it straight out from the camera. This starts development.

4. Allow to develop for recommended time (45 seconds at room temperature). Do not disturb print while developing.

5. After full development time has elapsed, separate print from negative by peeling back at end nearest yellow tab.

6. Wait to see the result of the first photo before making other exposures.

RESULTS AND DISCUSSION

Each type II restriction enzyme has a corresponding methylase that recognizes the same nucleotide sequence. However, rather than cutting the DNA at this point, the methylase adds a methyl group within the recognition sequence. This "modification" blocks the recognition site and prevents the restriction enzyme from binding. Within the bacterium, methylation protects the host DNA from cleavage by its own restriction enzyme. Unmethylated foreign DNA is not protected from cleavage.

The methylation reaction requires *S*-adenosyl methionine (SAM), the common methyl-donating molecule in both prokaryotes and eukaryotes. As its name implies, SAM is composed of the nucleoside adenosine and the amino acid methionine. The donation of a methyl from the methionine portion of the molecule converts it into *S*-adenosyl homoserine.

One common occurrence in this laboratory is partial methylation, where methyl groups are added to only a fraction of the *Eco*RI sites within the λ DNA molecules. Cleavage at the unprotected sites produces a partial digest, yielding restriction fragments of varying lengths. These fragments are evidenced as lower-molecular-weight bands in an agarose gel. The intensity of bands is inversely proportional to the level of DNA methylation.

1. Examine the photograph of your gel (or view on a light box or overhead projector). Compare your gel to the ideal gel shown below. How can you account for differences in separation and band intensity?

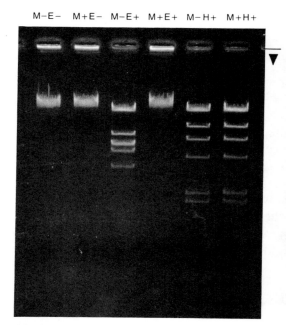

Ideal Gel

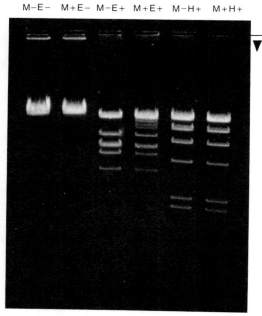

Incomplete Methylation
Faint bands in Lane M+E+ DNA partially cut by *Eco*RI

2. What does the M+H+ control tell you about M.*Eco*RI methylation?

3. What does the M+E− control tell you about methylation?

4. What biological function do methylases perform in bacteria? What adaptive value do they have for a bacterium?

5. Experimental constraints demand that a plasmid be constructed in two major steps, whose order cannot be reversed. In Step 1, a *Bam*HI fragment is inserted into the *Bam*HI site of plasmid pAMP. In Step 2, an *Eco*RI fragment must be cloned into an *Eco*RI site *within* the *Bam*HI insert. Experimental constraints demand that the order of the steps not be reversed. Unfortunately, the pAMP "backbone" also

contains an *Eco*RI site, which is not the intended cloning site for the *Eco*RI fragment in Step 2.

a. Draw a diagram of this cloning experiment.

b. Explain how M.*Eco*RI methylase could be used to solve this experimental problem.

FOR FURTHER RESEARCH

1. Design and execute a series of experiments to study the kinetics of a methylation reaction.

    a. Determine the approximate percentage of sites protected at various time points.

    b. Repeat experiments with several methylase dilutions and several DNA dilutions.

    c. In each case, at what time point does protection appear to be complete?

2. Design and execute experiments to use M.*Eco*RI methylase to map the locations of *Eco*RI restriction sites in the λ genome.

3. Research the use of methylases in constructing a genomic library.

4. Research the role of DNA methylation in gene regulation in higher organisms.

5. Research the role of methylation in controlling the movement of transposable elements in maize (corn).

Laboratory 5

Rapid Colony Transformation of *E. coli* with Plasmid DNA

This Laboratory demonstrates a rapid method to transform *E. coli* with a foreign gene. The bacterial cells are rendered "competent" to uptake plasmid DNA containing a gene for resistance to the antibiotic ampicillin (pAMP). Transformants are detected by their antibiotic-resistant phenotype.

Samples of *E. coli* cells are scraped off a nutrient agar plate (LB agar) and suspended in two tubes containing a solution of calcium chloride. Plasmid pAMP is added to one cell suspension, and both tubes are incubated at 0°C for 15 minutes. Following a brief heat shock at 42°C, cooling, and addition of LB broth, samples of the cell suspensions are plated on two types of media: LB agar and LB agar plus ampicillin (LB/amp).

The plates are incubated for 12–24 hours at 37°C and then checked for bacterial growth. Only cells that have been transformed by taking up the plasmid DNA with the ampicillin-resistant gene will grow on the LB/amp plate. Subsequent division of a single antibiotic-resistant cell produces a colony of resistant clones. Thus, each colony seen on an ampicillin plate represents a single transformation event.

LABORATORY 5

Rapid Colony Transformation of *E. coli* with Plasmid DNA

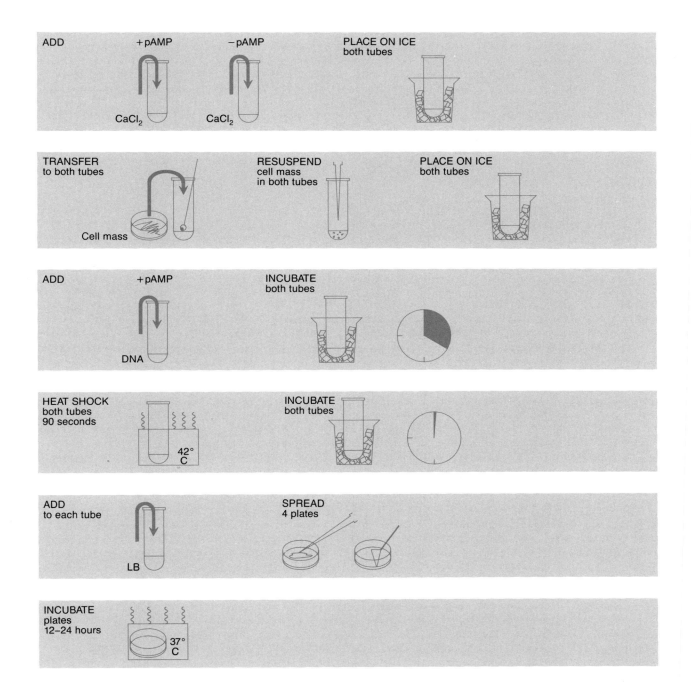

PRELAB NOTES

Review Prelab Notes in Laboratories 1 and 2 regarding sterile technique and *E. coli* culture.

Colony transformation is a simplification of the classic transformation protocol of Laboratory 8, which requires mid-log phase cells grown in liquid culture. This abbreviated protocol begins with *E. coli* colonies scraped from an agar plate. Since liquid culturing is not used, equipment for shaking incubation is not required. The procedure entails minimal preparation time and is virtually foolproof.

Transformation Scheme

Most transformation protocols can be conceptualized as four major steps.

1. *Preincubation:* Cells are suspended in a solution of cations and incubated at 0°C. The cations are thought to complex with exposed phosphates of lipids in the *E. coli* cell membrane. The low temperature freezes the cell membrane, stabilizing the distribution of charged phosphates.

2. *Incubation:* DNA is added, and the cell suspension is further incubated at 0°C. The cations are thought to neutralize negatively charged phosphates in DNA and the cell membrane.

3. *Heat shock:* The cell/DNA suspension is briefly incubated at 42°C and then returned to 0°C. The rapid temperature change creates a thermal imbalance on either side of the *E. coli* membrane, which is thought to create a draft that sweeps plasmids into the cell.

4. *Recovery:* LB broth is added to the DNA/cell suspension and incubated at 37°C (ideally with shaking) prior to plating on selective media. Transformed cells recover from the treatment, amplify the transformed plasmid, and begin to express the antibiotic resistance protein.

The incubation and heat-shock steps are critical. Since the preincubation and recovery steps do not consistently improve the efficiency of colony transformation, they have been omitted from this protocol. If time permits, a preincubation of 5–15 minutes and/or a recovery of 5–30 minutes may be included.

Relative Inefficiency of Colony Transformation

The transformation efficiencies achieved with the colony protocol (5×10^3 to 5×10^4 colonies per microgram of plasmid) are 2–200 times less than those of the classic protocol (5×10^4 to 10^6 colonies per microgram). Colony transformation is perfectly suitable for transforming *E. coli* with purified, intact plasmid DNA. However, it will give poor results with ligated DNA, which is composed of relaxed circular plasmid and linear plasmid DNA. These forms yield 5–100-fold fewer transformants than an equivalent mass of intact, supercoiled plasmid.

Maintenance of *E. coli* Strains for Colony Transformation

Prolonged reculturing (passaging) of *E. coli* can result in a loss of competence that makes the bacterium virtually impossible to transform using the colony method. There is some evidence that loss of transforming ability in MM294 may result from exposure of cells to temperatures below 4°C. Therefore, take care to store stab/slant cultures and streaked plates at room temperature. If there is a severe drop in number of transformants—from the expected 50–500 colonies per plate to essentially zero—discard the culture and obtain a fresh one.

Plasmid pAMP

Almost any plasmid containing a selectable antibiotic resistance marker can be substituted for pAMP for the purpose of demonstrating transformation of *E. coli* to an antibiotic resistance phenotype. However, pAMP was constructed specifically as a teaching molecule and offers the following advantages in other contexts:

1. pAMP is derived from a pUC expression vector that replicates to a high number of copies per cell. Therefore, yields from plasmid preparations are significantly greater than those obtained with pBR322 and other less highly amplified plasmids.

2. pAMP was designed for use with another teaching plasmid, pKAN. Each produces unique and readily recognizable restriction fragments when separated on an agarose gel. Thus, recombinant molecules formed by ligating these fragments can be easily characterized.

Antibiotic Selection

Ampicillin is the most practical antibiotic resistance marker for demonstration purposes, especially in the rapid transformation protocol described here. Ampicillin interferes with construction of the peptidoglycan layer and only kills replicating cells that are assembling new cell envelopes. It does not kill outright preexisting *E. coli* with intact cell envelopes. Thus, cells can be plated onto ampicillin-containing medium directly following heat shock, omitting the recovery step. Kanamycin selection, on the other hand, is less amenable to rapid transformation. A recovery step prior to plating is essential, because the antibiotic acts quickly to kill *any* preexisting cells that are not actively expressing the resistance protein.

Test Tube Selection

The type of test tube used is a critical factor in achieving high-efficiency transformation and may also be important in the colony protocol. Therefore, we recommend using a presterilized 15-ml (17 mm × 100 mm) polypropylene culture tube. The critical heat-shock step has been optimized for the thermal properties of a 15-ml polypropylene tube. Tubes of a different material (such as polycarbonate) or thickness conduct heat differently. Also, the small volume of cell suspension forms a thin layer across the bottom of a 15-ml tube, allowing heat to be quickly transferred to all cells. A smaller tube (such as 1.5-ml) increases the depth of cell suspension through which heat must be conducted. Thus, *any* change in tube specifications requires recalibrating the duration of the heat shock. The Becton Dickinson Falcon 2059 tube is the standard for transformation experiments. Although other quality brands may be completely acceptable, batches of tubes are occasionally contaminated with surfactants or other chemicals that inhibit transformation. Unfortunately, this can only be determined by hard luck.

Purified Water

Extraneous salts and minerals in the transformation buffer can also affect results. Use the most highly purified water available; pharmacy-grade distilled water is recommended. It might pay to obtain from a local research center or hospital several liters of water purified through a multistage ion-exchange system, such as Milli-Q.

Presterilized Supplies

Presterilized supplies can be used to good effect in transformations; 15-ml culture tubes and individually packaged 100–1000-μl micropipet tips are handy. A 3-ml transfer pipet, marked in 250-μl gradations, can be substituted for a 100–1000-μl micropipettor with no loss of speed or accuracy. A presterilized inoculating loop, calibrated to deliver 10 μl, can be substituted for a 1–10-μl micropipettor. Eliminate flaming when using individually wrapped plasticware.

Technically, everything used in this experiment should be sterilized. However, it is acceptable to use clean, but nonsterile, 1.5-ml tubes for aliquots of calcium chloride, LB broth, and plasmid DNA, *provided they will be used within a day or two*. Clean, nonsterile 1–10-μl micropipet tips can be used for adding DNA to cells in Step 9. Plastic supplies, if not handled extensively, are rarely contaminated. Antibiotic selection covers such minor lapses of sterile technique.

For Further Information

The protocol presented here is based on the following published methods:

Cohen, S.N., A.C.Y. Chang, and L. Hsu. 1972. Nonchromosomal antibiotic resistance in bacteria: Genetic transformation of *Escherichia coli* by R-factor DNA. *Proc. Natl. Acad. Sci.* 69: 2110.

Hanahan, D. 1983. Studies on transformation of *Escherichia coli* with plasmids. *J. Mol. Biol.* 166: 557.

Hanahan, D. 1987. Techniques for transformation of *E. coli*. In *DNA Cloning: A Practical Approach* (ed. D.M. Glover), vol. 1. IRL Press, Oxford.

Mandel, M. and A. Higa. 1970. Calcium-dependent bacteriophage DNA infection. *J. Mol. Biol.* 53: 159.

PRELAB PREPARATION

1. Within 5 days of the laboratory, streak out several fresh "starter plates" of MM294 or other *E. coli* host strain. Follow procedure in Laboratory 2A, Isolation of Single Colonies. Following an initial overnight incubation at 37°C, continue to incubate plates at room temperature for 1–3 additional days. During this time, colonies grow to large size and become tacky, making them easier to scrape up with inoculating loop.

2. PLAN AHEAD. Be sure to have a streaked plate or stab/slant culture of viable *E. coli* cells from which to streak starter plates. Also, streak the *E. coli* strain on an LB/amp plate to ensure that an ampicillin-resistant strain has not been used by mistake.

3. Sterilize 50 mM calcium chloride ($CaCl_2$) solution and LB broth by autoclaving or filtering through a 0.45- or 0.22-μm filter (Nalgene or Corning). Be sure to prewash the filter by drawing through 50–100 ml of deionized water; this removes any surfactants that may be used in the manufacture of the filter. To eliminate autoclaving completely, store filtered solutions in presterilized 50-ml conical tubes.

4. Prepare for each experiment:
   two LB agar plates.
   two LB + ampicillin plates (LB/amp).

5. Aliquot for each experiment:
   1 ml of sterile 50 mM $CaCl_2$ in 1.5-ml tube (store on ice).
   1 ml of sterile LB broth in 1.5-ml tube.
   12 μl of 0.005 μg/μl pAMP in 1.5-ml tube (store on ice).

6. Adjust water bath to 42°C. A constant-temperature water bath can be made by maintaining a trickle flow of 42°C tap water into a Styrofoam box. Monitor temperature with a thermometer. An aquarium heater can be used to maintain temperature.

7. Prewarm incubator to 37°C.

8. To retard evaporation, keep ethanol in beaker covered with Parafilm or plastic wrap prior to use. Retrieve and reuse ethanol exclusively for flaming.

Rapid Colony Transformation of *E. coli* with Plasmid DNA

CULTURE, MEDIA, AND REAGENTS

MM294 starter culture
50 mM $CaCl_2$
0.005 µg/µl of pAMP
LB broth
2 LB plates
2 LB/amp plates

SUPPLIES AND EQUIPMENT

inoculating loop
100–1000-µl micropipettor + tips (or 3-ml transfer pipets)
0.5–10-µl micropipettor + tips (or 10-µl calibrated loop)
2 15-ml culture tubes
beaker of crushed or cracked ice
beaker of 95% ethanol
beaker for waste/used tips
"bio-bag" or heavy-duty trash bag
10% bleach or disinfectant
Bunsen burner
cell spreader
37°C incubator
permanent marker
test tube rack
37°C water bath (optional)
42°C water bath

Prepare *E. coli* Colony Transformation (40 minutes)

This entire experiment *must be performed under sterile conditions.* Review sterile techniques in Laboratory 1, Measurements, Micropipetting, and Sterile Techniques.

1. Use permanent marker to label one sterile 15-ml tube +pAMP. Label another 15-ml tube −pAMP. Plasmid DNA will be added to the +pAMP tube; none will be added to −pAMP tube.

2. Use a 100–1000-µl micropipettor and sterile tip (or sterile transfer pipet) to add 250 µl of $CaCl_2$ solution to each tube.

3. Place both tubes on ice.

If there are no separate colonies on the starter plate, scrape up a *small* cell mass from a streak. Transformation efficiency decreases if too many cells are added to the calcium chloride.

Optimally, flame the mouth of the 15-ml tube after removing and before replacing cap.

Cells become difficult to resuspend if allowed to clump together in the CaCl$_2$ solution for several minutes. Resuspending cells in +pAMP tube first allows cells to preincubate for several minutes at 0°C while the −pAMP tube is being prepared. If time permits, both tubes can be preincubated on ice for 5–15 minutes.

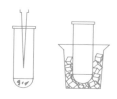

Double check both tubes for complete resuspension of cells, which is probably the most important variable in obtaining good results.

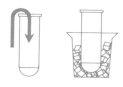

To save plates, different experimenters may omit *either* the +LB *or* the −LB plate.

4. Use sterile inoculating loop to transfer one or two large (3-mm) colonies from starter plate to +pAMP tube:

   a. Sterilize loop in Bunsen burner flame until it glows red hot. Then pass the lower half of the shaft through flame.

   b. Stab loop several times into side of agar to cool.

   c. Scrape up a visible cell mass, but be careful not to transfer any agar. (Impurities in agar can inhibit transformation.)

   d. Immerse loop tip in CaCl$_2$ solution and *vigorously* tap against wall of tube to dislodge cell mass. Hold tube up to light to observe the cell mass drop off into the CaCl$_2$ solution. Make sure cell mass is not left on loop or on side of tube.

   e. Reflame loop before placing it on lab bench.

5. Immediately resuspend cells in +pAMP tube by repeatedly pipetting in and out, using a 100–1000-μl micropipettor with sterile tip (or sterile transfer pipet).

> **CAUTION**
>
> Keep nose and mouth away from tip end when pipetting suspension culture to avoid inhaling any aerosol that might be created.

   a. Pipet carefully to avoid making bubbles in suspension or splashing suspension far up sides of tube.

   b. Hold tube up to light to check that suspension is homogeneous. No visible clumps of cells should remain.

6. Return +pAMP tube to ice.

7. Transfer a second mass of cells to −pAMP tube as described in Steps 4 and 5 above.

8. Return −pAMP tube to ice. Both tubes should be on ice.

9. Use a 1–10-μl micropipettor (or calibrated inoculating loop) to add 10 μl of 0.005 μg/μl pAMP solution *directly into cell suspension* in +pAMP tube. Tap tube with finger to mix. Avoid making bubbles in suspension or splashing suspension up the sides of tube.

10. Return +pAMP tube to ice. Incubate both tubes on ice for an additional 15 minutes.

11. While cells are incubating on ice, use permanent marker to label two LB plates and two LB/amp plates with your name and the date.

Label one LB/amp plate +. This is the experimental plate.
Label the other LB/amp plate −. This is the negative control.
Label one LB plate +. This is a positive control.
Label one LB plate −. This is a positive control.

12. Following a 15-minute incubation, heat shock the cells in +pAMP and −pAMP tubes. *It is critical that cells receive a sharp and distinct shock.*

   a. Carry ice beaker to water bath. Remove tubes from ice, and *immediately* immerse in 42°C water bath for 90 seconds.

   b. Immediately return both tubes to ice for at least 1 additional minute.

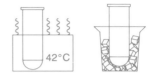

STOP POINT An extended period on ice following the heat shock will not affect the transformation. If necessary, store +pAMP and −pAMP tubes on ice in refrigerator (0°C) for up to several days, until there is time to plate cells. Do not put cell suspensions in the freezer.

13. Place +pAMP and −pAMP tubes in test tube rack at room temperature.

    If time permits, allow +pAMP and −pAMP cells to recover at 37°C for 5–30 minutes. Gentle shaking is also helpful.

14. Use a 100–1000-μl micropipettor and sterile tip (or sterile transfer pipet) to add 250 μl of LB broth to each tube. Gently tap tubes with finger to mix.

15. Use matrix below as a checklist as +pAMP and −pAMP cells are spread on each type of plate:

	Transformed cells +pAMP	Nontransformed cells −pAMP
LB/amp	100 μl	100 μl
LB	100 μl	100 μl

16. Use micropipettor with sterile tip (or transfer pipet) to add 100 μl of cell suspension from −pAMP tube onto −LB plate and another 100 μl onto −LB/amp plate. *Do not allow suspensions to sit on plates too long before proceeding to Step 17.*

    If too much liquid is absorbed by agar, cells will not be evenly distributed.

    The object is to evenly distribute and separate cells on agar so that each gives rise to a distinct colony of clones. *It is essential not to overheat spreader in burner flame and to cool it before touching cell suspensions.* A hot spreader will kill *E. coli* cells on the plate.

Sterile Spreading Technique (Steps 16 and 17)

17. Sterilize cell spreader, and spread cells over surface of each − plate in succession.

   a. Dip spreader into the ethanol beaker and *briefly* pass it through Bunsen flame to ignite alcohol. Allow alcohol to burn off *away from* Bunsen flame; spreading rod will become too hot if left in flame.

   > CAUTION
   >
   > Be extremely careful not to ignite ethanol in beaker.

   b. Lift lid of one − plate only enough to allow spreading; do not place lid on lab bench.

   c. Cool spreader by gently rubbing it on surface of the agar *away* from cell suspension or by touching it to condensed water on plate lid.

   d. Touch spreader to cell suspension, and gently drag it back and forth several times across surface of agar. Rotate plate one-quarter turn, and repeat spreading motion. Be careful not to gouge agar.

   e. Replace plate lid. Return cell spreader to ethanol *without flaming*.

18. Use micropipettor with sterile tip (or transfer pipet) to add 100 µl of cell suspension from +pAMP tube onto +LB plate and to add another 100 µl of cell suspension onto +LB/amp plate.

19. Repeat Step 17a–e to spread cell suspension on +LB and +LB/amp plates.

20. Reflame spreader one last time before placing it on lab bench.

21. Let plates set for several minutes to allow suspension to become absorbed into agar. Then wrap plates together with tape.

22. Place plates upside down in 37°C incubator, and incubate for 12–24 hours.

23. After initial incubation, store plates at 4°C to arrest *E. coli* growth and to slow the growth of any contaminating microbes.

24. If planning to do Laboratory 6, Purification and Identification of Plasmid DNA, save +LB/amp plate as source of a colony to begin an overnight suspension culture.

25. Take time for responsible cleanup.

    a. Segregate for proper disposal—culture plates and tubes, pipets, and micropipettor tips that have come in contact with *E. coli*.

    b. Disinfect cell suspensions, tubes, and tips with 10% bleach or disinfectant.

    c. Wipe down lab bench with soapy water, 10% bleach solution, or disinfectant (such as Lysol).

    d. Wash hands before leaving lab.

RESULTS AND DISCUSSION

Count the number of individual colonies on the +LB/amp plate. Observe colonies through the bottom of the culture plate, and use a permanent marker to mark each colony as it is counted. If the transformation worked well, between 50 and 500 colonies should be observed on the +LB/amp plate. One hundred colonies is equal to a transformation efficiency of 10^4 colonies per microgram of plasmid DNA. (Question 3 explains how to compute transformation efficiency.)

If plates have been overincubated or left at room temperature for several days, tiny "satellite" colonies may be observed that radiate from the edges of large, well-established colonies. Nonresistant satellite colonies grow in an "antibiotic shadow" where ampicillin has been broken down by the large resistant colony. Do not include satellite colonies in the count of transformants. A "lawn" should be observed on positive controls, where the bacteria cover nearly the entire agar surface and individual colonies cannot be discerned.

1. Record your observation of each plate in the matrix below. If cell growth is too dense to count individual colonies, record "lawn." Were results as expected? Explain possible reasons for variations from expected results.

	Transformed cells +pAMP	Nontransformed cells −pAMP
LB/amp	experiment	negative control
LB	positive control	positive control

2. Compare and contrast the growth on each of the following pairs of plates. What does each pair of results tell you about the experiment?

    a. +LB and −LB

    b. −LB/amp and −LB

    c. +LB/amp and −LB/amp

    d. +LB/amp and +LB

3. Transformation efficiency is expressed as the number of antibiotic resistant colonies per microgram of pAMP DNA. The object is to determine the mass of pAMP that was spread on the experimental plate and was therefore responsible for the transformants observed.

    a. Determine total mass (in micrograms) of pAMP used in Step 9: concentration × volume = mass.

    b. Determine fraction of cell suspension spread onto +LB/amp plate (Step 18): volume suspension spread/*total* volume suspension (Steps 2 and 14) = fraction spread.

    c. Determine mass of pAMP in cell suspension spread onto +LB/amp plate: total mass pAMP (*a*) × fraction spread (*b*) = mass pAMP spread.

    d. Determine number of colonies per microgram of pAMP. Express answer in scientific notation: colonies observed/mass pAMP spread (*c*) = transformation efficiency.

4. What factors might influence transformation efficiency?

5. Your Favorite Gene (YFG) is cloned into pAMP, and *E. coli* is transformed with 0.2 µg of intact pAMP/YFG recombinant according to the protocol above. Using the information below, calculate the number of molecules of YFG that have been cloned *in the entire culture* 200 minutes after inoculation—when the culture enters stationary phase.

    a. A transformation efficiency equal to 10^6 colonies per microgram of intact pAMP is achieved.

    b. pAMP has an average copy number of 100 molecules per transformed cell.

c. Following heat shock (Step 12), the entire 250 μl of cell suspension is used to inoculate 25 ml of fresh LB broth. The culture is incubated, with shaking, at 37°C.

d. Transformed cells enter log phase 60 minutes after inoculation and then begin to replicate an average of once every 20 minutes.

6. The transformation protocol above is used with 10 μl of intact plasmid DNA at four different concentrations. The following numbers of colonies are obtained when 100 μl of transformed cells are plated on selective medium:

0.00001 μg/μl	4 colonies
0.00005 μg/μl	12 colonies
0.0001 μg/μl	32 colonies
0.0005 μg/μl	125 colonies
0.001 μg/μl	442 colonies
0.005 μg/μl	542 colonies
0.01 μg/μl	507 colonies
0.05 μg/μl	475 colonies
0.1 μg/μl	516 colonies

a. Calculate transformation efficiencies at each concentration.

b. Plot a graph of DNA mass *versus* colonies.

c. Plot a graph of DNA mass *versus* transformation efficiency.

d. What is the relationship between mass of DNA transformed and transformation efficiency?

e. At what point does the transformation reaction appear to be saturated?

f. What is the true transformation efficiency?

FOR FURTHER RESEARCH

Interpretable experimental results can only be achieved when the colony transformation can be repeated with reproducible results. Only attempt experiments below when you are able to routinely achieve 100–500 colonies on the +LB/amp plate.

1. Design and execute an experiment to compare transformation efficiencies of linear *versus* circular plasmid DNAs. Keep molecular weight constant.

2. Design and execute a series of experiments to test the relative importance of each of the four major steps of most transformation protocols: (1) preincubation, (2) incubation, (3) heat shock, and (4) recovery. Which steps are absolutely necessary?

3. Design and execute a series of experiments to compare the transforming effectiveness of $CaCl_2$ versus the salts of other monovalent (+), divalent (++), and trivalent (+++) cations.

   a. Make up 50 mM solutions of each salt.

   b. Check pH of each solution, and buffer to approximately pH 7 when necessary.

   c. Is $CaCl_2$ unique in its ability to facilitate transformation?

   d. Is there any consistent difference in the transforming ability of monovalent versus divalent versus trivalent cations?

4. Do a series of experiments to determine saturating conditions for transformation reactions.

   a. Transform E. coli using DNA concentrations listed in Question 6 above.

   b. Plot a graph of DNA mass versus colonies per plate.

   c. Plot a graph of DNA mass versus transformation efficiency.

   d. At what mass does the reaction appear to become saturated?

   e. Repeat experiment with concentrations clustered on either side of the presumed saturation point to produce a fine saturation curve.

5. Repeat Experiment 4 above, but transform with a 1:1 mixture of pAMP and pKAN at each concentration. Plate transformants on LB/amp, LB/kan, and LB/amp+kan plates. *Be sure to include a 40–60-minute recovery, with shaking.*

   a. Calculate percentage of double transformations at each mass.

   $$\frac{\text{colonies LB/amp+kan plate}}{\text{colonies LB/amp plate + colonies LB/kan plate}}$$

   b. Plot a graph of DNA mass versus colonies per plate.

   c. Plot a graph of DNA mass versus percentage double transformations. Under saturating conditions, what percentage of bacteria are doubly transformed?

6. Plot a recovery curve for E. coli transformed with pKAN. Allow cells to recover for 0–120 at 20-minute intervals.

   a. Plot a graph of recovery time versus colonies per plate.

   b. At what time point is antibiotic expression maximized?

   c. Can you discern a point at which the cells began to replicate?

Laboratory 6

Purification and Identification of Plasmid DNA

Growth of *E. coli* on ampicillin plates demonstrates transformation to an antibiotic-resistant phenotype. Laboratory 5 showed that the observed phenotype is due to uptake of plasmid pAMP, a DNA molecule that is well-characterized.

In experimental situations where numerous recombinant plasmids are generated by joining two or more DNA fragments, the antibiotic resistance marker only functions to indicate which cells have taken up a plasmid bearing the resistance gene. It does not indicate anything about the structure of the new plasmid. Therefore, it is standard procedure to isolate plasmid DNA from transformed cells and to identify the molecular genotype using DNA restriction analysis. In cases where the recombinant molecules are formed by combining well-characterized fragments, restriction analysis is sufficient to confirm the structure of a hybrid plasmid. In other cases, the nucleotide sequence of the insert must be determined.

In Part A, Plasmid Minipreparation of pAMP, a small-scale protocol is used to purify from transformed *E. coli* enough plasmid DNA for restriction analysis. Cells taken from an ampicillin-resistant colony are grown to stationary phase in suspension culture. The cells from 1 ml of culture are harvested and lysed, and plasmid DNA is separated from the cellular proteins, lipids, and chromosomal DNA. This procedure yields 2–5 µg of relatively crude plasmid DNA, in contrast to large-scale preparations that yield 1 mg or more of pure plasmid DNA from a 1-liter culture.

In Part B, Restriction Analysis of Purified pAMP, a sample of plasmid DNA isolated in Part A and a control sample of pAMP are cut with the restriction enzymes *Bam*HI and *Hin*dIII. The two samples are coelectrophoresed on an agarose gel, and the restriction patterns are stained and visualized. The purified DNA is shown to have a restriction "fingerprint" identical to that of pAMP. *Bam/Hin*d restriction fragments of the miniprep DNA comigrate with the 784- and 3755-bp *Bam/Hin*d fragments of pAMP. This provides genotypic proof that pAMP molecules were successfully transformed into *E. coli* in Laboratory 5.

LABORATORY 6/PART A

Plasmid Minipreparation of pAMP

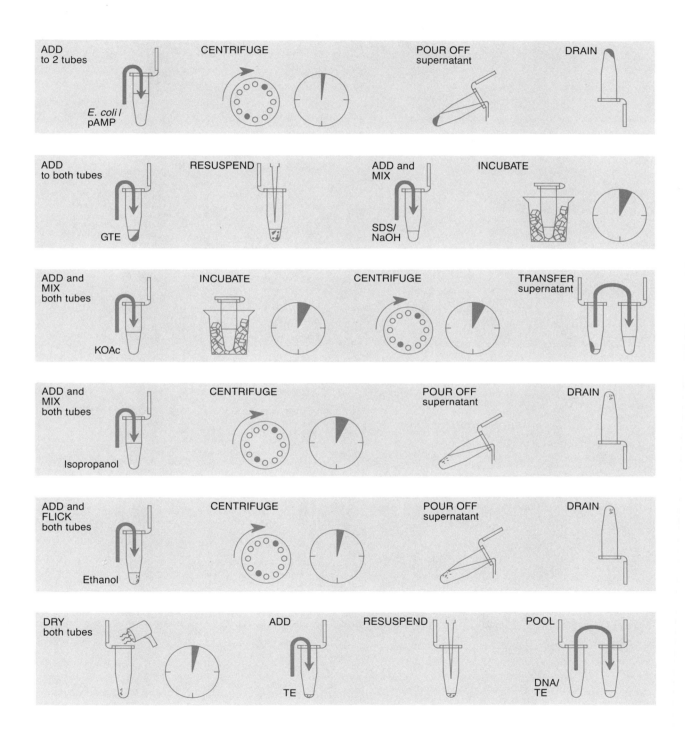

PRELAB NOTES

Optimally, minipreps should be done on cells that have been recently manipulated for transformation. This completes a conceptual stream that firmly cements the genotype-phenotype relationship. Alternately, use streaked plates of transformed *E. coli* to prepare overnight cultures.

Plasmid Selection

pAMP gives superior yields on minipreps compared to pBR322. A derivative of a pUC expression vector, pAMP is highly amplified—more than 100 copies are present per *E. coli* cell. If substituting a different plasmid for miniprep purposes, select a commercially available member of the pUC family, such as pUC18 or pUC19.

Centrifuge Requirements

A microfuge that generates approximately 12,000 times the force of gravity ($12,000g$) is needed for efficient and rapid purification of plasmid DNA. A slower-spinning clinical or preparatory centrifuge cannot be substituted.

Supplies

Sterile supplies are not required for this protocol. Standard 1-ml pipets, transfer pipets, and/or microcapillary pipets can be used instead of micropipettors. Use good-quality, colorless 1.5-ml tubes, beginning with Step 11. The walls of poor-quality tubes, especially colored ones, often contain tiny air bubbles that can be mistaken for ethanol droplets in Step 19. We have observed workers drying DNA pellets for 15 minutes or more, trying to rid their tubes of these phantom droplets. (Typical drying time is actually several minutes.)

Fine Points of Technique

Be careful not to overmix reagents; excessive manipulation shears both plasmid and chromosomal DNA. The success of this protocol in large part depends on maintaining chromosomal DNA in large pieces that can be differentially separated from intact plasmid DNA. Mechanical shearing reduces the yield of high-quality, supercoiled plasmid DNA and increases contamination with short-sequence chromosomal DNA. Make sure that microfuge will be immediately available for Step 13. If sharing a microfuge, coordinate with other experminenters to begin Steps 12 and 13 together.

For Further Information

The protocol presented here is based on the following published methods:

Birnboim, H.C. and J. Doly. 1979. A rapid alkaline extraction method for screening recombinant plasmid DNA. *Nucleic Acids Res.* 7: 1513.

Ish-Horowicz, D. and J.F. Burke. 1981. Rapid and efficient cosmid cloning. *Nucleic Acids Res.* 9: 2989.

PRELAB PREPARATION

1. The day before the laboratory, prepare an *E. coli* culture according to the protocol in Laboratory 2B, Overnight Suspension Culture. Inoculate the culture with a cell mass scraped from one colony selected from the +LB/amp plate from Laboratory 5, Rapid Colony Transformation of *E. coli* with Plasmid DNA. Maintain antibiotic selection with LB broth plus ampicillin. Alternately, the culture can be prepared 2–3 days in advance and stored at 4°C or incubated at 37°C without shaking for 24–48 hours. In either case, the cells will settle at the bottom of the culture tube. Shake tube to resuspend cells before beginning procedure.

2. The SDS/sodium hydroxide solution must be fresh; prepare within a few days of lab. Store solution at room temperature; a soapy precipitate may form at lower temperature. Warm solution by placing tube in a beaker of hot tap water, and shake gently to dissolve the precipitate.

3. Aliquot for each experiment:

    250 μl of glucose/Tris/EDTA (GTE) solution (store on ice).
    500 μl of SDS/sodium hydroxide (SDS/NaOH) solution.
    400 μl of potassium acetate/acetic acid (KOAc) solution (store on ice).
    1000 μl of isopropanol.
    500 μl of 95% ethanol.
    50 μl of Tris/EDTA (TE) solution.

4. Review Part B, Restriction Analysis of Purified pAMP.

Plasmid Minipreparation of pAMP

CULTURE AND REAGENTS

E. coli/pAMP overnight culture
glucose/Tris/EDTA (GTE)
SDS/sodium hydroxide (SDS/NaOH)
potassium acetate/acetic acid (KOAc)
isopropanol
95–100% ethanol
Tris/EDTA (TE)

SUPPLIES AND EQUIPMENT

100–1000-µl micropipet + tips
0.5–10-µl micropipet + tips
1.5-ml tubes
beaker of crushed ice
beaker for waste/used tips
10% bleach or disinfectant
clean paper towels
hair dryer
microfuge
permanent marker
test tube rack

Prepare Duplicate Minipreps (50 minutes)

The instructions below are for making duplicate minipreps, which provide balance in the microfuge and insurance if a critical mistake is made.

1. Shake culture tube to resuspend *E. coli* cells.

2. Label two 1.5-ml tubes with your initials. Use a micropipettor to transfer 1000 µl of *E. coli*/pAMP overnight suspension into each tube.

3. Close caps, and place tubes in a *balanced* configuration in microfuge rotor. Spin for 1 minute to pellet cells.

4. Pour off supernatant from both tubes into waste beaker for later disinfection. *Be careful not to disturb cell pellets.* Invert tubes, and tap gently on surface of clean paper towel to drain thoroughly.

5. Add 100 µl of GTE solution to each tube. Resuspend pellets by pipetting solution in and out several times. Hold tubes up to light to check that suspension is homogeneous and that no visible clumps of cells remain.

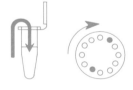
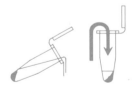

The cell pellet will appear as a small off-white smear on the bottom-side of the tube. Although the cell pellets are readily seen, the DNA pellets in Step 14 are very difficult to observe. Get into the habit of aligning tube with cap hinges facing up in the microfuge rotor. Then, the pellets should always be located at the tube bottom beneath the hinge.

Accurate pipetting is essential to good plasmid yield. The volumes of reagents are precisely calibrated so that sodium hydroxide added in Step 6 is neutralized by acetic acid in Step 8.

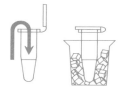

In Step 11, the supernatant is saved and the precipitate is discarded. The situation is reversed in Steps 14 and 17, where the precipitate is saved and the supernatant is discarded.

Do Step 12 quickly, and make sure that microfuge will be immediately available for Step 13.

The pellet may appear as a tiny smear or small particles on the bottom-side of each tube. Do not be concerned if pellet is not visible; pellet size is not a predictor of plasmid yield. A large pellet is composed primarily of RNA and cellular debris carried over from the original precipitate. A smaller pellet often means a cleaner preparation.

Nucleic acid pellets are not soluble in ethanol and will not resuspend during washing.

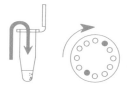

6. Add 200 μl of SDS/NaOH solution to each tube. Close caps, and mix solutions by rapidly inverting tubes five times.

7. Stand tubes on ice for 5 minutes. Suspension will become relatively clear.

8. Add 150 μl of *ice-cold* KOAc solution to each tube. Close caps, and mix solutions by rapidly inverting tubes five times. A white precipitate will immediately appear.

9. Stand tubes on ice for 5 minutes.

10. Place tubes in a *balanced* configuration in microfuge rotor, and spin for 5 minutes to pellet precipitate along side of tube.

11. Transfer 400 μl of supernatant from each tube into two clean 1.5-ml tubes. *Avoid pipetting precipitate,* and wipe off any precipitate clinging to outside of tip prior to expelling supernatant. Discard old tubes containing precipitate.

12. Add 400 μl of isopropanol to each tube of supernatant. Close caps, and mix vigorously by rapidly inverting tubes five times. *Stand at room temperature for only 2 minutes.* (Isopropanol preferentially precipitates nucleic acids rapidly; however, proteins remaining in solution also begin to precipitate with time.)

13. Place tubes in a *balanced* configuration in microfuge rotor, and spin for 5 minutes to pellet the nucleic acids. Align tubes in rotor so that cap hinges point outward. The nucleic acid residue, visible or not, will collect under hinge during centrifugation.

14. Pour off supernatant from both tubes. *Be careful not to disturb nucleic acid pellets.* Invert tubes, and tap gently on surface of clean paper towel to drain thoroughly.

15. Add 200 μl of 100% ethanol to each tube, and close caps. Flick tubes several times to wash pellets.

STOP POINT   Store DNA in ethanol at −20°C until ready to continue.

---

16. Place tubes in a *balanced* configuration in microfuge rotor, and spin for 2–3 minutes.

17. Pour off supernatant from both tubes. *Be careful not to disturb nucleic acid pellets.* Invert tubes, and tap gently on surface of clean paper towel to drain thoroughly.

18. Dry nucleic acid pellets by one of following methods:

    a. Direct a stream of warm air from a hair dryer into open ends of tubes for about 3 minutes. *Be careful not to blow pellets out of tubes.*

    or

    b. Close caps, and pulse tubes in microfuge to pool remaining ethanol. *Carefully* draw off drops of ethanol using a 1–10-µl micropipettor. Allow pellets to air dry at room temperature for 10 minutes.

19. At the end of the drying period, hold each tube up to light to check that no ethanol droplets remain. If ethanol is still evaporating, an alcohol odor can be detected by sniffing mouth of tube. All ethanol must be evaporated before proceeding to Step 20.

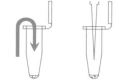

20. Add 15 µl of TE to each tube. Resuspend pellets by smashing with the pipet tip and pipetting in and out vigorously. Rinse down the side of tube several times, concentrating on the area where the pellet should have formed during centrifugation (beneath cap hinge). Check that all DNA is dissolved and that no particles remain in tip or on side of tube.

If you are using a 0.5–10-µl micropipettor, set to 7.5 µl and pipet twice.

21. Pool DNA/TE solution into one tube.

STOP POINT   Freeze DNA/TE solution at −20°C until ready to continue. Thaw before using.

---

22. Take time for responsible cleanup.

    a. Segregate for proper disposal—culture tubes and micropipettor tips that have come in contact with *E. coli*.

    b. Disinfect overnight culture, tips, and supernatant from Step 4 with 10% bleach or disinfectant.

    c. Wipe down lab bench with soapy water, 10% bleach solution, or disinfectant (such as Lysol).

    d. Wash hands before leaving lab.

RESULTS AND DISCUSSION

The minipreparation is a simple and efficient procedure for isolating plasmid DNA. You should be familiar with the molecular and biochemical effects of each reagent used in the protocol.

- *Glucose-Tris-EDTA:* The Tris buffers the cells at pH 7.9. EDTA binds divalent cations in the lipid bilayer, thus weakening the cell envelope.

- *SDS-sodium hydroxide:* This alkaline mixture lyses the bacterial cells. The detergent SDS dissolves the lipid components of the cell envelope, as well as cellular proteins. The sodium hydroxide denatures the chromosomal and plasmid DNA into single strands. The intact circles of plasmid DNA remain intertwined.

- *Potassium acetate–acetic acid:* The acetic acid returns the pH to neutral, allowing DNA strands to renature. The large, disrupted chromosomal strands cannot rehybridize perfectly, but instead collapse into a partially hybridized tangle. At the same time, the potassium acetate precipitates the SDS from the cell suspension, along with proteins and lipids with which it has associated. The renaturing chromosomal DNA is trapped in the SDS/lipid/protein precipitate. Only smaller plasmid DNA and RNA molecules escape the precipitate and remain in solution.

- *Isopropanol:* The alcohol rapidly precipitates nucleic acids, but only slowly precipitates proteins. Thus, a quick precipitation preferentially brings down nucleic acids.

- *Ethanol:* A wash with ethanol removes some remaining salts and SDS from the preparation.

- *Tris-EDTA:* Tris buffers the DNA solution. EDTA protects the DNA from degradation by DNases by binding divalent cations that are necessary cofactors for DNase activity.

1. Consider the three major classes of biologically important molecules: proteins, lipids, and nucleic acids. Which steps of the miniprep procedure act on proteins? On lipids? On nucleic acids?

2. What aspect of plasmid DNA structure allows it to renature efficiently in Step 8?

3. What other kinds of molecules, in addition to plasmid DNA, would you expect to be present in the final miniprep sample? How could you find out?

FOR FURTHER RESEARCH

Determine the approximate mass of plasmid DNA you isolated per milliliter of cells.

a. Set up 20-μl *Hin*dIII restriction reactions using 15 μl of your pAMP preparation and a known mass of λ DNA as a control.

b. Make 1:10, 1:50, and 1:100 dilutions of the digested pAMP and λ DNAs.

c. Electrophorese equal volumes of each dilution in an agarose gel, and stain with ethidium bromide.

> **CAUTION**
> Review Responsible Handling of Ethidium Bromide (page 256). Wear rubber gloves when staining, viewing, and photographing gel and during cleanup. Confine all staining to a restricted sink area.

d. Identify a lane of the λ digest where the 4361-bp fragment is *just* visible, and identify a lane of pAMP (4539 bp) that is of equal intensity. These bands should have a nearly equivalent mass of DNA.

e. Determine the mass of λ DNA in the selected fragment, using the formula below. Be sure to take into account the dilution factor.

$$\frac{\text{fragment bp (conc. DNA) (vol. DNA)}}{\lambda \text{ bp}}$$

f. Multiply the mass from Step 5 by the dilution factor of the selected pAMP lane.

LABORATORY 6/PART B

Restriction Analysis of Purified pAMP

I. **Set Up Restriction Digest**

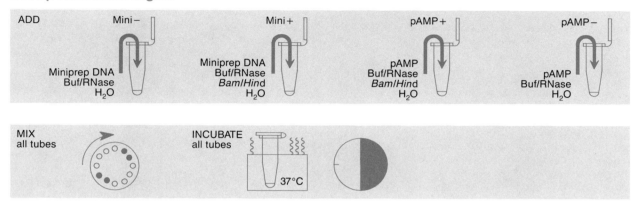

II. **Cast 0.8% Agarose Gel**

III. **Load Gel and Electrophorese**

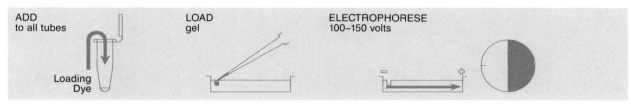

IV. **Stain Gel and View**   V. **Photograph Gel**

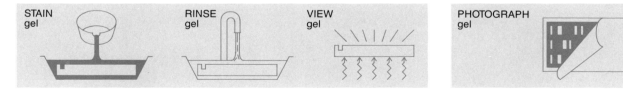

PRELAB NOTES

Review Prelab Notes in Laboratory 3, DNA Restriction Analysis.

Limiting DNase Activity

Unlike highly purified plasmid DNA available from commercial vendors, miniprep DNA is impure. A significant percentage of nucleic acid in the preparation is, in fact, RNA and fragmented chromosomal DNA. Typically, miniprep DNA is contaminated with nucleases (DNases) that cleave DNA into small fragments. Residual DNases will degrade plasmid DNA if minipreps are left for long periods of time at room temperature or even on ice. For this reason, store minipreps at $-20°C$, and thaw just prior to use.

The situation is further complicated during restriction digestion. DNases and restriction endonucleases both require divalent cations, such as Mg^{++}. Included in TE buffer at a low concentration of 1 mM, EDTA chelates (binds) a large proportion of divalent cations in the DNA preparation. Although it is likely that some divalent cations remain free to activate DNases, a higher concentration of EDTA would chelate Mg^{++} in the restriction buffer, inhibiting restriction enzyme activity. Thus, a balance is struck at an EDTA concentration that inhibits most of the contaminating DNases without significantly reducing the activity of the restriction enzymes.

Another balance must be struck. On the one hand, contaminants in the miniprep limit restriction enzyme activity—a 20-minute incubation is not usually sufficient for complete digestion. On the other hand, DNases are activated by Mg^{++} in the restriction buffer and will significantly degrade plasmid DNA if the restriction reaction is incubated too long. Experience has shown that a 30-minute incubation gives optimal results.

RNase

Miniprep DNA is contaminated by large amounts of ribosomal RNA and smaller amounts of messenger RNA and transfer RNA. If not removed from the preparation, this RNA will obscure the DNA bands in the agarose gel. Therefore, RNase is added to the restriction digest; during incubation, the RNase digests RNA into very small fragments (several hundred nucleotides or less). These RNA fragments run well ahead of the DNA fragments of interest. Use only RNase A from bovine pancreas. Boil RNase solution for 15 minutes to destroy contaminating DNases that may be present in the preparation.

For Further Information

The protocol presented here is based on the following published methods:

Helling, R.B., H.M. Goodman, and H.W. Boyer. 1974. Analysis of *Eco*RI fragments of DNA from lambdoid bacteriophages and other viruses by agarose-gel electrophoresis. *J. Virol. 14:* 235.

Sharp, P.A., B. Sugden, and J. Sambrook. 1973. Detection of two restriction endonuclease activities in *Haemophilus parainfluenzae* using analytical agarose-ethidium bromide electrophoresis. *Biochemistry 12:* 3055.

PRELAB PREPARATION

1. Mix in 1:1 proportion, BamHI + HindIII (6 μl per experiment).

2. Aliquot for each experiment:

    12 μl of 0.1 μg/μl pAMP (store on ice).
    12 μl of 5× restriction buffer/RNase (store on ice).
    6 μl of BamHI/HindIII (store on ice).
    500 μl of distilled water.
    500 μl of loading dye.

    If another plasmid was substituted for pAMP in the transformation, use that plasmid as a control in the restriction digest.

3. Prepare 0.8% agarose solution (approximately 40–50 ml per experiment). Keep agarose liquid in a hot-water bath (at about 60°C) throughout experiment. Cover solution with aluminum foil to retard evaporation.

4. Prepare 1× Tris/Borate/EDTA (TBE) buffer for electrophoresis (400–500 ml per experiment).

5. Prepare ethidium bromide or methylene blue staining solution (100 ml per experiment).

6. Adjust water bath to 37°C.

Restriction Analysis of Purified pAMP

REAGENTS

miniprep DNA/TE
0.1 µg/µl pAMP
BamHI/HindIII
5× restriction buffer/RNase
distilled water
loading dye
0.8% agarose
1× Tris/Borate/EDTA (TBE) buffer
1 µg/µl ethidium bromide
  (or 0.025% methylene blue)

SUPPLIES AND EQUIPMENT

0.5–10-µl micropipettor +
  tips
1.5-ml tubes
aluminum foil
beaker for agarose
beaker for waste/used tips
10% bleach
camera and film (optional)
electrophoresis box
masking tape
microfuge (optional)
Parafilm or wax paper
  (optional)
permanent marker
plastic wrap (optional)
power supply
rubber gloves
test tube rack
transilluminator (optional)
37°C water bath

I. Set Up Restriction Digest (40 minutes, including incubation)

Refer to Laboratory 3, DNA Restriction Analysis, for more detailed instructions.

1. Use permanent marker to label four 1.5-ml tubes, in which restriction reactions will be performed:

    Mini −   = miniprep, no enzymes
    Mini +   = miniprep + BamHI/HindIII
    pAMP +   = pAMP + BamHI/HindIII
    pAMP −   = pAMP, no enzymes

2. Use matrix on following page as a checklist while adding reagents to each reaction. Read down each column, adding the same reagent to all appropriate tubes. *Use a fresh tip for each reagent.* Refer to detailed directions that follow.

Tube	Miniprep DNA	pAMP	Buffer/ RNase	BamHI/ HindIII	H$_2$O
Mini−	5 μl	—	2 μl	—	3 μl
Mini+	5 μl	—	2 μl	2 μl	1 μl
pAMP+	—	5 μl	2 μl	2 μl	1 μl
pAMP−	—	5 μl	2 μl	—	3 μl

3. Collect reagents, and place in test tube rack on lab bench.

4. Add 5 μl of miniprep DNA to tubes labeled Mini− and Mini+.

5. Use *fresh tip* to add 5 μl of pAMP to tubes labeled pAMP+ and pAMP−.

6. Use *fresh tip* to add 2 μl of restriction buffer/RNase to a clean spot on each reaction tube.

7. Use *fresh tip* to add 2 μl of BamHI/HindIII to tubes labeled Mini+ and pAMP+.

8. Use *fresh tip* to add proper volumes of distilled H$_2$O to each tube.

9. Close tube tops. Pool and mix reagents by pulsing in a microfuge or by sharply tapping the tube bottom on lab bench.

Do not overincubate. During longer incubation, DNases in miniprep may degrade plasmid DNA.

10. Place reaction tubes in 37°C water bath, and incubate for 30 minutes only.

STOP POINT Following incubation, freeze reactions at −20°C until ready to continue. Thaw reactions before continuing to Section III, Step 1.

Gel is cast directly in box in some electrophoresis apparatuses.

II. Cast 0.8% Agarose Gel (15 minutes)

1. Seal ends of gel-casting tray with tape, and insert well-forming comb. Place gel-casting tray out of the way on lab bench so that agarose poured in next step can set undisturbed.

2. Carefully pour enough agarose solution into casting tray to fill to a depth of about 5 mm. Gel should cover only about one-third the height of comb teeth. Use a pipet tip to move large bubbles or solid debris to sides or end of tray, while gel is still liquid.

3. Gel will become cloudy as it solidifies (about 10 minutes). *Be careful not to move or jar casting tray while agarose is solidifying.* Touch corner of agarose *away* from comb to test whether gel has solidified.

4. When agarose has set, unseal ends of casting tray. Place tray on platform of gel box, so that comb is at negative (black) electrode.

5. Fill box with TBE buffer to a level that just covers entire surface of gel.

6. Gently remove comb, taking care not to rip wells.

7. Make certain that sample wells left by comb are completely submerged. If "dimples" are noticed around wells, slowly add buffer until they disappear.

STOP POINT Cover electrophoresis tank and save gel until ready to continue. Gel will remain in good condition for at least several days if it is completely submerged in buffer.

Too much buffer will channel current over the top rather than through the gel, increasing the time required to separate DNA. TBE buffer can be used several times; do not discard. If using buffer remaining in electrophoresis box from a previous experiment, rock chamber back and forth to remix ions that have accumulated at either end.

Buffer solution helps to lubricate the comb. Some gel boxes are designed such that the comb must be removed prior to inserting the casting tray into the box. In this case, flood casting tray and gel surface with running buffer *before* removing the comb. Combs removed from a dry gel can cause tearing of the wells.

III. Load Gel and Electrophorese (30–50 minutes)

1. Add loading dye to each reaction. Either

   a. Add 1 μl of loading dye to each reaction tube. Close tube tops, and mix by tapping tube bottom on lab bench, pipetting in and out, or pulsing in a microfuge. Make sure tubes are placed in a *balanced* configuration in rotor.

   or

   b. Place four individual droplets of loading dye (1 μl each) on a small square of Parafilm or wax paper. Withdraw contents from reaction tube, and mix with a loading dye droplet by pipetting in and out. Immediately load dye mixture according to Step 2. Repeat successively, *with clean tip,* for each reaction.

2. Use micropipettor to load entire contents of each reaction tube into separate well in gel, as shown on following page. Use *fresh tip* for each reaction.

   a. Steady pipet over well using two hands.

   b. If there is air in the end of the tip, carefully depress plunger to push the sample to the end of the tip. (If air bubble forms "cap" over well, DNA/loading dye will flow into buffer around edges of well.)

   c. Dip pipet tip through surface of buffer, center it over the well, and gently depress pipet plunger to slowly expel sample. Sucrose in the loading dye weighs down the sample, causing it to sink to the bottom of the well. *Be careful not to punch tip of pipet through bottom of gel.*

A piece of dark construction paper beneath the gel box will make the wells more visible.

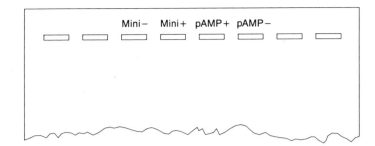

Alternately, set power supply on lower voltage, and run gel for several hours. When running two gels off the same power supply, current is double that for a single gel at the same voltage.

The *Bam*HI/*Hin*dIII digest yields two bands containing small fragments of 784 bp and 3755 bp, which are easily resolved during a short electrophoresis run. The 784-bp fragment runs directly behind the purplish band of bromophenol blue (equivalent to approximately 300 bp), while the 3755-bp fragment runs in front of the aqua band of xylene cyanol (equivalent to approximately 9000 bp).

Staining may be performed by an instructor in a controlled area when students not present.

3. Close top of electrophoresis box, and connect electrical leads to a power supply, anode to anode (red-red) and cathode to cathode (black-black). Make sure both electrodes are connected to same channel of power supply.

4. Turn power supply on, and set to 100–150 volts. The ammeter should register approximately 50–100 milliamperes. If current is not detected, check connections and try again.

5. Electrophorese for 20–40 minutes. Good separation will have occurred when the bromophenol blue band has moved 4–6 cm from wells. If time allows, electrophorese until the bromophenol blue band nears the end of the gel. *Stop* electrophoresis before bromophenol blue band runs off end of gel.

6. Turn off power supply, disconnect leads from the inputs, and remove top of electrophoresis box.

7. Carefully remove casting tray from electrophoresis box, and slide gel into disposable weigh boat or other shallow tray. Label staining tray with your name.

STOP POINT Cover electrophoresis tank and save gel until ready to continue. Gel can be stored in a zip-lock plastic bag and refrigerated overnight for viewing/photographing the next day. However, over longer periods of time, the DNA will diffuse through the gel, and the bands will become indistinct or disappear entirely.

8. Stain and view gel using one of the methods described in Sections IVA and IVB.

IVA. Stain Gel with Ethidium Bromide and View (10–15 minutes)

CAUTION

Review Responsible Handling of Ethidium Bromide (page 256). Wear rubber gloves when staining, viewing, and photographing gel and during cleanup. Confine all staining to a restricted sink area.

1. Flood gel with ethidium bromide solution (1 μg/ml), and allow to stain for 5–10 minutes.

2. Following staining, use funnel to decant as much ethidium bromide solution as possible from staining tray back into storage container.

Staining time increases markedly for thicker gels.

Ethidium bromide solution may be reused to stain 15 or more gels. When staining time increases markedly, disable ethidium bromide solution as explained on page 256.

3. Rinse gel and tray under running tap water.

4. If desired, gel can be destained in tap water or distilled water for 5 minutes or more to help remove background ethidium bromide from the gel.

STOP POINT Staining intensifies dramatically if rinsed gels set overnight at room temperature. Stack staining trays, and cover top gel with plastic wrap to prevent desiccation.

5. View under ultraviolet transilluminator or other UV source.

> **CAUTION**
>
> Ultraviolet light can damage eyes. Never look at unshielded UV light source with naked eyes. View only through a filter or safety glasses that absorb harmful wavelengths.

6. Take time for responsible cleanup.
   a. Wipe down camera, transilluminator, and staining area.
   b. Wash hands before leaving lab.

IVB. Stain Gel with Methylene Blue and View (30+ minutes)

1. Wear rubber gloves during staining and cleanup.

2. Flood gel with 0.025% methylene blue, and allow to stain for 20–30 minutes.

3. Following staining, use funnel to decant as much methylene blue solution as possible from staining tray back into storage container.

4. Rinse gel in running tap water. Let gel soak for several minutes in several changes of fresh water. DNA bands will become increasingly distinct as gel destains.

Destaining time is decreased by rinsing the gel in warm water, with agitation.

STOP POINT For best results, continue to destain overnight in a *small volume* of water. (Gel may destain too much if left overnight in large volume of water.) Cover staining tray to retard evaporation.

5. View gel over light box; cover surface of light box with plastic wrap to prevent staining.

V. Photograph Gel (5 minutes)

Exposure times vary according to mass of DNA in lanes, level of staining, degree of background staining, thickness of gel, and density of filter. Experiment to determine best exposure. When possible, stop lens down (to higher f/number) to increase the depth of field and sharpness of bands.

1. *For ultraviolet (UV) photography of ethidium-bromide-stained gels:* Use Polaroid high-speed film Type 667 (ASA 3000). Set camera aperture to f/8 and shutter speed to B. Depress shutter for a 2–3-second time exposure.

   For white-light photography of methylene-blue-stained gels: Use Polaroid Type 667 film, with an aperture of f/8 and a shutter speed of 1/125 second.

   > **CAUTION**
   >
   > Avoid getting caustic developing jelly on skin or clothes. If jelly gets on skin, wash immediately with plenty of soap and water.

2. Place left hand firmly on top of camera to steady. Firmly grasp small white tab, and pull straight out from camera. This causes a large yellow tab to appear.

3. Grip yellow tab in center, and in one steady motion, pull it straight out from the camera. This starts development.

4. Allow to develop for recommended time (45 seconds at room temperature). Do not disturb print while developing.

5. After full development time has elapsed, separate print from negative by peeling back at end nearest yellow tab.

6. Wait to see the result of the first photo before making other exposures.

RESULTS AND DISCUSSION

Interpreting gels containing plasmid DNA is not always straightforward and is complicated by impurities in miniprep DNA. Examine the gel and determine which lanes contain cut and uncut control pAMP and which lanes contain miniprep DNA. Even if you have confused the prescribed loading order, the miniprep lanes can be distinguished by the following characteristics:

1. A background "smear" of degraded and partially digested chromosomal DNA, plasmid DNA, and RNA is typically seen running much of the length of the miniprep lanes. The smear is composed of faint bands of virtually every nucleotide length that grade together. A heavy background smear, along with high-molecular-weight DNA at the top of the undigested lane, indicates the miniprep is contaminated with large amounts of chromosomal DNA.

2. Frequently, undissolved material and high-molecular-weight DNA are seen "trapped" at the front edge of the well. These anomalies are not seen in commercial preparations, where plasmid DNA is separated from degraded nucleic acids by ultracentrifugation in a cesium chloride gradient.

3. A "cloud" of low-molecular-weight RNA is often seen in both the cut and uncut miniprep lanes at a position corresponding to 100–200 bp. Again, variously sized molecules are represented, which are the remnants of larger molecules that have been partially digested by the RNase. However, the majority of RNA is usually digested into fragments that are too small to intercalate the ethidium bromide dye or that electrophorese off the end of the gel.

4. Only two bands (784 bp and 3755 bp) are expected to be seen in the cut miniprep lane. However, it is common to see one or more faint bands higher up on the gels that comigrate with the uncut plasmid forms described below. Incomplete digestion is usually due to contaminants in the preparation that inhibit restriction enzyme activity or may occur when the miniprep solution contains a very high concentration of plasmid DNA.

It is especially confusing to see several bands in a lane known to contain only uncut plasmid DNA. This occurs because the migration of plasmid DNA in an agarose gel depends on its molecular conformation, as well as its molecular weight (base-pair size). Plasmid DNA exists in one of three major conformations:

- *Form I, supercoiled:* Although a plasmid is usually pictured as an open circle, within the *E. coli* cell (in vivo), the DNA strand is wound around histone-like proteins to make a compact structure. Adding these coils to the coiled DNA helix produces a *super*coiled molecule. The enzyme DNA gyrase (topoisomerase II) rotates one strand of the DNA molecule to introduce the supercoils around the supporting proteins. The extraction procedure strips proteins from plasmid DNA, causing the molecule to coil about itself. (Supercoiling can be demonstrated by adding twists to one end of a loop of rope.) Under most gel conditions, the supercoiled plasmid DNA is the fastest-moving form. Its compact molecular shape allows it to move most easily through the agarose matrix. Therefore, the fastest moving band of uncut plasmid DNA is assumed to be supercoiled.

- *Form II, relaxed or nicked circle:* During DNA replication, the enzyme topoisomerase I introduces a nick into one strand of the DNA helix and rotates the strand to release the torsional strain that holds the molecule in a supercoil. The relaxed section of the DNA uncoils, allowing access to the replicating enzymes. Introducing nicks into supercoiled plasmid DNA produces the open circular structure with which we are familiar. Physical shearing and enzymatic cleavage during plasmid isolation introduce nicks in the supercoiled plasmid DNA. Thus, the percentage of supercoiled plasmid DNA is an indicator of the care with which the DNA is extracted from the *E. coli* cell. The relaxed circle is the slowest-migrating form of plasmid DNA; its "floppy" molecular shape impedes movement through the agarose matrix.

- *Form III, linear:* Linear DNA is produced when a restriction enzyme cuts the plasmid at a single recognition site or when damage results in strand nicks directly opposite each other on the DNA helix. Under most gel conditions, linear plasmid DNA migrates at a rate intermediate between supercoiled and relaxed circle. The presence of linear DNA in a plasmid preparation is a sign of contamination with nucleases or of sloppy lab procedure (overmixing or mismeasuring SDS/NaOH and KOAc).

MM294 and other strains of *E. coli*, termed *recA*$^+$, have an enzyme system that recombines plasmids to form large concatemers of two or more plasmid units. A general mechanism for shuffling DNA strands, homologous recombination occurs when identical regions of nucleotides are exchanged between two DNA molecules. Homologous recombination occurs frequently between plasmids, which exist as multiple identical copies within the cell.

The *recA* enzyme binds to single-strand regions of nicked plasmids, promoting crossover and rejoining of homologous sequences. This results in multimeric ("super") plasmids that appear as a series of slow-migrating bands near the top of the gel (see facing page [top]). Since the concatemers form head-to-tail, they produce the identical restriction fragments as a monomer (single plasmid) when cut with restriction enzymes. To confuse matters further, multimers can exist in any of the three forms mentioned above. Supercoiled multimers may appear further down on the gel than relaxed or linear plasmids with fewer nucleotides.

1. Examine the photograph of your stained gel (or view on a light box or overhead projector). Compare your gel with the ideal gel on facing page (bottom). Label the size of fragments in each lane of your gel.

2. Compare the two gel lanes containing miniprep DNA with the two lanes containing control pAMP. Explain possible reasons for variations.

Restriction Analysis of Purified pAMP 339

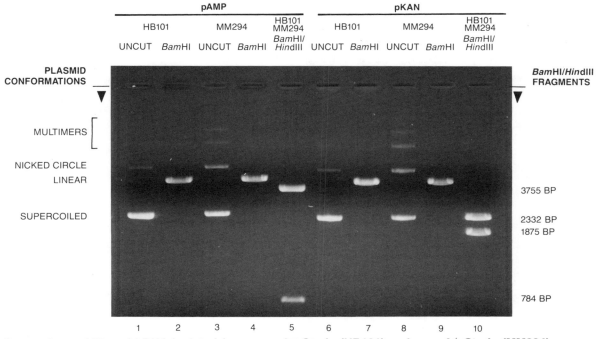

Comparison of Plasmid DNA Isolated from a recA⁻ Strain (HB101) and a recA⁺ Strain (MM294)

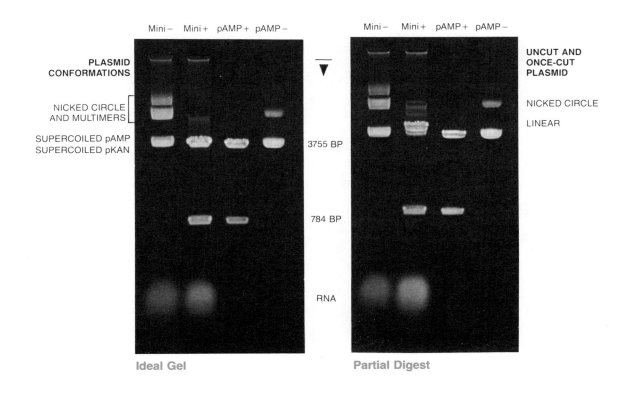

Ideal Gel   Partial Digest

3. A plasmid preparation of pAMP is composed entirely of dimeric molecules (pAMP/pAMP). The two molecules are joined *head-to-head* at a "hot spot" for recombination located 655 bp from the *Hin*dIII site near the origin of replication.

   a. Draw a map of the dimeric plasmid described above.

   b. Draw a map of the dimeric pAMP that actually forms by head-to-tail recombination at the site described above.

   c. Now draw the gel-banding patterns that would result from double digestion of each of these plasmids with *Bam*HI and *Hin*dIII, and label the base-pair size of fragments in each band.

4. Explain why EDTA is an important component of TE buffer in which the miniprep DNA is dissolved.

FOR FURTHER RESEARCH

1. Isolate and characterize an unknown plasmid. Make overnight cultures of *E. coli* strains containing any of several commercially available plasmids (such as pAMP, pKAN, pUC19, and pBR322). Digest miniprep and control samples of each plasmid with *Bam*HI/*Hin*dIII, and electrophorese in an agarose gel.

2. Transform pAMP and/or other plasmids into a $recA^+$ strain (MM294) and a $recA^-$ strain (HB101). Do minipreps from overnight cultures of each strain, and incubate samples of each with no enzyme, *Hin*dIII, and *Bam*HI + *Hin*dIII. Electrophorese samples as far as possible in an agarose gel. Compare the banding patterns of the two strains, especially in the uncut lanes.

3. Research the potential use of homologous recombination in targeted gene therapy.

Laboratory 7

Recombination of Antibiotic Resistance Genes

This laboratory begins an experimental stream to construct and analyze a recombinant DNA molecule. The starting reactants are the plasmids pAMP and pKAN, each of which carries a single antibiotic resistance gene—ampicillin in pAMP and kanamycin in pKAN. The goal is to construct a recombinant plasmid that contains both ampicillin and kanamycin resistance genes.

In Part A, Restriction Digest of Plasmids pAMP and pKAN, samples of both plasmids are digested in separate restriction reactions with *Bam*HI and *Hin*dIII. Following incubation at 37°C, samples of digested pAMP and pKAN are electrophoresed in an agarose gel to confirm proper cutting. Each plasmid contains a single recognition site for each enzyme, yielding only two restriction fragments. Cleavage of pAMP yields fragments of 784 bp and 3755 bp, and cleavage of pKAN yields fragments of 1875 bp and 2332 bp.

In Part B, Ligation of pAMP and pKAN Restriction Fragments, the restriction digests of pAMP and pKAN are heated to destroy *Bam*HI and *Hin*dIII activity. A sample from each reaction is mixed with DNA ligase plus ATP and incubated at room temperature. Complementary *Bam*HI and *Hin*dIII "sticky ends" hydrogen-bond to align restriction fragments. Ligase catalyzes the formation of phosphodiester bonds that covalently link the DNA fragments to form stable recombinant DNA molecules.

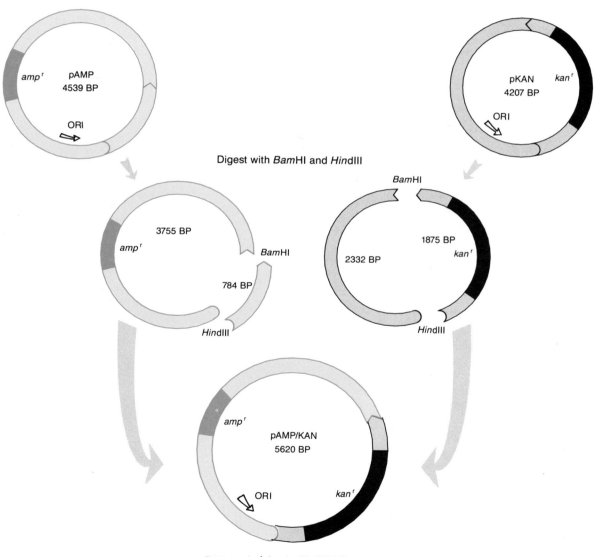

Formation of the "Simple Recombinant" pAMP/KAN

LABORATORY 7/PART A

Restriction Digest of Plasmids pAMP and pKAN

I. **Prepare Restriction Digest**

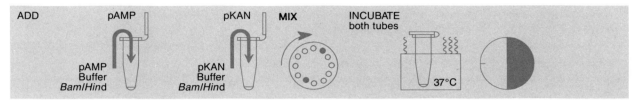

II. **Cast 0.8% Agarose Gel**

III. **Load Gel and Electrophorese**

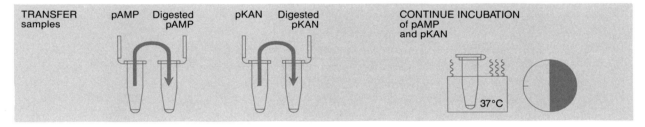

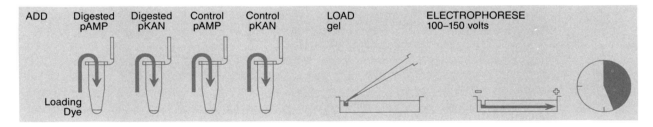

IV. **Stain Gel and View**

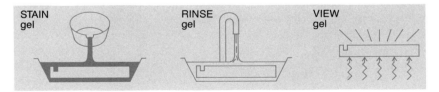

V. **Photograph Gel**

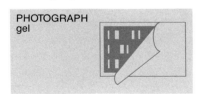

PRELAB NOTES

Review Prelab Notes in Laboratory 3, DNA Restriction Analysis.

Plasmid Substitution

The process of constructing and analyzing recombinant molecules in Laboratories 7 through 10 is not trivial. However, good results can be expected if the directions are followed carefully. The protocols have been optimized for the teaching plasmids pAMP and pKAN, and the extensive analysis of results is based *entirely* on recombinant molecules derived from these parent molecules. Substituting other plasmids in this experimental stream would be self-defeating.

The Prudent Control

In Section III, samples of the restriction digests are electrophoresed, prior to ligation, to confirm complete cutting by the endonucleases. This prudent control is standard experimental procedure. If pressed for time, omit electrophoresis and ligate DNA directly following the restriction digest. However, be sure to pretest the activity of the *Bam*HI and *Hin*dIII to determine the incubation time needed for complete digestion.

It is fairly impractical to use methylene blue staining for this step, which demands a rapid and sensitive assay to check for complete digestion of the plasmid DNAs. Methylene blue destaining requires *at least* 30 minutes, and it could fail to detect a small but significant amount of uncut DNA. However, if you prefer to use methylene blue staining for this lab, refer to the staining procedure in Step IVB of Laboratory 6 (part B) on page 335.

Saving DNA

Restriction reactions and controls in this experiment use a relatively large amount of plasmid DNA, which is the most expensive reagent used in the course. *To minimize expense, the protocol directs that the lab be prepared by aliquoting exactly the required volumes of pAMP and pKAN into 1.5-ml tubes. Then restriction reactants are added directly to the prealiquoted DNA.*

PRELAB PREPARATION

1. Mix in 1:1 proportion, *Bam*HI + *Hin*dIII (6 μl per experiment).

2. Aliquot for each experiment:

   5.5 μl of 0.20 μg/μl pAMP (store on ice).
   5.5 μl of 0.20 μg/μl pKAN (store on ice).
   5 μl of 0.10 μg/μl pAMP (store on ice).
   5 μl of 0.10 μg/μl pKAN (store on ice).
   20 μl of 2× restriction buffer (store on ice).
   6.0 μl of *Bam*HI/*Hin*dIII.
   500 μl of distilled water.
   500 μl of loading dye.

3. Prepare 0.8% agarose solution (40–50 ml per experiment). Keep agarose liquid in a hot-water bath (at about 60°C) throughout lab. Cover with aluminum foil to retard evaporation.

4. Prepare 1× Tris/Borate/EDTA (TBE) buffer for electrophoresis (400–500 ml per experiment).

5. Prepare ethidium bromide staining solution (100 ml per experiment).

6. Adjust water bath to 37°C.

7. Review Part B, Ligation of pAMP and pKAN Restriction Fragments.

Restriction Digest of Plasmids pAMP and pKAN

REAGENTS

For digest:
  0.20 μg/μl pAMP
  0.20 μg/μl pKAN

For control electrophoresis:
  0.1 μg/μl pAMP
  0.1 μg/μl pKAN
  BamHI/HindIII
  2× restriction buffer
  distilled water
  loading dye
  0.8% agarose
  1× Tris/Borate/EDTA (TBE) buffer
  1 μg/μl ethidium bromide

SUPPLIES AND EQUIPMENT

0.5–10-μl micropipettor + tips
1.5-ml tubes
aluminum foil
beaker for agarose
beaker for waste/used tips
10% bleach
camera and film (optional)
electrophoresis box
masking tape
microfuge (optional)
Parafilm or wax paper (optional)
permanent marker
plastic wrap (optional)
power supply
rubber gloves
test tube rack
transilluminator
37°C water bath

I. Set Up Restriction Digest

(40–60 minutes, including incubation through Section III)

Refer to Laboratory 3, DNA Restriction Analysis, for more detailed instructions.

1. Use matrix below as a checklist while adding reagents to each reaction. Read down each column, adding the same reagent to all appropriate tubes. *Use a fresh tip for each reagent.* Refer to detailed directions that follow.

Tube	pAMP 0.2 μg/μl	pKAN 0.2 μg/μl	2× Buffer	BamHI/HindIII
pAMP	5.5 μl	—	7.5 μl	2 μl
pKAN	—	5.5 μl	7.5 μl	2 μl

To avoid confusion, keep 0.1 μg/μl control pAMP and pKAN aside until needed in Section III.

To minimize waste and expense, plasmid DNA can be prealiquoted into tubes labeled pAMP and pKAN. *Add reagents directly to these tubes.*

2. Collect 2× buffer and *Bam*HI/*Hin*dIII, and place in test tube rack on lab bench.

3. Add 7.5 µl of 2× restriction buffer to each tube.

4. Use *fresh tip* to add 2 µl of *Bam*HI/*Hin*dIII to each tube.

5. Close tube tops. Pool and mix reagents by pulsing in a microfuge or by sharply tapping the tube bottom on lab bench.

6. Place reaction tubes in 37°C water bath, and incubate for a minimum of 30 minutes. Reactions can be incubated for a longer period of time.

STOP POINT After a full 30-minute incubation (or longer), freeze reactions at −20°C until ready to continue. Thaw reactions before proceeding to Section III, Step 1.

II. Cast 0.8% Agarose Gel (15 minutes)

1. Seal ends of gel-casting tray with tape, and insert well-forming comb. Place gel-casting tray out of the way on lab bench so that agarose poured in next step can set undisturbed.

2. Carefully pour enough agarose solution into casting tray to fill to a depth of about 5 mm. Gel should cover only about one-third the height of comb teeth. Use a pipet tip to move large bubbles or solid debris to sides or end of tray, while gel is still liquid.

3. Gel will become cloudy as it solidifies (about 10 minutes). *Be careful not to move or jar casting tray while agarose is solidifying.* Touch corner of agarose *away* from comb to test whether gel has solidified.

4. When agarose has set, unseal ends of casting tray. Place tray on platform of gel box, so that comb is at negative (black) electrode.

Too much buffer will channel current over the top rather than through the gel, increasing the time required to separate DNA. TBE buffer can be used several times; do not discard. If using buffer remaining in electrophoresis box from a previous experiment, rock chamber back and forth to remix ions that have accumulated at either end.

5. Fill box with TBE buffer, to a level that just covers surface of gel.

6. Gently remove comb, taking care not to rip wells.

7. Make certain that sample wells left by comb are completely submerged. If "dimples" are noticed around wells, slowly add buffer until they disappear.

STOP POINT Cover electrophoresis tank and save gel until ready to continue. Gel will remain in good condition for at least several days if it is completely submerged in buffer.

III. Load Gel and Electrophorese (20–30 minutes)

Only a fraction of the *Bam*HI/*Hin*dIII digests of pAMP and pKAN are electrophoresed to check whether plasmids are completely cut. These restriction samples are electrophoresed along with uncut pAMP and pKAN as controls.

1. Use a permanent marker to label two clean 1.5-ml tubes:

   Digested pAMP
   Digested pKAN

2. Remove tubes labeled pAMP and pKAN from 37°C water bath. Transfer 5 µl of plasmid from pAMP tube into Digested pAMP tube. Transfer 5 µl of plasmid from pKAN tube into Digested pKAN tube.

3. *Immediately return the pAMP and pKAN tubes to the water bath, and continue incubating at 37°C during electrophoresis.*

4. Collect 1.5-ml tubes containing 5 µl each of purified plasmid at 0.1 µg/µl; label tubes:

   Control pAMP
   Control pKAN

5. Add 1 µl of loading dye to each tube of Digested and Control pAMP and pKAN. Close tube tops, and mix by tapping tube bottom on lab bench, pipetting in and out, or pulsing in a microfuge.

6. Load entire contents of each sample tube into separate well in gel, as shown in diagram below. *Use fresh tip for each sample. Expel any air in tip before loading, and be careful not to punch tip of pipet through bottom of the gel.*

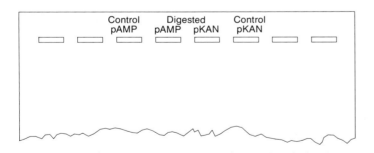

7. Electrophorese at 100–150 volts for 15–30 minutes. Adequate separation will have occurred when the bromophenol blue band has moved 2–4 cm from wells.

8. Turn off power supply, disconnect leads from the inputs, and remove top of electrophoresis box.

The 784-bp *Bam*HI/*Hin*dIII fragment of pAMP migrates just behind the bromophenol blue marker. Stop electrophoresis before bromophenol blue band runs off end of gel or this fragment may be lost.

9. Carefully remove casting tray from electrophoresis chamber, and slide gel into disposable weigh boat or other shallow tray. Label staining tray with your name.

STOP POINT Cover electrophoresis tank and save gel until ready to continue. Gel can be stored in a zip-lock plastic bag and refrigerated overnight for viewing/photographing the next day. However, over longer periods of time, the DNA will diffuse through the gel and the bands will become indistinct or disappear entirely.

Staining may be performed by an instructor in a controlled area when students are not present.

10. Stain and view gel as described in Section IV.

IV. Stain Gel with Ethidium Bromide and View (10–15 minutes)

> **CAUTION**
>
> Review Responsible Handling of Ethidium Bromide (page 256). Wear rubber gloves when staining, viewing, and photographing gels and during cleanup. Confine all staining to a restricted sink area.

1. Flood gel with ethidium bromide solution (1 μg/ml), and allow to stain for 5–10 minutes.

Ethidium bromide solution may be reused to stain 15 or more gels. Disable spent staining solution as explained on page 256.

2. Following staining, use funnel to decant as much ethidium bromide solution as possible from staining tray back into storage container.

3. Rinse gel and tray under running tap water.

4. If desired, gel can be destained in tap water or distilled water for 5 minutes or more to remove background ethidium bromide.

5. View under ultraviolet transilluminator or other UV source.

> **CAUTION**
>
> Ultraviolet light can damage eyes. Never look at unshielded UV light source with naked eyes. View only through a filter or safety glasses that absorb harmful wavelengths.

6. Take time for responsible cleanup.

   a. Wipe down camera, transilluminator, and staining area.

   b. Wash hands before leaving lab.

V. Photograph Gel (5 minutes)

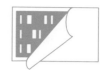

Exposure times vary according to mass of DNA in lanes, level of staining, degree of background staining, thickness of gel, and density of filter. Experiment to determine best exposure. When possible, stop lens down (to higher f/number) to increase depth of field and sharpness of bands.

1. *For ultraviolet (UV) photography of ethidium-bromide-stained gels:* Use Polaroid high-speed film Type 667 (ASA 3000). Set camera aperture to f/8 and shutter speed to B. Depress shutter for a 2–3-second time exposure.

   > **CAUTION**
   > Avoid getting caustic developing jelly on skin or clothes. If jelly gets on skin, wash immediately with plenty of soap and water.

2. Place left hand firmly on top of camera to steady. Firmly grasp small white tab, and pull straight out from camera. This causes a large yellow tab to appear.

3. Grip yellow tab in center, and in one steady motion, pull it straight out from the camera. This starts development.

4. Allow to develop for recommended time (45 seconds at room temperature). Do not disturb print while developing.

5. After full development time has elapsed, separate print from negative by peeling back at end nearest yellow tab.

6. Wait to see the result of the first photo before making other exposures.

RESULTS AND DISCUSSION

Examine the photograph of your stained gel (or view on a light box). Compare your gel with the ideal gel on following page, and check whether both plasmids have been completely digested by *Bam*HI and *Hin*dIII.

1. The Digested pAMP lane should show two distinct fragments: 784 bp and 3755 bp.

2. The Digested pKAN lane should show two distinct fragments: 1875 bp and 2332 bp.

3. Additional bands that comigrate with bands in the uncut Control pAMP and Control pKAN should be faint or absent, indicating that most or all of the pAMP and pKAN plasmid has been completely digested by both enzymes.

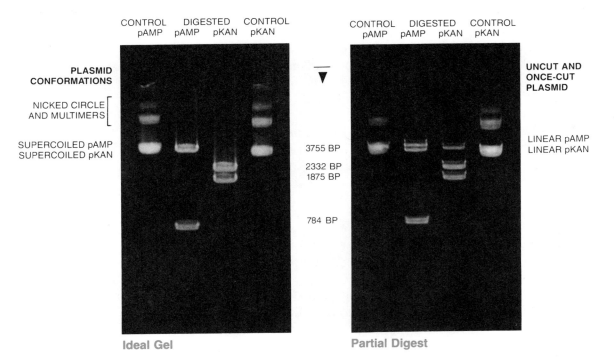

Ideal Gel | Partial Digest

4. If both digests look complete, or nearly so, continue on to Part B, Ligation of pAMP and pKAN Restriction Fragments. The reaction will have gone to completion with the additional incubation during electrophoresis.

5. If either or both digests look very incomplete, add another 1 µl of BamHI/HindIII solution and incubate for an additional 20 minutes. Then continue on to Part B, Ligation of pAMP and pKAN Restriction Fragments.

STOP POINT   Freeze BamHI/HindIII reactions at −20°C until ready to continue. Thaw reactions before proceeding to Part B, Ligation of pAMP and pKAN Restriction Reactions.

LABORATORY 7/PART B

Ligation of pAMP and pKAN Restriction Fragments

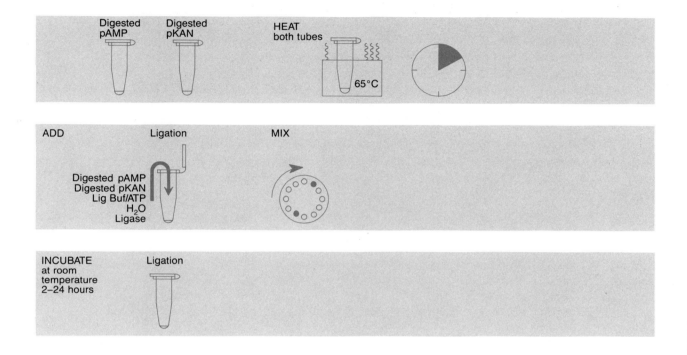

PRELAB NOTES

DNA Ligase

Use only T4 DNA ligase. *E. coli* DNA ligase requires different reaction conditions and cannot be substituted in this experiment. Two different units are used to calibrate ligase activity:

- *Weiss Unit (Pyrophosphate Exchange Unit):* One unit of enzyme catalyzes the exchange of 1 nmole of ^{32}P from pyrophosphate into $[\gamma,\beta-^{32}P]ATP$ in 20 minutes at 37°C. This unit is used by Bethesda Research Laboratories (BRL); International Biotechnologies, Inc. (IBI), and Boehringer Mannheim.

- *Cohesive-end Unit:* One unit of enzyme ligates 50% of *Hin*dIII fragments of λ DNA (6 μg in 20 μl) in 30 minutes at 16°C. This unit is used by New England Biolabs (NEB) and Carolina Biological Supply Company (CBS).

1 Weiss Unit = 67 Cohesive-end Units

Researchers typically incubate ligation reactions overnight at room temperature. *For brief ligations, down to a minimum of 2 hours, it is essential to choose a high-concentration T4 DNA ligase with at least 5 Weiss Units/μl or 100–500 Cohesive-end Units/μl.*

For Further Information

The protocol presented here is based on the following published method:

Cohen, S.N., A.C.Y. Chang, and H.W. Boyer. 1973. Construction of biologically functional bacteria plasmids in vitro. *Proc. Natl. Acad. Sci.* 70: 3240.

PRELAB PREPARATION

1. Prepare fresh 2× ligation buffer/ATP solution just prior to use. ATP is somewhat unstable in solution.

2. T4 DNA ligase is critical to the experiment and rather expensive. Make one aliquot of ligase sufficient for all experiments, and hold on ice during laboratory. We suggest that the instructor dispense ligase directly into each experimenter's reaction tube.

3. Prewarm water bath to 65°C.

4. Dispose of 2× restriction buffer from Part A, Restriction Digest of Plasmids pAMP and pKAN, to avoid mistaking it for 2× ligation buffer/ATP.

Ligation of pAMP and pKAN Restriction Fragments

REAGENTS	SUPPLIES AND EQUIPMENT
digested pAMP	0.5–10-µl micropipettor + tips
digested pKAN	1.5-ml tube
2× ligation buffer/ATP	beaker for waste/used tips
T4 DNA ligase	microfuge (optional)
distilled water	test tube rack
	65°C water bath

Ligate pAMP and pKAN   (30 minutes, including incubation)

1. Incubate tubes labeled Digested pAMP and Digested pKAN (from Part A) in 65°C water bath for 10 minutes.

2. Label a clean 1.5-ml tube Ligation.

3. Use matrix below as a checklist while adding reagents to the Ligation tube. *Use a fresh tip for each reagent.* Refer to detailed directions that follow.

Tube	Digested pAMP	Digested pKAN	2× Ligation Buffer/ATP	Water	Ligase
Ligation	3 µl	3 µl	10 µl	3 µl	1 µl

Step 1 is critical. Heat denatures protein, thus inactivating the restriction enzymes.

4. Collect reagents (except ligase), and place in test tube rack on lab bench.

5. Add 3 µl of Digested pAMP.

6. Use *fresh tip* to add 3 µl of Digested pKAN.

7. Use *fresh tip* to add 10 µl of 2× ligation buffer/ATP.

8. Use *fresh tip* to add 3 µl of distilled water.

9. Use *fresh tip* to add 1 µl of DNA ligase. Carefully check that droplet of ligase is on the *inside wall* of tube.

Ligase may be dispensed by your instructor.

10. Close tube top. Pool and mix reagents by pulsing in a microfuge or by sharply tapping tube bottom on lab bench.

For brief ligations of 2–4 hours, it is essential to use a high-concentration T4 DNA ligase with at least 5 Weiss Units/μl or 100—500 Cohesive-end Units/μl.

11. Incubate reaction at *room temperature* for 2–24 hours.

12. If time permits, ligation may be confirmed by electrophoresing 5 μl of the ligation reaction, along with *Bam*HI/*Hin*dIII digests of pAMP and pKAN. None of the parent *Bam*HI/*Hin*dIII fragments should be observed in the lane of ligated DNA, which should show multiple bands of high-molecular-weight DNA high up on the gel.

STOP POINT Freeze reaction at −20°C until ready to continue. Thaw reaction before proceeding to Laboratory 8.

RESULTS AND DISCUSSION

Ligation of the four *Bam*HI/*Hin*dIII restriction fragments of pAMP and pKAN produces many types of hybrid molecules, including plasmids composed of more than two fragments. However, only those constructs possessing an origin of replication will be maintained and expressed. Three different replicating plasmids, with selectable antibiotic resistance, are created by ligating combinations of two *Bam*HI/*Hin*dIII fragments:

- Ligation of the 784-bp fragment to the 3755-bp fragment regenerates pAMP.

- Ligation of the 1875-bp fragment to the 2332-bp fragment regenerates pKAN.

- Ligation of the 1875-bp fragment to the 3755-bp fragment produces the "simple recombinant" plasmid pAMP/KAN, in which the kanamycin resistance gene has been fused into the pAMP backbone.

1. Make a scale drawing of the simple recombinant molecule pAMP/KAN described above. Include fragment sizes, locations of *Bam*HI and *Hin*dIII restriction sites, location of origin(s), and location of antibiotic resistance gene(s).

2. Make scale drawings of other two-fragment recombinant plasmids having the following properties.

   a. Three kinds of plasmids having two origins.

   b. Three kinds of plasmids having no origin.

   Whenever possible, include fragment sizes, locations of *Bam*HI and *Hin*dIII restriction sites, location of origin(s), and location of antibiotic resistance gene(s).

3. Ligation of the 784-bp fragment, 3755-bp fragment, 1875-bp fragment, and 2332-bp fragment produces a "double plasmid" pAMP/pKAN. Make a scale drawing of the double plasmid pAMP/pKAN.

4. Make scale drawings of several recombinant plasmids composed of any three of the four *Bam*HI/*Hin*dIII fragments of pAMP and pKAN. Include fragment sizes, locations of *Bam*HI and *Hin*dIII restriction sites, location of origin(s), and location of antibiotic resistance gene(s). What rule governs the construction of plasmids from three kinds of restriction fragments?

5. What kind of antibiotic selection would identify *E. coli* cells that have been transformed with each of the plasmids drawn in Questions 1–4?

6. Explain what is meant by "sticky ends." Why are they so useful in creating recombinant DNA molecules?

7. Why is ATP essential for the ligation reaction?

Laboratory 8

Transformation of *E. coli* with Recombinant DNA

In Part A, Classic Procedure for Preparing Competent Cells, *E. coli* cells are rendered competent to uptake plasmid DNA using a method essentially unchanged since its publication in 1970 by M. Mandel and A. Higa. The procedure begins with vigorous *E. coli* cells grown in suspension culture. Cells are harvested in mid-log phase by centrifugation and incubated at 0°C with two successive changes of calcium chloride solution.

This procedure is more involved than the rapid colony protocol introduced in Laboratory 5. However, the classic procedure typically achieves transformation efficiencies ranging from 5×10^4 to 10^6 colonies per microgram of plasmid—a 10–200-fold increase over the colony procedure. The enhanced efficiency is important when transforming ligated DNA (composed primarily of relaxed circular plasmids and linear DNA), which produces 5–100 times fewer transformants than plasmid DNA purified from *E. coli* (containing a high proportion of the supercoiled form).

In Part B, Transformation of *E. coli* with Recombinant DNA, the competent *E. coli* cells are transformed with the ligation products from Laboratory 7, Recombination of Antibiotic Resistance Genes. Ligated plasmid DNA is added to one sample of competent cells, and purified pAMP and pKAN plasmids are added as controls to two other samples. The cell suspensions are incubated with the plasmid DNAs for 20 minutes at 0°C. Following a brief heat shock at 42°C, the cells recover in LB broth for 40–60 minutes at 37°C. Unlike ampicillin selection in Laboratory 5, Rapid Colony Transformation of *E. coli* with Plasmid DNA, the recovery step is essential for the kanamycin selection in this lab. Samples of transformed cells are plated onto three types of LB agar: with ampicillin (LB/amp), with kanamycin (LB/kan), and with both ampicillin and kanamycin (LB/amp+kan).

The ligation reaction produces many kinds of recombinant molecules composed of *Bam*HI/*Hin*dIII fragments, including the religated parental plasmids pAMP and pKAN. The object is to select for transformed cells with dual antibiotic resistance, which must contain a 3755-bp fragment from pAMP containing the ampicillin resistance gene (plus the origin of replication) and a 1875-bp fragment from pKAN containing the kanamycin resistance gene. Bacteria transformed with a single plasmid containing these sequences, or those doubly transformed with both pAMP and pKAN plasmids, form colonies on the LB/amp+kan plate.

LABORATORY 8/PART A

Classic Procedure for Preparing Competent Cells

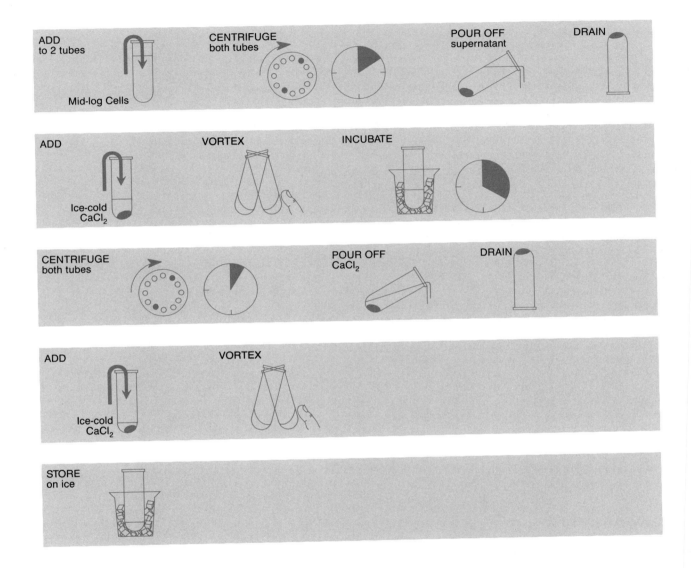

PRELAB NOTES

Review Prelab Notes in Laboratories 1, 2, and 5 regarding sterile techniques, *E. coli* culture, and transformation.

Seasoning Cells for Transformation

If possible, schedule experiments so that competent cells (Part A) are prepared one day prior to transformation with recombinant DNA (Part B). "Seasoning" cells for 12–24 hours at 0°C (an ice bath inside the refrigerator) generally increases transformation efficiency five- to tenfold. This enhanced efficiency will help to assure successful cloning of the recombinant molecules produced in Laboratory 7.

For Further Information

The protocol presented here is based on the following published methods:

Cohen, S.N., A.C.Y. Chang, and L. Hsu. 1972. Nonchromosomal antibiotic resistance in bacteria: Genetic transformation of *Escherichia coli* by R-factor DNA. *Proc. Natl. Acad. Sci.* 69: 2110.

Dagert, M. and S.D. Ehrlich. 1979. Prolonged incubation in calcium chloride improves the competence of *Escherichia coli* cells. *Gene* 6: 23.

Mandel, M. and A. Higa. 1970. Calcium-dependent bacteriophage DNA infection. *J. Mol. Biol.* 53: 159.

PRELAB PREPARATION

1. On the day before this lab, begin an *E. coli* culture from a streaked plate of MM294, according to the protocol in Laboratory 2B, Overnight Suspension Culture.

2. PLAN AHEAD. Be sure to have a streaked plate of viable *E. coli* cells from which to inoculate overnight. Also, streak the *E. coli* strain on LB/amp and LB/kan plates to ensure that a resistant strain has not been used by mistake.

3. Approximately 2–4 hours before the lab, begin an *E. coli* culture according to the protocol in Laboratory 2C, Mid-log Suspension Culture. Cells are optimal for transformation when the culture reaches an OD_{550} of 0.3–0.40 (1.75–2.0 hours after inoculation). Hold cells in mid-log phase by storing culture on ice for up to 2 hours prior to beginning calcium chloride treatment. *Each experiment requires 20 ml of mid-log suspension culture.*

4. Sterilize 50 mM calcium chloride ($CaCl_2$) solution and LB broth by autoclaving or filtering through a 0.45- or 0.22-μm filter (Nalgene or Corning). Be sure to prewash the filter by drawing through 50–100 ml of deionized water; this removes any surfactants that may be used in the manufacturing of the filter. To eliminate autoclaving completely, store filtered solutions in presterilized 50-ml conical tubes.

5. Aliquot for each experiment:

    15 ml of 50 mM $CaCl_2$ in sterile 50-ml tube (store on ice).
    2–10 ml of mid-log phase *E. coli* cells in sterile 15-ml culture tubes (store on ice).
    12 μl of 0.005 μg/μl pAMP in 1.5-ml tube (store on ice).
    12 μl of 0.005 μg/μl pKAN in 1.5-ml tube (store on ice).

6. Review Part B, Transformation of *E. coli* with Recombinant DNA.

Classic Procedure for Preparing Competent Cells

CULTURE AND REAGENTS

2–10 ml of mid-log MM294 cells
50 mM $CaCl_2$

SUPPLIES AND EQUIPMENT

5- or 10-ml pipets, sterile pipet aid or bulb
100–1000-µl micropipettor + tips (or 1-ml pipet)
beaker of crushed or cracked ice
beaker for waste
10% bleach or disinfectant
Bunsen burner
clean paper towels
clinical centrifuge (2000–4000 rpm)
test tube rack

Prepare Competent Cells (40–50 minutes)

This entire experiment *must be performed under sterile conditions.* Review sterile techniques in Laboratory 1, Measurements, Micropipetting, and Sterile Techniques.

1. Place sterile tube of $CaCl_2$ solution on ice.

2. Obtain two 15-ml tubes, each with 10 ml of mid-log cells, and label with your name.

3. *Securely close caps,* and place both tubes of cells in a *balanced* configuration in rotor of clinical centrifuge. Centrifuge at 2000–4000 rpm for 10 minutes to pellet cells on the bottom-side of the culture tube.

4. Sterilely pour off supernatant from each tube into waste beaker for later disinfection. *Be careful not to disturb cell pellet.*

    a. Remove cap from culture tube, and briefly flame mouth. Do not place cap on lab bench.

    b. Carefully pour off supernatant. Invert culture tube, and tap gently on surface of clean paper towel to drain thoroughly.

    c. *Reflame mouth of culture tube, and replace cap.*

A tight pellet of cells should be easily seen at the bottom of the tube. If the pellet does not appear to be consolidated, recentrifuge for an additional 5 minutes.

Plan out manipulations for Step 4. Organize lab bench, and work quickly. Locate Bunsen burner in a central position on lab bench to avoid reaching over flame.

If working as a team, one person handles pipets, and the other removes and replaces tube caps.

The cell pellet becomes increasingly difficult to suspend the longer it sits in the CaCl₂ solution.

Double check both tubes for complete resuspension of cells, which is probably the most important variable in obtaining good results.

CaCl₂ treatment alters the adhering properties of the *E. coli* membranes. The cell pellet is much more dispersed after the second centrifugation.

5. Use a 5- or 10-ml pipet to sterilely add 5 ml of ice-cold $CaCl_2$ solution to each culture tube:

   a. Briefly flame pipet cylinder.

   b. Remove cap from $CaCl_2$ tube, and flame mouth. Do not place cap on lab bench.

   c. Withdraw 5 ml of $CaCl_2$. Reflame mouth of tube, and replace cap.

   d. Remove cap of culture tube, and flame mouth. Do not place cap on lab bench.

   e. Expel $CaCl_2$ into culture tube. Reflame mouth of culture tube, and replace cap.

6. Immediately finger vortex to resuspend pelleted cells in each tube.

   a. Close cap tightly.

   b. Hold upper part of tube securely with thumb and index finger.

   c. With your other hand, vigorously hit the bottom end of tube with index finger or thumb to create a vortex that lifts the cell pellet off the bottom of tube. Continue "finger vortexing" until all traces of cell mass are completely resuspended. This may take a couple of minutes, depending on technique.

   d. Hold tube up to light to check that suspension is homogeneous. No visible clumps of cells should remain.

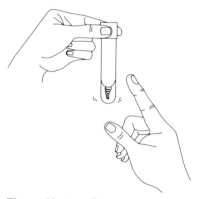

Finger Vortex (Steps 6 and 11)

7. Return both tubes to ice, and incubate for 20 minutes.

8. Following incubation, re-spin cells in clinical centrifuge for 5 minutes at 2000–4000 rpm.

9. Sterilely pour off $CaCl_2$ from each tube into waste beaker. *Be careful not to disturb cell pellet.*

   a. Remove cap from culture tube, and briefly flame mouth. Do not place cap on lab bench.

b. Carefully pour off supernatant. Invert culture tube, and tap gently on surface of clean paper towel to drain thoroughly.

c. Reflame mouth of culture tube, and replace cap.

10. Use a 100–1000-μl micropipettor (or 1-ml pipet) to sterilely add 1000 μl (1 ml) of fresh, ice-cold CaCl$_2$ to each tube.

    a. Remove cap from CaCl$_2$ tube, and flame mouth. Do not place cap on lab bench.

    b. Withdraw 1000 μl (1 ml) of CaCl$_2$. Reflame mouth of tube, and replace cap.

    c. Remove cap of culture tube, and flame mouth. Do not place cap on lab bench.

    d. Expel CaCl$_2$ into culture tube. Reflame mouth of culture tube, and replace cap.

Do not flame micropipet tip.

11. Close caps tightly, and immediately finger vortex to resuspend pelleted cells in each tube. Hold tube up to light to check that suspension is homogeneous. No visible clumps of cells should remain.

STOP POINT Store cells in beaker of ice in refrigerator (approximately 0°C) until ready for use. "Seasoning" at 0°C for up to 24 hours increases the competency of cells five- to tenfold.

Cell pellet may appear more diffuse than at the beginning of the procedure and will resuspend more easily. Double check both tubes for complete resuspension of cells.

12. Take time for responsible cleanup:

    a. Segregate for proper disposal—culture plates and tubes, pipets, and micropipettor tips that have come in contact with *E. coli*.

    b. Disinfect mid-log culture, tips, and supernatant from Steps 4 and 9 with 10% bleach or disinfectant.

    c. Wipe down lab bench with soapy water, 10% bleach solution, or disinfectant (such as Lysol).

    d. Wash hands before leaving lab.

LABORATORY 8/PART B
Transformation of *E.coli* with Recombinant DNA

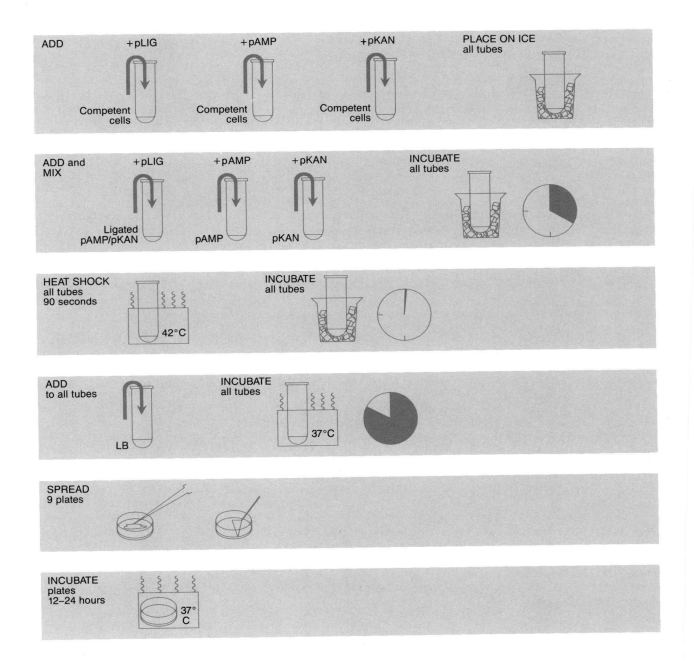

PRELAB NOTES

Review Prelab Notes in Laboratory 5, Rapid Colony Transformation of *E. coli* with Plasmid DNA.

Equipment Substitutions

A standard 1-ml pipet or transfer pipet can be substituted for a 100–1000-μl micropipettor. A calibrated inoculating loop can be substituted for a 1–10-μl micropipettor.

Recovery Period

A 40–60-minute postincubation recovery at 37°C, with shaking, is essential prior to plating transformed cells on kanamycin, which acts quickly to kill any cell that is not actively expressing the resistance protein.

For Further Information

The protocol presented here is based on the following published method:

Cohen, S.N., A.C.Y. Chang, and H.W. Boyer. 1973. Construction of biologically functional bacteria plasmids in vitro. *Proc. Natl. Acad. Sci.* 70: 3240.

PRELAB PREPARATION

1. Prepare for each experiment:

    three LB + ampicillin plates (labeled LB/amp).
    three LB + kanamycin plates (labeled LB/kan).
    three LB + ampicillin + kanamycin plates (labeled LB/amp + kan).

    If only one control transformation, with *either* pAMP *or* pKAN, is done, then one less plate of each type is required.

2. Adjust water baths to 65°C, 42°C, and 37°C.

3. Prewarm incubator to 37°C.

4. To retard evaporation, keep beaker of ethanol covered with Parafilm or plastic wrap prior to use. Retrieve and reuse ethanol exclusively for flaming.

Transformation of *E. coli* with Recombinant DNA

CULTURE, MEDIA, AND REAGENTS

competent *E. coli* cells
 (from Part A)
ligation tube (from Laboratory 7)
0.005 µg/µl pAMP
0.005 µg/µl pKAN
LB broth
three LB/amp plates
three LB/kan plates
three LB/amp+kan plates

SUPPLIES AND EQUIPMENT

100–1000-µl micropipettor
 + tips
0.5–10-µl micropipettor
 + tips
3 15-ml culture tubes
beaker of crushed or
 cracked ice
beaker of 95% ethanol
beaker for waste/used tips
"bio-bag" or heavy-duty
 trash bag
10% bleach or disinfectant
Bunsen burner
cell spreader
37°C incubator
permanent marker
37°C shaking water bath
test tube rack
42°C water bath

Perform *E. coli* Transformation (70–90 minutes)

This entire experiment must be performed under sterile conditions. Review sterile techniques in Laboratory 1, Measurements, Micropipetting, and Sterile Techniques.

1. Use permanent marker to label three *sterile* 15-ml culture tubes:

   +pLIG = ligated DNA
   +pAMP = pAMP control
   +pKAN = pKAN control

2. Use a 100–1000-µl micropipettor and *sterile tip* to add 200 µl of competent cells to each tube.

   Optimally, flame mouths of tubes after removing and before replacing caps.

3. Place all three tubes on ice.

Store remainder of ligated DNA at 4°C. Electrophorese with cut pAMP and pKAN controls in Laboratory 10 to observe the products of the ligation reaction.

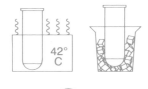

To save plates, experimenters may omit *either* Set b *or* Set c.

Optimally, flame mouths of tubes after removing and before replacing caps.

If shaking water bath is not available, warm cells for several minutes in 37°C water bath, then transfer to dry shaker inside 37°C incubator. Alternately, occasionally swirl tubes by hand in non-shaking 37°C water bath.

4. Use a 1–10-µl micropipettor to add 10 µl of ligated pAMP/KAN solution *directly into cell suspension* in tube labeled +pLIG.

5. Use *fresh tip* to add 10 µl of 0.005 µg/µl pAMP solution *directly into cell suspension* in tube labeled +pAMP.

6. Use *fresh tip* to add 10 µl of 0.005 µg/µl pKAN solution *directly into cell suspension* in tube labeled +pKAN.

7. Close caps, and tap tubes with finger to mix. Avoid making bubbles in suspension or splashing suspension up sides of tubes.

8. Return all three tubes to ice for 20 minutes.

9. While cells are incubating on ice, use permanent marker to label all nine LB agar plates with your name and the date. Divide plates into three sets of three plates each and mark as follows:

   Set a
   Mark L on one LB/amp, one LB/kan, and one LB/amp+kan plate

   Set b
   Mark A on one LB/amp, one LB/kan, and one LB/amp+kan plate

   Set c
   Mark K on one LB/amp, one LB/kan, and one LB/amp+kan plate

10. Following 20-minute incubation, heat shock the cells in all three tubes. *It is critical that cells receive a sharp and distinct shock.*

    a. Carry ice beaker to water bath. Remove tubes from ice, and *immediately* immerse in 42°C water bath for 90 seconds.

    b. Immediately return all three tubes to ice for at least 1 additional minute.

11. Use a 100–1000-µl micropipettor with a sterile tip to add 800 µl of LB broth to each tube. Gently tap tubes with finger to mix.

12. Allow cells to recover by incubating all three tubes at 37°C in a shaking water bath (with moderate agitation) for 40–60 minutes.

STOP POINT Cells may be allowed to recover for up to several hours. A longer recovery period assures the growth of as many kanamycin-resistant recombinants as possible and can help to compensate for a poor ligation or cells of low competence.

13. Use matrix below as a checklist as +pLIG, +pAMP, and +pKAN cells are spread on each type of antibiotic plate in Steps 14–19:

	Ligated DNA L	pAMP control A	pKAN control K
LB/amp	100 µl	100 µl	100 µl
LB/kan	100 µl	100 µl	100 µl
LB/amp+kan	100 µl	100 µl	100 µl

14. Use micropipettor with sterile tip to add 100 µl of cell suspension from tube labeled pLIG onto three plates marked L. *Do not allow suspensions to sit on plates too long before proceeding to Step 15.*

    If too much liquid is absorbed by agar, cells will not be evenly distributed.

15. Sterilize cell spreader, and spread cells over surface of each L plate in succession.

    a. Dip spreader into the ethanol beaker and *briefly* pass it through Bunsen flame to ignite alcohol. Allow alcohol to burn off *away from* Bunsen flame; spreading rod will become too hot if left in flame.

       The object is to evenly distribute and separate cells on agar so that each gives rise to a distinct colony of clones. *It is essential not to overheat spreader in burner flame and to cool it before touching cell suspensions. A hot spreader will kill E. coli cells on the plate.*

CAUTION
Be extremely careful not to ignite ethanol in beaker.

Sterile Spreading Technique (Steps 14 and 15)

b. Lift lid of first L plate only enough to allow spreading; do not place lid on lab bench.

c. Cool spreader by gently rubbing it on surface of the agar *away* from cell suspension or by touching it to condensed water on plate lid.

d. Touch spreader to cell suspension, and gently drag it back and forth several times across surface of agar. Rotate plate one-quarter turn, and repeat spreading motion. Be careful not to gouge agar.

e. Replace plate lid. Return the cell spreader to ethanol *without flaming*.

f. Repeat Steps a through e in succession for the remaining two L plates.

16. Use *fresh sterile tip* to add 100 µl of cell suspension from tube labeled +pAMP onto three plates marked A.

17. Repeat Step 15a–f to sterilize cell spreader and spread cells over the surface of each A plate in succession.

18. Use *fresh sterile tip* to add 100 µl of cell suspension from tube labeled +pKAN onto three plates marked K.

19. Repeat Step 15a–f to sterilize cell spreader and spread cells over the surface of each K plate in succession.

20. Reflame spreader one last time before placing it on lab bench.

21. Let plates set for several minutes to allow suspension to become absorbed into agar. Then wrap plates together with tape.

22. Place plates upside down in 37°C incubator, and incubate for 12–24 hours.

23. After initial incubation, store plates at 4°C to arrest *E. coli* growth and to slow the growth of any contaminating microbes.

Save L LB/amp and L LB/kan plates if planning to do Laboratory 9. Save L LB/amp+kan as a source of colonies to begin overnight suspension cultures if planning to do Laboratory 10.

24. Take time for responsible cleanup:

a. Segregate for proper disposal—culture plates and tubes, pipets, and micropipettor tips that have come in contact with *E. coli*.

b. Disinfect overnight cell suspensions, tubes, and tips with 10% bleach or disinfectant.

c. Wipe down lab bench with soapy water, 10% bleach solution, or disinfectant (such as Lysol).

d. Wash hands before leaving lab.

Observe colonies through the bottom of the culture plate, using a permanent marker to mark each colony as it is counted. If the experiment worked well, 5–50 colonies should be observed on the L LB/amp+kan experimental plate, 500–5000 colonies on the A LB/amp control plate, and 200–2000 colonies on the K LB/kan control plate. Approximately tenfold fewer colonies should be observed on the corresponding L LB/amp plate and L LB/kan plate. An extended recovery period would inflate these numbers. (Question 3 explains how to compute transformation efficiency.) If plates have been overincubated or left at room temperature for several days, "satellite" colonies may be observed on the LB/amp plates. Satellite colonies are never observed on the LB/kan or LB/amp+kan plates.

RESULTS AND DISCUSSION

1. Record your observation of each plate in the matrix below. If cell growth is too dense to count individual colonies, record "lawn." Were results as expected? Explain possible reasons for variations from expected results.

	Ligated DNA L	pAMP control A	pKAN control K
LB/amp			
LB/kan			
LB/amp+kan			

2. Compare and contrast the growth on each of the following pairs of plates. What does each pair of results tell you about transformation and/or antibiotic selection?

   L LB/amp and A LB/amp
   L LB/kan and A LB/kan
   A LB/amp and K LB/kan
   L LB/amp and L LB/kan
   L LB/amp and L LB/amp+kan
   L LB/kan and L LB/amp+kan

3. Calculate transformation efficiencies of A LB/amp and K LB/kan positive controls. Remember that transformation efficiency is expressed as the number of antibiotic resistant colonies per microgram of intact plasmid DNA. The object is to determine the mass of pAMP or pKAN that was spread on each plate and was therefore responsible for the transformants observed.

   a. Determine total mass (in micrograms) of pAMP used in Step 5 and of pKAN used in Step 6: concentration × volume = mass.

   b. Determine fraction of cell suspension spread onto A LB/amp plate (Step 16) and K LB/kan plate (Step 18): volume suspension spread/*total* volume suspension (Steps 2 and 11) = fraction spread.

c. Determine mass of plasmid pAMP and pKAN in cell suspension spread onto A LB/amp plate and K LB/kan plate: total mass plasmid (a) × fraction spread (b) = mass plasmid spread.

d. Determine number of colonies per microgram of pAMP and pKAN. Express answer in scientific notation: colonies observed/mass plasmid spread (c) = transformation efficiency.

4. Calculate transformation efficiencies of the L LB/amp, L LB/kan, and the L LB/amp + kan plates.

   a. Calculate the mass of pAMP and pKAN used in the restriction reactions of Laboratory 7 (see matrix in Part A.1). Then calculate the concentration of plasmid in each restriction reaction.

   b. Calculate mass of pAMP and pKAN used in ligation reaction (Laboratory 7, Part B.3). Then calculate the *total* concentration of plasmid in the ligation mixture.

   c. Use this concentration in calculations following Steps a–d of Question 3 above.

5. Compare the transformation efficiencies you calculated for the A LB/amp plate in this laboratory and the +AMP plate in Laboratory 5. By what factor is the classic procedure more or less efficient than colony transformation? What differences in the protocols contribute to the increase in efficiency?

6. Compare the transformation efficiencies you calculated for control pAMP and pKAN *versus* the ligated pAMP and pKAN. How can you account for the differences in efficiency? Take into account the formal definition of transformation efficiency.

FOR FURTHER RESEARCH

Interpretable experimental results can only be achieved when the classic transformation protocol can be repeated with reproducible results. Only attempt experiments below when you are able to routinely achieve 500–2000 colonies on the A LB/amp plate.

1. Design and execute an experiment to compare the transformation efficiencies of linear *versus* circular plasmid DNAs. Keep molecular weight constant.

2. Design and execute a series of experiments to test the relative importance of each of the four major steps of most transformation protocols: (1) preincubation, (2) incubation, (3) heat shock, and (4) recovery. Which steps are absolutely necessary?

3. Design and execute a series of experiments to compare the transforming effectiveness of $CaCl_2$ *versus* salts of other monovalent (+),

divalent (++), and trivalent (+++) cations.

   a. Make up 50 mM solutions of each salt.

   b. Check pH of each solution, and buffer to pH 7 when necessary.

   c. Is $CaCl_2$ unique in its ability to facilitate transformation?

   d. Is there any consistent difference in the transforming ability of monovalent *versus* divalent *versus* trivalent cations?

4. Design a series of experiments to determine saturating conditions for transformation reactions.

   a. Transform *E. coli* using the following DNA concentrations:

      0.00001 µg/µl
      0.00005 µg/µl
      0.0001 µg/µl
      0.0005 µg/µl
      0.001 µg/µl
      0.005 µg/µl
      0.01 µg/µl
      0.05 µg/µl
      0.1 µg/µl

   b. Plot a graph of DNA mass *versus* colonies per plate.

   c. Plot a graph of DNA mass *versus* transformation efficiency.

   d. At what mass does the reaction appear to become saturated?

   e. Repeat experiment with concentrations clustered on either side of the presumed saturation point to produce a fine saturation curve.

5. Repeat Experiment 4 above, but transform with a 1:1 mixture of pAMP and pKAN at each concentration. Plate transformants on LB/amp, LB/kan, and LB/amp+kan plates. *Be sure to include a 40–60-minute recovery, with shaking.*

   a. Calculate percentage of double transformations at each mass.

$$\frac{\text{colonies amp+kan plate}}{\text{colonies amp plate + colonies kan plate}}$$

   b. Plot a graph of DNA mass *versus* colonies per plate.

   c. Plot a graph of DNA mass *versus* percentage double transformations. Under saturating conditions, what percentage of bacteria are doubly transformed?

6. Plot a recovery curve for *E. coli* transformed with pKAN. Allow cells to recover for 0–120 minutes at 20-minute intervals.

   a. Plot a graph of recovery time *versus* colonies per plate.

b. At what time point is antibiotic expression maximized?

c. Can you discern a point at which the cells began to replicate?

7. Attempt to isolate pAMP/KAN recombinants using the colony transformation protocol in Laboratory 5. What trick would increase the likelihood of retrieving ampicillin/kanamycin-resistant colonies?

Laboratory 9

Replica Plating to Identify Mixed *E. coli* Populations

Ligation of *Bam*HI/*Hin*dIII fragments of pAMP and pKAN in Laboratory 7 created a number of molecules containing either an ampicillin resistance gene (amp^r), a kanamycin resistance gene (kan^r), or both genes together (amp^r/kan^r). Samples of this ligated DNA (L) were then transformed into competent *E. coli* cells, which were plated onto LB/amp, LB/kan, and LB/amp+kan plates in Laboratory 8. The results clearly indicate that colonies growing on the L LB/amp+kan plate (having an amp^r/kan^r phenotype) contain both resistance genes (the amp^r/kan^r genotype).

It is not possible, however, to be certain of the amp^r/kan^r genotypes of bacteria growing on the L LB/amp or L LB/kan plates. Although a colony growing on the L LB/amp plate possesses an amp^r gene, it is not possible to say whether or not it also possesses a kan^r gene. Conversely, although a colony growing on the L LB/kan must possess a kan^r gene, it is not possible to know whether or not it also possesses an amp^r gene. A conclusion can be made about the presence of an antibiotic resistance gene only when the organism has been challenged with that antibiotic.

In this laboratory, replica plating provides a rapid means to distinguish between single- and dual-resistant colonies growing on the L LB/amp and L LB/kan plates. Cells from 12 colonies on the L LB/amp plate and from 12 colonies on the L LB/kan plate are transferred onto one fresh LB/amp plate and one fresh LB/kan plate to which numbered grids have been attached. An L colony is scraped with a sterile toothpick (or inoculating loop), and a sample of cells is streaked successively into the same-numbered squares of the fresh LB/amp and LB/kan plates. Following overnight incubation at 37°C, colonies that grow in the same squares of both the LB/amp and LB/kan plates have the amp^r/kan^r genotype.

LABORATORY 9

Replica Plating to Identify Mixed *E. coli* Populations

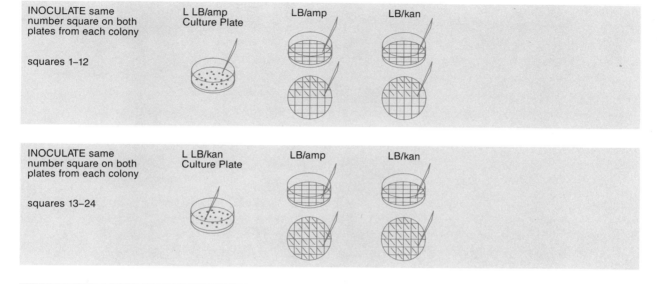

INOCULATE same number square on both plates from each colony

squares 1–12

L LB/amp Culture Plate   LB/amp   LB/kan

INOCULATE same number square on both plates from each colony

squares 13–24

L LB/kan Culture Plate   LB/amp   LB/kan

INCUBATE 12–24 hours

37° C

PRELAB NOTES

Review Prelab Notes in Laboratory 2A, Overnight Suspension Culture.
Replica plating provides a rapid means to screen L LB/amp and L LB/kan plates for dual-resistant colonies that potentially contain pAMP/KAN recombinant plasmids. If colonies were not obtained on the L LB/amp + kan plate, replica plating provides another chance to identify dual-resistant colonies from which to isolate recombinant plasmids for Laboratory 10, Purification and Identification of Recombinant DNA.

For Further Information

The protocol presented here is based on the following published method:

Lederberg, J. and E.M. Lederberg. 1952. Replica plating and indirect selection of bacterial mutants. *J. Bact. 63:* 399.

PRELAB PREPARATION

1. Prepare for each experiment:

    one LB + ampicillin plate (LB/amp).
    one LB + kanamycin plate (LB/kan).

2. Sterilize 30 toothpicks per experiment. Place toothpicks in 50-ml beaker, cover with aluminum foil, and autoclave for 15 minutes at 121°C. (Although much less rapid, a flamed and cooled inoculating loop can be used to transfer colonies.)

3. Xerox two replica-plating grids (below) per experiment.

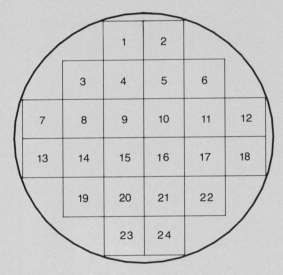

4. Prewarm incubator to 37°C.

Replica Plating to Identify Mixed *E. coli* Populations

CULTURES AND MEDIA

L LB/amp plate w/colonies
L LB/kan plate w/colonies
 (from Laboratory 8)
1 LB/amp plate
1 LB/kan plate

SUPPLIES AND EQUIPMENT

2 replica-plating grids
sterile toothpicks
 (or inoculating loop +
 Bunsen burner)
beaker for waste
"bio-bag" or heavy-duty trash
 bag
10% bleach or disinfectant
37°C incubator
permanent marker

Prepare Replica Plates (10 minutes)

1. Attach a replica-plating grid to the *bottom* of an LB/amp plate and to the *bottom* of an LB/kan plate. Use permanent marker to label each plate with your name and the date.

   A 24-square grid may be drawn on plate bottom with permanent marker.

2. Replica plate a sample of cells from one colony on the L LB/amp plate onto the fresh LB/amp and LB/kan plates.

   Lift plate lids only enough to select colony and streak. Do not place lids on lab bench.

   a. Use a sterile toothpick (or inoculating loop) to scrape up a cell mass from a well-defined colony on the L LB/amp plate.

   b. Immediately drag the *same* toothpick (or loop) gently across agar surface to make a short diagonal (/) streak *within Square 1* of the LB/amp plate.

   c. Immediately use the *same* toothpick (or loop) to make a diagonal (/) streak *within Square 1* of the LB/kan plate.

   d. Discard toothpick in waste beaker (or reflame and cool inoculating loop).

3. Repeat Step 2a–d *using fresh toothpicks* (or *flamed* and *cooled* inoculating loop) to streak cells from 11 *different* L LB/amp colonies onto Squares 2–12 of both LB/amp and LB/kan plates.

4. Repeat Step 2a–d *using fresh toothpicks* (or *flamed* and *cooled* inoculating loop) to streak cells from 12 *different* L LB/kan colonies onto Squares 13–24 of both LB/amp and LB/kan plates.

   If you have fewer than 12 colonies on either plate, obtain a plate from another experimenter.

Save replica plates as a source of colonies from which to isolate plasmid DNA in Laboratory 10 if your L LB/amp + kan plate has fewer than two colonies.

5. Place plates upside down in 37°C incubator, and incubate for 12–24 hours.

6. After initial incubation, store plates at 4°C to arrest *E. coli* growth and to slow the growth of any contaminating microbes.

7. Take time for responsible cleanup:

   a. Segregate for proper disposal—bacterial cultures *and* used toothpicks.

   b. Disinfect overnight culture, tips, and supernatant from Step 4 with 10% bleach or disinfectant.

   c. Wipe down lab bench with soapy water, 10% bleach solution, or disinfectant (such as Lysol).

   d. Wash hands before leaving lab.

RESULTS AND DISCUSSION

In general, the results of replica plating indicate the success of the ligation in Laboratory 7 and parallel the results observed on the L LB/amp + kan plate from Laboratory 8. Thus, if there were a large number of colonies on the LB/amp + kan plate, it is likely that there will be a high percentage of dual-resistant colonies that grow on both the LB/amp and LB/kan replica plates. Experience indicates that 30–70% of transformants selected with only ampicillin *or* kanamycin actually have dual resistance. Roughly equal numbers of dual-resistant colonies are identified from L LB/amp and L LB/kan plates.

1. Observe the LB/amp and LB/kan plates. Use the matrix below to record as + the squares in which new bacterial growth has expanded

Colony source *L LB/amp*	Replica plates		*Colony source* *L LB/kan*	Replica plates	
	LB/amp	LB/kan		LB/amp	LB/kan
1			13		
2			14		
3			15		
4			16		
5			17		
6			18		
7			19		
8			20		
9			21		
10			22		
11			23		
12			24		

the width of the initial streak. Record as − the squares in which no new growth has expanded the initial streak. Remember that nonresistant cells may survive, separated from the antibiotic on top of a heavy initial streak; however, no *new growth* will be observed.

2. On the basis of your observations:

   a. Calculate the percentage of dual-resistant colonies taken from the L LB/amp plate (Squares 1–12).

   b. Calculate the percentage of dual-resistant colonies taken from the L LB/kan plate (Squares 13–24).

   c. Give an explanation for the similarity or difference in the percentages of dual-resistant colonies taken from the two source plates.

3. Draw restriction maps for different plasmid molecules that could be responsible for the dual resistance phenotype.

FOR FURTHER RESEARCH

1. We have discussed the phenomenon of satellite colonies on LB/amp plates—small nonresistant colonies that grow in a halo around large resistant colonies. To prove this, replica plate satellite colonies from an LB/amp plate onto fresh LB and LB/amp plates.

2. The *amp*^{r} protein, β-lactamase, is not actively secreted into the medium but is believed to "leak" through the cell envelope of *E. coli*. Satellite colonies do not form on kanamycin plates because the antibiotic kills all nonresistant cells outright. The following experiment tests whether resistance protein escapes from ampicillin- and kanamycin-resistant cells.

   a. Grow separate overnight cultures of an *amp*^{r} colony from the A LB/amp plate and a *kan*^{r} colony from the K LB/kan plate (from Laboratory 8, or use other *amp*^{r} and *kan*^{r} strains). Inoculate 5 ml of *plain* LB broth, according to the protocol in Laboratory 2B, Overnight Suspension Culture.

   b. Pass each overnight culture through a 0.22- or 0.45-μm filter, and collect filtrate in a clean, sterile 15-ml tube. Filtering removes all *E. coli* cells.

   c. Use permanent marker to mark one LB/amp plate and one LB/kan plate. Draw a line on plate bottom to divide each plate into two equal parts; mark one half +.

   d. Sterilely spread 100 μl of the A filtrate onto the + *half only* of the LB/amp plate. Sterilely spread 100 μl of the K filtrate onto the + *half only* of the LB/kan plate. Allow filtrates to soak into plates for 10–15 minutes.

e. Sterilely streak wild-type (nontransformed) *E. coli* cells on each filtrate-treated plate, taking care to streak back and forth across the dividing line.

f. Incubate plates at 37°C for 12–24 hours. Compare growth on the treated *versus* untreated sides of each plate.

3. The *amp*^{r} protein is believed to leak primarily from stationary-stage cells. The following experiment tests the hypothesis that leakage of β-lactamase is growth-phase-dependent.

   a. Grow an overnight culture of an ampicillin-resistant colony from the A LB/amp plate in Laboratory 8 (or other *amp*^{r} strain). Inoculate 1 ml of *plain* LB broth, according to the protocol in Laboratory 2B, Overnight Suspension Culture.

   b. Use overnight culture to inoculate 100 ml of fresh LB broth, and grow according to protocol in Laboratory 2C, Mid-log Suspension Culture.

   c. Sterilely withdraw 10-ml aliquots from the culture after 1, 2, and 4 hours, holding aliquots on ice.

   d. Take the OD_{550} of each aliquot.

   Note: The objective is to test resistance-protein "leakage" as a function of culture age, *not as a function of cell number*. Because cell number increases over time, the cell number must be equalized by diluting the 2- and 4-hour samples with sterile LB to the *E. coli* concentration of the 1-hour sample. Since the OD_{550} values are proportional to the cell number, they can be used to compute the dilution factor.

   e. Pass each of the three samples through a 0.22- or 0.45-μm filter to remove the bacteria.

   f. Prepare a 10- and 100-fold dilution for each filtrate, using sterile LB broth.

   g. Use a permanent marker to draw a line dividing each of six LB/amp plates into equal parts.

   h. Sterilely spread 100 μl of undiluted filtrate from the 1-hour sample over half of the first plate. Label the plate with time point and dilution factor. Allow filtrate to soak into plate for 10–15 minutes.

   i. Spread 100 μl of undiluted 2- and 4-hour filtrates over separate halves of the second plate, as described above.

   j. Repeat spreading procedure for the 10- and 100-fold dilutions, as described in Steps h and i.

   k. After filtrates have soaked into the plates, sterilely streak wild-type (nontransformed) *E. coli* cells on each half of each plate.

   l. Incubate plates at 37°C for 12–24 hours. Compare growth for each time point across each dilution.

Laboratory 10

Purification and Identification of Recombinant DNA

Growth of *E. coli* colonies on the L LB/amp + kan plate in Laboratory 8 confirms that they have been transformed to a dual resistance (Amp^r/Kan^r) phenotype. This resistance is expressed by one or more replicating plasmids, which were assembled in Laboratory 7 by ligating four *Bam*HI/*Hin*dIII restriction fragments of the parental plasmids pAMP and pKAN:

- a 784-bp pAMP fragment
- a 3755-bp pAMP fragment containing an origin of replication and an *amp*^{r} gene
- a 1875-bp pKAN fragment containing a *kan*^{r} gene
- a 2332-bp pKAN fragment containing an origin of replication

The goal of this laboratory is to determine the genotype responsible for dual resistance; that is, the number and probable arrangement of any two or more pAMP and pKAN fragments. In Part A, Plasmid Minipreparation of pAMP/KAN Recombinants, plasmid DNA is isolated from overnight cultures of two different colonies from an L LB/amp + kan plate (Laboratory 8) or from replica plates (Laboratory 9). In Part B, Restriction Analysis of Purified Recombinant DNA, samples of the plasmids isolated in Part A and a control sample of pAMP + pKAN are incubated with *Bam*HI and *Hin*dIII. The three digested samples and samples of uncut minipreps are coelectrophoresed in an agarose gel, along with uncut pAMP and λ/*Hin*dIII size markers. The comigration of *Bam*/*Hin*d fragments in the lanes of miniprep DNA and pAMP/pKAN controls, along with an evaluation of the relative sizes of uncut supercoiled DNAs, gives evidence of the structure, size, and number of plasmids present in each of the transformed strains.

LABORATORY 10/PART A

Plasmid Minipreparation of pAMP/pKAN Recombinants

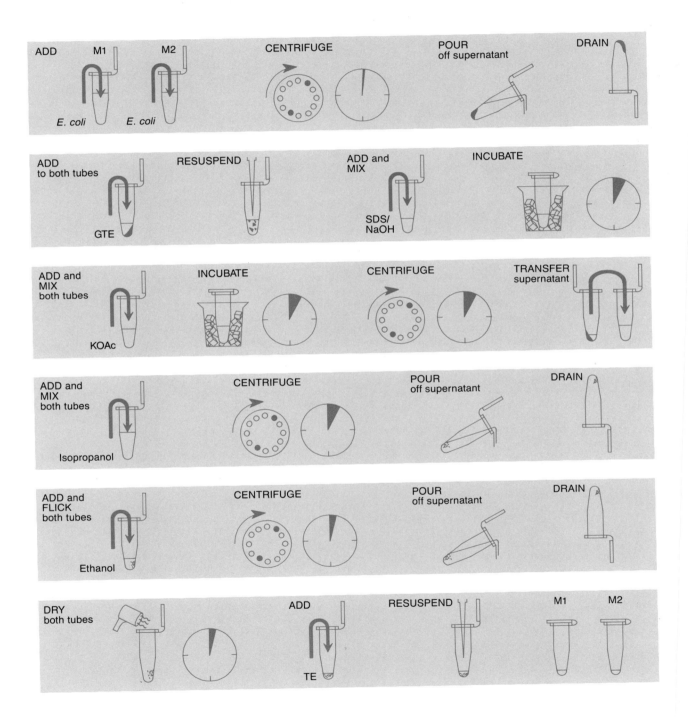

PRELAB NOTES

Review Prelab Notes in Laboratory 6A, Plasmid Minipreparation of pAMP.

Antibiotic Selection

Although plain LB broth may be used, it is safest to maintain antibiotic selection when culturing overnight cultures of the *E. coli* transformed with recombinant plasmid. Single selection in LB/amp is sufficient.

Alternate Sources of Dual-resistant Colonies

If colonies were not obtained on L LB/amp + kan plate:

1. Use dual-resistant colonies identified by replica plating in Laboratory 9. These may be picked from either LB/amp or LB/kan replica plate.

2. If there is not enough time to replica plate, prepare overnight cultures from five or more different colonies. The following selection strategies allow only dual-resistant colonies to grow; there should be at least two cloudy cultures in the morning.

    a. Inoculate LB/amp with colonies from L LB/kan plate.

    b. Inoculate LB/kan with colonies from L LB/amp plate.

    c. Inoculate LB/amp + kan with colonies from either L LB/amp plate or L LB/kan plate.

For Further Information

The protocol presented here is based on the following published methods:

Birnboim, H.C. and J. Doly. 1979. A rapid alkaline extraction method for screening recombinant plasmid DNA. *Nucleic Acids Res.* 7:1513.
Ish-Horowicz, D. and J.F. Burke. 1981. Rapid and efficient cosmid cloning. *Nucleic Acids Res.* 9:2989.

PRELAB PREPARATION

1. The day before the laboratory, prepare two *E. coli* cultures according to the protocol in Laboratory 2B, Overnight Suspension Culture. Inoculate each overnight culture with a cell mass scraped from a different colony on the L LB/amp+kan plate from Laboratory 8. Maintain antibiotic selection with LB broth plus ampicillin. Alternately, the culture can be prepared 2–3 days in advance and stored at 4°C or incubated at 37°C without shaking for 24–48 hours. In either case, the cells will settle at the bottom of the culture tube. Shake tube to resuspend cells, before beginning procedure. If colonies were not obtained on the L LB/amp+kan plate, see Alternate Sources of Dual-resistant Colonies in Prelab Notes.

2. Circle colonies on plate used to inoculate each overnight culture. Label one colony and its overnight culture M1 (miniprep 1). Label the other colony and its overnight culture M2 (miniprep 2). This will allow you to refer back to the original colony.

3. Prepare SDS/sodium hydroxide solution within a few days of lab. Store solution at room temperature; a soapy precipitate may form at lower temperature. Warm solution by placing tube in beaker of hot tap water, and shake gently to dissolve the precipitate.

4. Aliquot for each experiment:

    250 μl of glucose/Tris/EDTA (GTE) solution (store on ice).
    500 μl of SDS/sodium hydroxide (SDS/NaOH) solution.
    400 μl of potassium acetate/acetic acid (KOAc) solution (store on ice).
    1000 μl of isopropanol 500 μl of 100% ethanol.
    500 μl of Tris/EDTA (TE) solution.

5. Review Part B, Restriction Analysis of Purified Recombinant DNA.

Plasmid Minipreparation of pAMP/KAN Recombinants

CULTURES AND REAGENTS

two *E. coli*/pAMP/pKAN
  overnight cultures
glucose/Tris/EDTA (GTE)
SDS/sodium hydroxide (SDS/NaOH)
potassium acetate/acetic acid (KOAc)
isopropanol
95–100% ethanol
Tris/EDTA (TE)

SUPPLIES AND EQUIPMENT

100–1000-μl micropipet
  + tips
0.5–10-μl micropipet +
  tips
1.5-ml tubes
beaker of crushed ice
beaker for waste/used
  tips
10% bleach or
  disinfectant
clean paper towels
hair dryer
microfuge
permanent marker
test tube rack

Perform Plasmid Miniprep (50 minutes)

1. Shake culture tubes to resuspend *E. coli* cells.

2. Label two 1.5-ml tubes with your initials. Label one tube M1, and label the other tube M2. Use a micropipettor to transfer 1000 μl of overnight suspension M1 and 1000 μl of overnight suspension M2 into appropriate tubes.

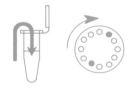

3. Close caps, and place tubes in a *balanced* configuration in microfuge rotor. Spin for 1 minute to pellet cells.

4. Pour off supernatant from both tubes into waste beaker for later disinfection. *Be careful not to disturb cell pellets.* Invert tubes, and tap gently on surface of clean paper towel to drain thoroughly.

The cell pellet will appear as a small off-white smear on the bottom-side of the tube. Although the cell pellets are readily seen, the DNA pellets in Step 14 are very difficult to observe. Get into the habit of aligning tube with cap hinges facing up in the microfuge rotor. Then, pellets should always be located at tube bottom beneath hinge.

5. Add 100 μl of GTE solution to each tube. Resuspend pellets by pipetting solution in and out several times. Hold tubes up to light to check that suspension is homogeneous and no visible clumps of cells remain.

Accurate pipetting is essential to good plasmid yield. The volumes of reagents are precisely calibrated so that sodium hydroxide added in Step 6 is neutralized by acetic acid in Step 8.

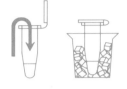

In Step 11, the supernatant is saved and precipitate is discarded. The situation is reversed in Steps 14 and 17, where the precipitate is saved and supernatant is discarded.

Do Step 12 quickly, and make sure that the microfuge will be immediately available for Step 13.

The pellet may appear as a tiny smear or small particles on the bottom-side of each tube. Do not be concerned if pellet is not visible; pellet size is not a predictor of plasmid yield. A large pellet is composed primarily of RNA and cellular debris carried over from the original precipitate. A smaller pellet often means a cleaner preparation.

Nucleic acid pellets are not soluble in ethanol and will not resuspend during washing.

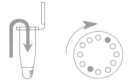

6. Add 200 µl of SDS/NaOH solution to each tube. Close caps, and mix solutions by rapidly inverting tubes five times.

7. Stand tubes on ice for 5 minutes. Suspension will become relatively clear.

8. Add 150 µl of *ice-cold* KOAc solution to each tube. Close caps and mix solutions by rapidly inverting tubes five times. A white precipitate will immediately appear.

9. Stand tubes on ice for 5 minutes.

10. Place tubes in a *balanced* configuration in microfuge rotor, and spin for 5 minutes to pellet precipitate along the side of the tube.

11. Transfer 400 µl of supernatant from M1 into a clean 1.5-ml tube labeled M1. Transfer 400 µl of supernatant from M2 into a clean 1.5-ml tube labeled M2. *Avoid pipetting precipitate,* and wipe off any precipitate clinging to the outside of the tip prior to expelling supernatant. Discard old tubes containing precipitate.

12. Add 400 µl of isopropanol to each tube of supernatant. Close caps, and mix vigorously by rapidly inverting tubes five times. *Stand at room temperature for only 2 minutes.* (Isopropanol preferentially precipitates nucleic acids rapidly; however, proteins remaining in solution also begin to precipitate with time.)

13. Place tubes in a *balanced* configuration in microfuge rotor, and spin for 5 minutes to pellet the nucleic acids. Align tubes in rotor so that cap hinges point outward. The nucleic acid residue, visible or not, will collect under hinge during centrifugation.

14. Pour off supernatant from both tubes. *Be careful not to disturb nucleic acid pellets.* Invert tubes, and tap gently on surface of clean paper towel to drain thoroughly.

15. Add 200 µl of 100% ethanol to each tube, and close caps. Flick tubes several times to wash pellets.

STOP POINT Store ethanol solution at −20°C until ready to continue.

16. Place tubes in a *balanced* configuration in microfuge rotor, and spin for 2–3 minutes.

17. Pour off supernatant from both tubes. *Be careful not to disturb nucleic acid pellets.* Invert tubes, and tap gently on surface of clean paper towel to drain thoroughly.

18. Dry nucleic acid pellets by one of following methods:

    a. Direct a stream of warm air from hair dryer into open ends of tube for about 3 minutes. *Be careful not to blow pellets out of tubes.*

    or

    b. Close caps, and pulse tubes in microfuge to pool remaining ethanol. *Carefully* draw off drops of ethanol using a 1–10-µl micropipettor. Allow pellets to air dry at room temperature for 10 minutes.

19. All ethanol must be evaporated before proceeding to Step 20. Hold each tube up to light to check that no ethanol droplets remain. If ethanol is still evaporating, an alcohol odor can be detected by sniffing mouth of tube.

20. Add 15 µl of TE to each tube. Resuspend pellets by smashing with the pipet tip and pipetting in and out vigorously. Rinse down the side of tube several times, concentrating on area where the pellet should have formed during centrifugation (beneath cap hinge). Check that all DNA is dissolved and that no particles remain in tip or on side of tube.

If using a 0.5–10-µl micropipettor, set to 7.5 µl and pipet twice.

21. Keep the two DNA/TE solutions *separate*. DO NOT pool into one tube.

STOP POINT   Freeze DNA/TE solution at −20°C until ready to continue. Thaw before using.

---

22. Take time for responsible cleanup.

    a. Segregate for proper disposal—culture tubes and micropipettor tips that have come in contact with *E. coli*.

    b. Disinfect overnight culture, tips, and supernatant from Step 4 with 10% bleach or disinfectant.

    c. Wipe down lab bench with soapy water, 10% bleach solution, or disinfectant (such as Lysol).

    d. Wash hands before leaving lab.

LABORATORY 10/PART B

Restriction Analysis of Purified Recombinant DNA

I. Set Up Restriction Digest

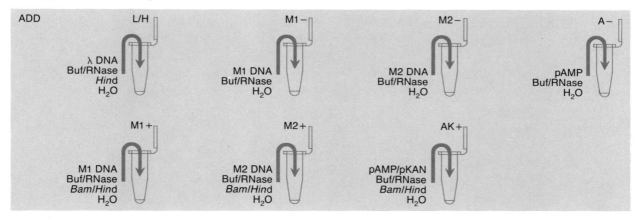

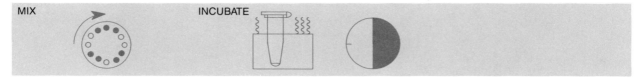

II. Cast 0.8% Agarose Gel

III. Load Gel and Electrophorese

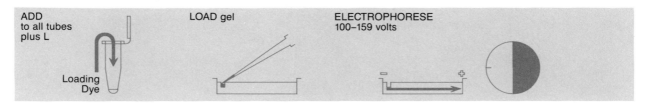

IV. Stain Gel and View

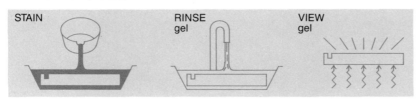

V. Photograph Gel

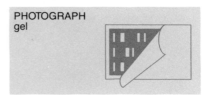

PRELAB NOTES

Review Prelab Notes in Laboratory 6B, Restriction Analysis of Purified pAMP.

For Further Information

The protocol presented here is based on the following published methods:

Cohen, S.N., A.C.Y. Chang, and H.W. Boyer. 1973. Construction of biologically functional bacteria plasmids in vitro. *Proc. Natl. Acad. Sci. 70:* 3240.

Helling, R.B., H.M. Goodman, and H.W. Boyer. 1974. Analysis of *Eco*RI fragments of DNA from lambdoid bacteriophages and other viruses by agarose-gel electrophoresis. *J. Virol. 14:* 1235.

Sharp, P.A., B. Sugden, and J. Sambrook. 1973. Detection of two restriction endonuclease activities in *Haemophilus parainfluenzae* using analytical agarose-ethidium bromide electrophoresis. *Biochemistry 12:* 3055.

PRELAB PREPARATION

1. Mix in 1:1 proportion:

   BamHI + HindIII (8 µl per experiment)
   pAMP + pKAN (6 µl per experiment)

2. Aliquot for each experiment:

   6 µl of 0.1 µg/µl pAMP/pKAN (store on ice).
   6 µl of 0.1 µg/µl pAMP (store on ice).
   6 µl of 0.1 µg/µl λ DNA (store on ice).
   16 µl of 5× restriction buffer/RNase (store on ice).
   8 µl of BamHI/HindIII (store on ice).
   2 µl of HindIII (store on ice).
   500 µl of deionized/distilled water.
   500 µl of loading dye.

   Precut λ/HindIII is readily available from commercial suppliers, or a large-scale λ/HindIII digest can be done in advance to provide size markers for a number of experiments. Omit HindIII aliquot if using predigested λ DNA.

3. Prepare 0.8% agarose solution (approximately 40–50 ml per experiment). Keep agarose liquid in a hot-water bath (at about 60°C) throughout experiment. Cover solution with aluminum foil to retard evaporation.

4. Prepare 1× Tris/Borate/EDTA (TBE) buffer for electrophoresis (400–500 ml per experiment).

5. Prepare ethidium bromide or methylene blue staining solution (100 µl per experiment).

6. Adjust water bath to 37°C.

Restriction Analysis of Purified Recombinant DNA

REAGENTS

minisprep DNA/TE (M1, M2)
0.1 µg/µl pAMP/pKAN
0.1 µg/µl pAMP
0.1 µg/µl λ DNA
*Hin*dIII
*Bam*HI/*Hin*dIII
5× restriction buffer/RNase
distilled water
loading dye
0.8% agarose
1× Tris/Borate/EDTA (TBE) buffer
1 µg/µl ethidium bromide (or 0.025% methylene blue)

SUPPLIES AND EQUIPMENT

0.5–10-µl micropipettor + tips
1.5-ml tubes
aluminum foil
beaker for agarose
beaker for waste/used tips
10% bleach
camera (optional)
electrophoresis box
microfuge (optional)
masking tape
Parafilm or wax paper (optional)
permanent marker
plastic wrap (optional)
power supply
rubber gloves
test tube rack
transilluminator (optional)
37°C water bath

I. Set Up Restriction Digest   (40 minutes, including incubation)

Refer to Laboratory 3, DNA Restriction Analysis, for more detailed instructions.

1. Use permanent marker to label seven 1.5-ml tubes, in which restriction reactions will be performed:

   L/H = λ DNA, *Hin*dIII
   M1− = miniprep 1, no enzyme
   M2− = miniprep 2, no enzyme
   A−   = pAMP, no enzyme
   M1+ = miniprep 1, *Bam*HI/*Hin*dIII
   M2+ = miniprep 2, *Bam*HI/*Hin*dIII
   AK+ = pAMP/pKAN, *Bam*HI/*Hin*dIII

   Skip L/H tube if using predigested λ/*Hin*dIII markers.

   Return unused miniprep DNA (M1 and M2) to freezer at −20°C for possible use in further experiments suggested at the end of this laboratory.

2. Use matrix below as a checklist while adding reagents to each reaction. Read down each column, adding the same reagent to all

appropriate tubes. *Use a fresh tip for each reagent.* Refer to detailed directions that follow.

Tube	λ DNA	M1	M2	pAMP	pAMP/ pKAN	Buffer/ RNase	HindIII	BamHI/ HindIII	H₂O
L/H	5 μl	—	—	—	—	2 μl	1 μl	—	2 μl
M1–	—	5 μl	—	—	—	2 μl	—	—	3 μl
M2–	—	—	5 μl	—	—	2 μl	—	—	3 μl
A–	—	—	—	5 μl	—	2 μl	—	—	3 μl
M1+	—	5 μl	—	—	—	2 μl	—	2 μl	1 μl
M2+	—	—	5 μl	—	—	2 μl	—	2 μl	1 μl
AK+	—	—	—	—	5 μl	2 μl	—	2 μl	1 μl

3. Collect reagents, and place in test tube rack on lab bench.

4. Add 5 μl of λ DNA to tube labeled L/H.

5. Use *fresh tip* to add 5 μl of M1 DNA to tubes labeled M1– and M1+.

6. Use *fresh tip* to add 5 μl of M2 DNA to tubes labeled M2– and M2+.

7. Use *fresh tip* to add 5 μl of pAMP to tube labeled A–.

8. Use *fresh tip* to add 5 μl of pAMP/pKAN to tube labeled AK+.

9. Use *fresh tip* to add 2 μl of restriction buffer/RNase to a clean spot on each reaction tube.

10. Use *fresh tip* to add 1 μl of *Hin*dIII to tube labeled L/H.

11. Use *fresh tip* to add 2 μl of *Bam*HI/*Hin*dIII to tubes labeled M1+, M2+, and AK+.

12. Use *fresh tip* to add proper volumes of distilled water to each tube.

13. Close tube tops. Pool and mix reagents by pulsing in a microfuge or by sharply tapping the tube bottom on lab bench.

14. Place reaction tubes in 37°C water bath, and incubate for 30 minutes only.

Do not overincubate. During longer incubation, DNases in miniprep may degrade plasmid DNA.

STOP POINT Following incubation, freeze reactions at −20°C until ready to continue. Thaw reactions before continuing to Section III, Step 1.

II. Cast 0.8% Agarose Gel (15 minutes)

1. Seal ends of gel-casting tray with tape, and insert well-forming comb. Place gel-casting tray out of the way on lab bench so that agarose poured in next step can set undisturbed.

2. Carefully pour enough agarose solution into casting tray to fill to a depth of about 5 mm. Gel should cover only about one-third the height of comb teeth. Use a pipet tip to move large bubbles or solid debris to sides or end of tray, while gel is still liquid.

3. Gel will become cloudy as it solidifies (about 10 minutes). *Be careful not to move or jar casting tray while agarose is solidifying.* Touch corner of agarose *away* from comb to test whether gel has solidified.

4. When agarose has set, unseal ends of casting tray. Place tray on platform of gel box, so that comb is at negative (black) electrode.

5. Fill box with TBE buffer, to a level that just covers surface of gel.

6. Gently remove comb, taking care not to rip wells.

7. Make certain that sample wells left by comb are completely submerged. If "dimples" are noticed around wells, slowly add buffer until they disappear.

Too much buffer will channel current over the top rather than through the gel, increasing the time required to separate DNA. TBE buffer can be used several times; do not discard. If using buffer remaining in electrophoresis box from a previous experiment, rock chamber back and forth to remix ions that have accumulated at either end.

STOP POINT Cover electrophoresis tank and save gel until ready to continue. Gel will remain in good condition for at least several days if it is completely submerged in buffer.

III. Load Gel and Electrophorese (30–50 minutes)

1. Add 1 µl of loading dye to each reaction tube. Close tube tops, and mix by tapping tube bottom on lab bench, pipetting in and out, or pulsing in a microfuge.

2. Load entire contents of each reaction tube into separate well in gel, as shown in diagram below. Use *fresh tip* for each reaction. *Expel any air in tip before loading, and be careful not to punch tip of pipet through bottom of the gel.*

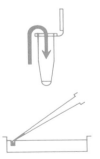

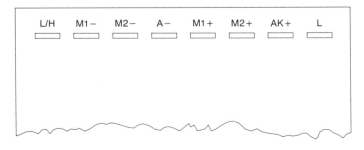

3. Add loading dye to ligated DNA saved from Laboratory 7, Recombination of Antibiotic Resistance Genes. Load entire contents of L tube (5–10 μl) into well 8.

4. Electrophorese at 100–150 volts for 20–40 minutes, or longer. Good separation will have occurred when the bromophenol blue band has moved 4–6 cm from wells.

   If time allows, electrophorese until the bromophenol blue band nears the end of the gel. This will allow maximum separation of uncut DNA, which is important in differentiating a large "superplasmid" from a double transformation of two smaller plasmids.

   Stop electrophoresis before bromophenol blue band runs off end of gel or the 784-bp *Bam*HI/*Hin*dIII fragment of pAMP, which migrates just behind the bromophenol blue marker, may be lost.

5. Turn off power supply, disconnect leads from the inputs, and remove top of electrophoresis box.

6. Carefully remove casting tray from electrophoresis box, and slide gel into disposable weigh boat or other shallow tray. Label the staining tray with your name.

STOP POINT Cover electrophoresis tank, and save gel until ready to continue. Gel can be stored in a zip-lock plastic bag and refrigerated overnight for viewing/photographing the next day. However, over longer periods of time, the DNA will diffuse through the gel, and the bands will become indistinct or disappear entirely.

Staining may be performed by an instructor in a controlled area when students are not present.

7. Stain and view gel using one of the methods described in Sections IVA and IVB.

IVA. Stain Gel with Ethidium Bromide and View (10–15 minutes)

> **CAUTION**
>
> Review Responsible Handling of Ethidium Bromide (page 256). Wear rubber gloves when staining, viewing, and photographing gel and during cleanup. Confine all staining to a restricted sink area.

1. Flood gels with ethidium bromide solution (1 μg/ml), and allow to stain for 5–10 minutes.

2. Following staining, use funnel to decant as much of the ethidium bromide solution as possible from staining tray back into storage container.

Ethidium bromide solution may be reused to stain 15 or more gels. Disable spent staining solution as explained on page 256.

3. Rinse gel and tray under running tap water.

4. If desired, the gel can be destained in tap water or distilled water for 5 minutes or more to help remove the background ethidium bromide.

STOP POINT Staining intensifies dramatically if rinsed gels set overnight at room temperature. Stack staining trays, and cover top gel with plastic wrap to prevent desiccation.

5. View under ultraviolet transilluminator or other UV source.

> **CAUTION**
> Ultraviolet light can damage eyes. Never look at unshielded UV light source with naked eyes. View only through a filter or safety glasses that absorb harmful wavelengths.

6. Take time for responsible cleanup.
   a. Wipe down camera, transilluminator, and staining area.
   b. Wash hands before leaving lab.

IVB. Stain Gel with Methylene Blue and View (30+ minutes)

1. Wear rubber gloves during staining and cleanup.

2. Flood gel with 0.025% methylene blue, and allow to stain for 20–30 minutes.

3. Following staining, use funnel to decant as much of the methylene blue solution as possible from staining tray back into storage container.

4. Rinse gel in running tap water. Let gel soak for several minutes in several changes of fresh water. DNA bands will become increasingly distinct as gel destains.

Destaining time is decreased by rinsing the gel in warm water, with agitation.

STOP POINT For best results, continue to destain overnight in a *small volume* of water. (Gel may destain too much if left overnight in large volume of water.) Cover staining tray to retard evaporation.

5. View gel over light box; cover surface with plastic wrap to prevent staining.

V. Photograph Gel (5 minutes)

Exposure times vary according to mass of DNA in lanes, level of staining, degree of background staining, thickness of gel, and density of filter. Experiment to determine best exposure. When possible, stop lens down (to higher f/number) to increase the depth of field and sharpness of bands.

1. *For ultraviolet (UV) photography of ethidium-bromide-stained gels:* Use Polaroid high-speed film Type 667 (ASA 3000). Set camera aperture to f/8 and shutter speed to B. Depress shutter for a 2–3-second time exposure.

   For white-light photography of methylene-blue-stained gels: Use Polaroid Type 667 film, with an aperture of f/8 and a shutter speed of 1/125 second.

   > **CAUTION**
   > Avoid getting caustic developing jelly on skin or clothes. If jelly gets on skin, wash immediately with plenty of soap and water.

2. Place left hand firmly on top of camera to steady. Firmly grasp small white tab, and pull straight out from camera. This causes a large yellow tab to appear.

3. Grip yellow tab in center, and in one steady motion, pull it straight out from the camera. This starts development.

4. Allow to develop for recommended time (45 seconds at room temperature). Do not disturb print while developing.

5. After full development time has elapsed, separate print from negative by peeling back at end nearest yellow tab.

6. Wait to see the result of the first photo before making other exposures.

Observe your gel and determine which lanes contain control pAMP/pKAN and which lanes contain minipreps M1 and M2. Even if you have confused the prescribed loading order, the miniprep lanes can be distinguished by the following characteristics:

RESULTS AND DISCUSSION

- a background "smear" of degraded and partially digested chromosomal DNA, plasmid DNA, and RNA
- undissolved material and high-molecular-weight DNA "trapped" at the front edge of the well
- a "cloud" of low-molecular-weight RNA at a position corresponding to 100–200 bp
- presence of high-molecular-weight bands of uncut plasmid in lanes of digested miniprep DNA

Refer to Results and Discussion section of Laboratory 6B, Restriction Analysis of Purified pAMP, for more details about interpreting miniprep gels and plasmid conformations.

Remember these three facts when considering possible constructions of the ligated plasmids in M1 and M2.

1. Every replicating plasmid must have an origin of replication. Recombinant plasmids with more than one origin also replicate normally; however, only one origin is active.

2. Each adjacent restriction fragment can only ligate at a like restriction site, *Bam*HI to *Bam*HI and *Hin*dIII to *Hin*dIII. Thus, an intact plasmid must be constructed of an *even* number of fragments (2, 4, 6, 8, etc.).

3. Repeated copies of a restriction fragment cannot exist adjacent to one another; that is, they must alternate with other fragments. Adjacent duplicate fragments form "inverted repeats" in which the sequences, one on either side of the restriction site, are complementary along the entire length of the duplicated fragment. Molecules with such inverted repeats cannot replicate properly. As the plasmid opens up to allow access to DNA polymerase, the single-strand regions on either side of the restriction site base pair to one another to form a large "hairpin loop," which fouls replication.

Follow Questions 1 through 8 to interpret each pair of miniprep results (M1 +/− and M2 +/−).

1. Examine the photograph of your stained gel (or view on a light box or overhead projector). Compare your gel with the ideal gel on following page. Label the size of fragments in each lane of your gel.

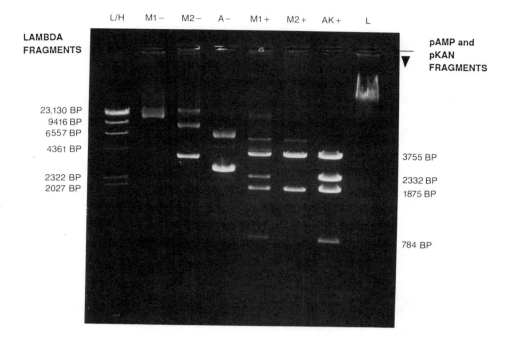

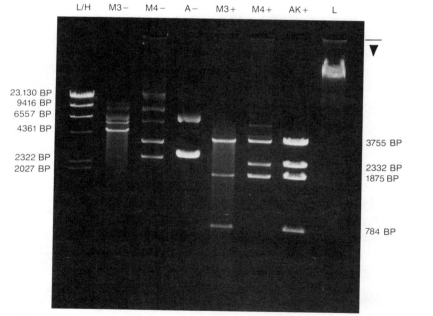

Restriction Analysis of Four pAMP/KAN Recombinants (M1, M2, M3, M4)

2. Label fragment sizes of the four bands in AK+ lane (cut control pAMP and pKAN) from top of gel to bottom: 3755 bp, 2332 bp, 1875 bp, and 784 bp. Every miniprep must contain the 3755-bp fragment containing the amp^r gene and the 1875-bp fragment containing the kan^r gene. Locate these bands by comparing the M+ lane (cut miniprep) with the AK+ lane (cut control).

3. Now look for evidence of any other bands in the M+ lane. Compare the M+ lane with the AK+ lane. The 2332-bp fragment and/or the 784-bp fragment may be present. If neither of these two additional bands is present, the molecule is termed a "simple recombinant."

4. If a third band of 784 bp is present, the molecule may be

   a. a "superplasmid" in which one of the three fragments is repeated

   or

   b. a double transformation of a simple recombinant *and* a religated pAMP

5. If a third band of 2332 bp is present, the molecule may be

   a. a superplasmid in which one of the three fragments is repeated

   or

   b. a double transformation of the simple recombinant *and* religated pKAN

   or

   c. a double transformation of the simple recombinant *and* ligated 3755-bp + 2332-bp fragments

   or

   d. a double transformation of religated pKAN *and* ligated 3755-bp + 2332-bp fragments

6. If all four bands are present, the molecule may be

   a. a superplasmid containing all four fragments

   or

   b. a double transformation of religated pAMP *and* religated pKAN

   or

   c. a double transformation of a simple recombinant *and* ligated 2332-bp + 784-bp fragments

7. To gauge the size of the miniprep plasmid, compare the M− lane (uncut miniprep) with the A− lane (uncut pAMP) and the L/H lane (λ markers). Remember that uncut plasmid can assume several conformations but that the fastest moving form is supercoiled.

   a. Locate the band that has migrated furthest in the A− lane; this is the supercoiled form of pAMP.

   b. Now examine the band(s) furthest down the M− lane. If this band and the pAMP band have comigrated similar distances, your miniprep is likely a double transformation. The possible molecules present in a double transformation range in size from 3106 bp (6c above) to 6077 bp (5c,d above), and thus may appear noticeably lower or higher on the gel than supercoiled pAMP.

c. If the fastest moving band of the uncut miniprep is very high on the gel, your molecule is likely a superplasmid. Compute the possible sizes of superplasmids composed of three or four fragments.

8. When bacteria are transformed with two different plasmids having related origins of replication, one of the two plasmids is preferentially replicated within the host cell. Over generations, one of the two plasmids is eventually lost. Thus, in double transformations with four different fragments, one pair of fragments should be fainter than the other pair.

9. Based on your evaluation above, make scale restriction maps of your M1 and M2 plasmids.

FOR FURTHER RESEARCH

Further research may reveal with certainty the structure of perplexing recombinant plasmids. To obtain additional plasmid for further experimentation, do double minipreps from your master colonies (M1 and M2) in Laboratory 8 or from colonies of interest from replica plates in Experiment 1 below.

1. Perform a series of experiments to distinguish between a superplasmid and a double transformation in Questions 4, 5, and 6 in Results and Discussion.

    a. Make a 1:10 dilution of your miniprep DNA.

    b. Use the dilute miniprep DNA to transform competent *E. coli* cells, and plate onto LB/amp and LB/kan.

    c. Replica plate colonies from each master plate onto fresh LB/amp and LB/kan plates.

    d. Examine the proportion of dual-resistant colonies.

    - If all the restriction fragments are contained in a single superplasmid, all transformants will have dual resistance.

    - If three or four restriction fragments are distributed among separate plasmids, the transformants will have mixed antibiotic resistance. Matching the observed pattern of antibiotic resistance with alternate two-gene recombinants can often reveal the structure of the two plasmids involved.

Digesting miniprep DNA with the restriction enzyme *Xho*I can elucidate some of the structures of superplasmids and plasmids in double transformations. This enzyme has a single recognition site within the 1875-bp pKAN fragment and *no sites* within any of the other three *Bam*HI/

HindIII fragments. Electrophorese XhoI digests of miniprep DNA with samples of uncut pAMP and λ/HindIII size markers.

2. If your miniprep DNA shows three fragments, including the 784-bp fragment (Question 4 in Results and Discussion), then you may have either a superplasmid with one repeated fragment or a double transformation.

   a. The results of a XhoI digest of a superplasmid will differ according to which fragment is repeated:
      - If the 784-bp fragment is repeated, a linear 7198-bp plasmid is produced.
      - If the 3755-bp fragment is repeated, a linear 10,169-bp plasmid is produced.
      - If the 1875-bp fragment containing the XhoI site is repeated, two fragments of 2659 bp and 5630 bp are produced.

   b. A double transformation of the simple recombinant and religated pAMP produces a linear 5630-bp plasmid *plus* an *uncut* pAMP plasmid.

3. If your miniprep DNA shows three bands, including the 2332-bp fragment (Question 5 in Results and Discussion), then you may have a superplasmid with one repeated fragment or a double transformation.

   a. The results of a XhoI digest of a superplasmid will differ according to which fragment is repeated:
      - If the 2332-bp fragment is repeated, a linear 10,274-bp plasmid is produced.
      - If the 3755-bp fragment is repeated, a linear 11,707-bp plasmid is produced.
      - If the 1875-bp fragment containing the XhoI site is repeated, two fragments of 4197 bp and 5130 bp are produced.

   b. A double transformation of the simple recombinant and religated pKAN produces two fragments of 5630 bp and 4197 bp.

   c. A double transformation of the simple recombinant and ligated 3755-bp + 2332-bp fragments produces a linear 5630-bp plasmid *plus* an *uncut* 6077-bp plasmid.

   d. A double transformation of religated pKAN and ligated 3755-bp + 2332-bp fragments produces a linear 4197-bp plasmid *plus* an *uncut* 6077-bp plasmid.

4. If your miniprep DNA contains all four fragments (Question 6 in Results and Discussion), the XhoI digest will discriminate between a superplasmid and double transformations.

a. A superplasmid produces a linear 8736-bp plasmid.

b. A double transformation of religated pAMP and religated pKAN produces a linear 4197-bp pKAN plasmid *plus* an *uncut* pAMP plasmid.

c. A double transformation of the simple recombinant and ligated 2332-bp + 784-bp fragments produces a linear 5630-bp plasmid *plus* an *uncut* 3106-bp plasmid.

5. Make a restriction map of the simple recombinant plasmid using *Bam*HI, *Hin*dIII, and *Pvu*I. Prior experiments showed that the *Bam*HI and *Hin*dIII sites are separated by 1875 bp. *Pvu*I cuts the recombinant plasmid at two positions.

   a. Do double minipreps to obtain additional plasmid from a master colony known to contain the simple recombinant.

   b. Digest aliquots of the miniprep DNA with

   *Pvu*I
   *Pvu*I + *Bam*HI
   *Pvu*I + *Hin*dIII
   *Bam*HI + *Hin*dIII
   *Bam*HI + *Hin*dIII + *Pvu*I

   c. Electrophorese the digested samples on a 1.2% agarose gel, stain, and photograph.

   d. The expected number of fragments and their sizes are shown in the diagram below:

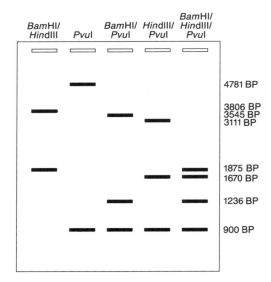

6. Using the data from Experiment 5 above and applying a little logic, the relative positions of the restriction sites can be positioned around

a circle to produce a restriction map of the simple recombinant plasmid.

a. The *Bam*HI/*Hin*dIII digest reveals that the *Bam*HI and *Hin*dIII sites are separated by 1875 bp.

b. The *Pvu*I digest reveals that the two *Pvu*I sites are separated by 900 bp.

c. The *Bam*HI/*Hin*dIII/*Pvu*I digest shows both 1875-bp and 900-bp fragments. This means that the 1670-bp and 1236-bp fragments separate the 900-bp *Pvu*I fragment from the 1875-bp *Bam*HI/*Hin*dIII fragment.

d. The *Pvu*I/*Bam*HI digest shows a 3545-bp fragment that must be composed of the 1875-bp fragment plus the 1670-bp fragment.

e. The *Pvu*I/*Hin*dIII digest shows a 3111-bp fragment that must be composed of the 1875-bp fragment plus the 1236-bp fragment.

f. Results d and e indicate that the 1236-bp fragment is adjacent to the *Bam*HI site and the 1670-bp fragment is adjacent to the *Hin*dIII site.

g. Complete the restriction map showing all restriction sites and the distances between them.

APPENDIX 1
Equipment, Supplies, and Reagents

I. Equipment

Materials needed	Laboratory number										CBS Catalog number*
	1	2	3	4	5	6	7	8	9	10	
Cell spreader					•			•			21-5820
Clinical centrifuge[a]								•			21-4075
Electrophoresis box[a]		•	•			•	•			•	21-3668
Electrophoresis power supply[a]		•	•			•	•			•	21-3673
Incubator[a]		•			•	•[b]		•	•	•[b]	21-5868
Inoculating loop[a]		•			•	•[b]		•[b]	•	•[b]	21-5826
Microfuge[a]	•		•	•		•	•			•	21-4050
Micropipet (adjustable volume)[a]											
0.5–10 μl	•		•	•	•	•	•	•		•	21-4650
10–100 μl			•[b]	•[b]	•[b]	•[b]	•[b]	•[b]		•[b]	21-4654
100–1000 μl	•				•	•		•		•	21-4656
(Micro) water bath			•	•	•	•	•	•		•	21-6250
Pipet aid[a]											
1 ml					•			•		•	21-4680
10 ml	•	•				•[b]		•		•[b]	21-4684
Shaking water bath		•				•[b]		•		•[b]	21-6258
Spectrophotometer								•[b]			21-4000
Test tube rack[a]	•	•	•	•	•	•	•	•		•	21-5572
Ultraviolet transilluminator/camera			•	•		•	•			•	21-3678
White light transilluminator/camera			•	•		•	•			•	21-3680

* Carolina Biological Supply (CBS) Company Catalog.
[a] Options available. See Carolina Biological Supply Company Catalog for more information.
[b] Used in laboratory preparation.

II. Supplies

Materials needed	Laboratory number										CBS Catalog number
	1	2	3	4	5	6	7	8	9	10	
Camera film (Polaroid)			•	•		•	•			•	21-3679
50-ml Conical tube (sterile, screwtop)	•	•				•[b]		•[b]		•[b]	21-5100
15-ml Culture tube (sterile, 2-position cap)[a]	•				•			•			21-5080
Gloves, rubber			•	•		•	•			•	70-6347
Micropipet tip[a]											
0.5–10 μl	•		•	•	•	•	•	•		•	21-5130
10–100 μl			•[b]	•[b]	•[b]	•[b]	•[b]	•[b]		•[b]	21-5120
100–1000 μl	•				•	•		•		•	21-5124
Petri dish (sterile, 100 × 15 mm)			•		•			•	•		21-4826
Pipet (sterile)[a]											
1 ml transfer					•			•			21-5840
10 ml	•	•				•[b]		•		•[b]	21-4626
Sterile filter (.45 μm)[a]					•			•			21-4232
1.5-ml Tube	•		•	•	•[b]	•	•	•		•	21-5222

[a] Options available. See Carolina Biological Supply Company Catalog for more information.
[b] Used in laboratory preparation.

III. Media (LB)

Materials needed	1	2	3	4	5	6	7	8	9	10	CBS Catalog number
A. Ready-to-pour. Sterile; 20 plates											
LB agar		•			•						21-6620
LB agar + ampicillin		•			•			•	•		21-6621
LB agar + kanamycin								•	•		21-6622
LB agar + amp + kan								•			21-6623
B. Ready-to-use. Prepoured; 10 plates											
LB agar		•			•						21-6610
LB agar + ampicillin		•			•			•	•		21-6611
LB agar + kanamycin								•	•		21-6612
LB agar + amp + kan								•			21-6613
C. Media solution. Sterile; 5–50-ml aliquots											
LB broth		•			•			•			21-6660
LB broth + ampicillin		•				•[b]				•[b]	21-6661

IV. Media Components

Materials needed	1	2	3	4	5	6	7	8	9	10	CBS Catalog number
LB agar base		•			•			•	•		21-6700
LB base		•			•	•		•	•	•	21-6710
Tryptone		•			•	•		•	•	•	21-6740
Yeast extract		•			•	•		•	•	•	21-6745
Sodium hydroxide, 4 N		•			•	•		•	•	•	21-8830
Ampicillin solution, 10 mg/ml		•			•	•		•	•	•	21-6860
powder		•			•	•		•	•	•	21-6880
Kanamycin solution, 10 mg/ml								•	•		21-6861
powder								•	•		21-6881

V. Biologicals and Enzymes

Materials needed	Laboratory number										CBS Catalog number
	1	2	3	4	5	6	7	8	9	10	
A. DNA											
λ DNA			•	•							21-1410
Plasmid pAMP					•	•	•	•		•	21-1430
Plasmid pKAN							•	•		•	21-1440
B. Bacterial Strains											
MM294		•			•			•			21-1529
MM294/pAMP		•				•					21-1539
C. Enzymes											
BamHI			•			•	•			•	21-1660
EcoRI			•	•							21-1670
M.EcoRI methylase (with 10× restriction buffer/SAM)				•							21-1671
HindIII			•	•		•	•			•	21-1690
T4 DNA ligase							•				21-1740
RNase A, pancreatic, 5 mg/ml						•				•	21-1745

VI. Reagents

Materials needed	\multicolumn{10}{c}{Laboratory number}	CBS Catalog number
	1	2	3	4	5	6	7	8	9	10	
Agarose, LE (low EEO)			•	•		•	•			•	21-7080
Calcium chloride solution, sterile, 50 mM					•			•			21-1320
Ethanol, 95%[a]					•	•		•		•	21-7410
Ethidium bromide solution, 5 mg/ml[b]			•	•		•	•			•	21-7420
Glucose-Tris-EDTA (GTE) solution						•				•	21-7710
Isopropanol						•				•	21-7910
$KMnO_4$, 0.5 M[c]			•	•		•	•			•	21-8601
Ligation buffer/ATP, 2×							•				included with T4 DNA ligase
Loading dye			•	•		•	•			•	21-8200
Methylene blue solution			•	•		•	•			•	21-8290
Potassium acetate–acetic acid solution						•				•	21-8602
Restriction buffer, 10× compromise			•	•		•	•			•	21-8770
Sodium dodecyl sulfate (SDS) 10% solution						•				•	21-8822
Sodium hydroxide, 4 N						•				•	21-8830
Tris/Borate/EDTA (TBE) buffer, 10× mix			•	•		•	•			•	21-9024
Tris-EDTA (TE) buffer						•				•	21-9026

[a] Be extremely careful not to ignite ethanol in beaker.
[b] Review Responsible Handling of Ethidium Bromide on page 256. Wear rubber gloves when staining, viewing, and photographing gels and during cleanup. Confine all staining to a restricted sink area.
[c] $KMnO_4$ is an irritant and is explosive. Solutions containing $KMnO_4$ should be handled in a chemical hood.

VII. Reagent Components

Materials needed	Laboratory number										CBS Catalog number
	1	2	3	4	5	6	7	8	9	10	
Acetic acid, glacial						•				•	21-7060
Adenosine triphosphate (ATP)							•				21-7066
Boric acid			•	•		•	•			•	21-7200
Bromophenol blue			•	•		•	•			•	21-7210
Calcium chloride solution, dihydrate					•			•			21-7260
Dithiothreitol (DTT)			•	•		•	•			•	21-7360
Dithiothreitol (DTT) solution 1 M			•	•		•	•			•	21-7362
Ethylenediaminetetra-acetate (EDTA) -disodium salt, dihydrate			•	•		•	•			•	21-7430
Ethylenediaminetetra-acetate (EDTA) solution, 0.5 M			•	•		•	•			•	21-7432
Glucose, anhydrous						•				•	21-7700
Magnesium chloride solution, 1 M			•	•		•	•			•	21-8262
Magnesium chloride, 6-hydrate			•	•		•	•			•	21-8260
β-Mercaptoethanol (BME), 14 M			•	•		•	•			•	21-8280
Potassium acetate						•				•	21-8600
Sodium chloride		•	•	•	•	•	•	•	•	•	21-8810
Sodium dodecyl sulfate (SDS)						•				•	21-8820
Sucrose			•	•		•	•			•	21-8840
Tris base			•	•		•	•			•	21-9020
Tris buffer, pH 8.0			•	•		•	•			•	21-9022
Xylene cyanol			•	•		•	•			•	21-9940

VIII. Ready-to-use Kits

Phenotypic Bacterial ID Kit
21-1130N Teacher Demonstration Kit
21-1132N Student Kit

Colony Transformation Kit
21-1140N Teacher Demonstration Kit
21-1142N Student Kit

Restriction Enzyme Cleavage of DNA Kit
21-1149N Teacher Demonstration or Student Kit

DNA Restriction Analysis Kit
21-1150N Kit with Ethidium Bromide
21-1151N Kit with Methylene Blue

APPENDIX 2
Recipes for Media, Reagents, and Stock Solutions

The success of the laboratories in this book depends on the use of uncontaminated reagents. Follow the recipes with care and pay scrupulous attention to cleanliness. Use a clean spatula for each ingredient or carefully pour each ingredient from its bottle.

The recipes are organized in six sections. Stock solutions that are used in more than one laboratory are listed once, according to their *first use* in the laboratories.

I. BACTERIAL CULTURE

4 N Sodium Hydroxide (NaOH)
10 mg/ml Ampicillin
10 mg/ml Kanamycin
Luria-Bertani (LB) Broth
LB Broth + Antibiotic
LB Agar Plates
LB Agar + Antibiotic
Stab Cultures

II. DNA RESTRICTION

1 M Tris (pH 8.0)
5 M Sodium Chloride (NaCl)
1 M Magnesium Chloride ($MgCl_2$)
1 M Dithiothreitol (DTT)
10× Compromise Restriction Buffer
2× Restriction Buffer
30 mM S-Adenosyl Methionine (SAM)
2× Restriction Buffer/SAM
0.05% Glacial Acetic Acid
5 mg/ml RNase A (Pancreatic RNase)
5× Restriction Buffer/RNase

III. GEL ELECTROPHORESIS

10× Tris/Borate/EDTA (TBE) Electrophoresis Buffer
1× Tris/Borate/EDTA (TBE) Electrophoresis Buffer
0.8% Agarose
Loading Dye
5 mg/ml Ethidium Bromide

1 μl/ml Ethidium Bromide Staining Solution
1% Methylene Blue
0.025% Methylene Blue Staining Solution

IV. BACTERIAL TRANSFORMATION

1 M Calcium Chloride ($CaCl_2$)
50 mM Calcium Chloride ($CaCl_2$)

V. PLASMID MINIPREPARATION

0.5 M Ethylenediaminetetraacetic Acid (EDTA)
Tris/EDTA (TE) Buffer
Glucose/Tris/EDTA (GTE)
5 M Potassium Acetate (KOAc)
Potassium Acetate/Acetic Acid
10% Sodium Dodecyl Sulfate (SDS)
1% SDS/0.2 N NaOH

VI. DNA LIGATION

1 M Tris (pH 7.6)
10× Ligation Buffer
0.1 M Adenosine Triphosphate (ATP)
2× Ligation Buffer + ATP

Notes on Buffers

1. Typically, solid reagents are dissolved in a volume of deionized or distilled water equivalent to 70–80% of the finished volume of buffer. This leaves room for the addition of acid or base to adjust the pH. Finally, water is added to bring the solution up to the final volume.

2. The final concentration of each liquid reagent is given in the right-hand column of the reagent list.

3. Buffers are used as 2×, 5×, or 10× solutions. Buffers are diluted when mixed with other reagents to produce a working concentration of 1×.

I. BACTERIAL CULTURE

4 N Sodium Hydroxide (NaOH)

Makes 100 ml.
Store at room temperature (indefinitely).

1. Slowly add 16 g of NaOH pellets (m.w. = 40.00) to 80 ml of deionized or distilled water, with stirring. The solution will get very hot.

2. When NaOH pellets are completely dissolved, add water to a final volume of 100 ml.

10 mg/ml Ampicillin

Makes 100 ml.
Store at $-20°C$ (1 year) or $4°C$ (3 months).

1. Add 1 g of ampicillin (sodium salt, m.w. = 371.40) to 100 ml of deionized or distilled water in a clean 250-ml flask. (The sodium salt dissolves readily; however, the free acid form is difficult to dissolve.)

2. Stir to dissolve.

3. Prewash a 0.45- or 0.22-μm sterile filter (Nalgene or Corning) by drawing through 50–100 ml of deionized or distilled water. Pass ampicillin solution through the washed filter.

4. Dispense 10-ml aliquots in sterile 15-ml tubes (Falcon 2059 or equivalent), and freeze at $-20°C$.

10 mg/ml Kanamycin

Makes 100 ml.
Store at $-20°C$ (1 year) or $4°C$ (3 months).

1. Add 1.0 g of kanamycin sulfate (m.w. = 582.60) to 100 ml of deionized or distilled water in a clean 250-ml flask.

2. Stir to dissolve.

3. Prewash a 0.45- or 0.22-μm sterile filter (Nalgene or Corning) by drawing through 50–100 ml of deionized or distilled water. Pass kanamycin solution through the washed filter.

4. Dispense 10-ml aliquots in sterile 15-ml tubes (Falcon 2059 or equivalent), and freeze at $-20°C$.

Luria-Bertani (LB) Broth

Makes 1 liter.
Store at room temperature (indefinitely).

1. Weigh out

   10 g of tryptone
   5 g of yeast extract
   10 g of NaCl (m.w. = 58.44)

   Alternately, use 25 g of premix containing all of these ingredients.

2. Add all ingredients to a clean 2-liter flask that has been rinsed with deionized or distilled water.

3. Add 1 liter of deionized or distilled water to flask.

4. Add 0.5 ml of 4 M NaOH.

5. Stir to dissolve the dry ingredients, preferably using a magnetic stir bar.

6. *If preparing for mid-log cultures,* split LB broth into two 500-ml aliquots in 2-liter flasks. Plug top with cotton or foam, and cover with aluminum foil. (Alternately, cover with aluminum foil only.) Autoclave for 15–20 minutes at 121°C.

   or

   If preparing for general use in transformations, dispense 100-ml aliquots into sterile 150–250-ml bottles using one of the following methods:

   a. Loosely put on the caps. Autoclave for 15–20 minutes at 121°C. (To help guard against breakage, autoclave the bottles in a shallow pan with a small amount of water.)

   or

   b. Prewash a 0.45- or 0.22-µm sterile filter (Nalgene or Corning) by drawing through 50–100 ml of deionized or distilled water. Pass LB broth through the filter, and aliquot into sterile bottles.

Note: LB broth can be considered sterile as long as the solution remains clear. Cloudiness is a sign of contamination by microbes. Always swirl solution to check for bacterial or fungal cells that may have settled at the bottom of the flask or bottle.

LB Broth + Antibiotic

Makes 100 ml.
Store at 4°C (3 months).

1. Sterilely add 1 ml of 10 mg/ml antibiotic to 100 ml of *cool* LB broth.

2. Swirl to mix.

LB Agar Plates

Makes 35–40 plates.
Store at 4°C (3 months) or room temperature (3 months).

1. Weigh out
    10 g of tryptone
    5 g of yeast extract
    10 g of NaCl (m.w. = 58.44)
    15 g of agar

    Alternately, use 40 g of premix containing all of these ingredients.

2. Add all ingredients to a clean 2-liter flask that has been rinsed with deionized or distilled water.

3. Add 1 liter of deionized or distilled water.

4. Add 0.5 ml of 4 N NaOH.

5. Stir to dissolve dry ingredients, preferably using a magnetic stir bar. Any undissolved material will dissolve during autoclaving.

6. Cover flask mouth with aluminum foil, and autoclave solution for 15 minutes at 121°C.

7. During autoclaving, the agar may settle to the bottom of flask. Swirl to mix agar evenly.

8. Allow solution to cool until flask can just be held in bare hands (55–60°C). (If solution cools too long and the agar begins to solidify, remelt by briefly autoclaving for 5 minutes or less.)

9. While agar is cooling, mark culture *plate bottoms* with the date and a description of the media (e.g., LB). If using presterilized poly-

styrene plates, carefully cut the end of plastic sleeves, and save the sleeves for storing the poured plates. Spread plates out on lab bench.

10. When agar flask is cool enough to hold, lift lid of culture plate only enough to pour solution. *Do not place lid on lab bench.* Quickly pour in agar to just cover plate bottom (approximately 3 mm). Tilt plate to spread agar, and immediately replace lid.

11. Continue pouring agar into plates. Occasionally flame mouth of flask to maintain sterility.

12. To remove bubbles in the surface of the poured agar, touch surface with Bunsen flame while agar is still liquid.

13. Allow agar to solidify undisturbed.

14. If possible, incubate plates *lidside down* for several hours at 37°C (overnight if convenient). This dries the agar, limiting condensation when plates are stored under refrigeration. It also allows the ready detection of any contaminated plates.

15. Stack plates in their original plastic sleeves for storage.

LB Agar + Antibiotic

Makes 30–45 plates.
Store at 4°C (3 months).

1. Follow recipe above for LB agar plates through Step 9.

2. When agar flask is cool enough to hold, sterilely add 10 ml of 10 mg/ml antibiotic. Ampicillin and kanamycin are destroyed by heat; therefore, *it is essential to cool agar before adding antibiotic.* (For LB/amp + kan plates, add 10 ml *each* of 10 mg/ml ampicillin and kanamycin.)

3. Swirl flask to mix antibiotic.

4. Resume recipe above with Step 10.

Note: In a pinch, antibiotic-containing plates can be quickly made by evenly spreading 200 µl of 10 mg/ml antibiotic on the surface of an LB agar plate. Allow the antibiotic to absorb into agar for 10–20 minutes before using. Outdated antibiotic plates can be refurbished in this manner.

Stab Cultures

Makes 30–40 stab cultures.
Store in dark at room temperature (1 year).

To prepare vials:

1. Weigh out

    1.0 g of tryptone
    0.5 g of yeast extract
    1.0 g of NaCl (m.w. = 58.44)
    0.7 g of agar

Note: This recipe is the same as LB agar recipe above, but the lower percentage of agar makes the stab culture easier to use. Standard LB agar or 4 g of premix can also be used.

2. Add all ingredients to a clean 250-ml flask that has been rinsed with deionized or distilled water.

3. Add 100 ml of deionized or distilled water.

4. Add 50 µl of 4 N NaOH.

5. Stir while heating to dissolve dry ingredients; preferably, use a magnetic stir bar and a hot plate.

6. Pour dissolved solution into 4-ml vials (15 × 45 mm) until two-thirds filled, and loosely replace caps.

7. Autoclave vials for 15 minutes at 121°C. (Alternately, sterilized agar can be poured into presterilized vials.)

8. Allow agar to solidify undisturbed.

9. Prior to storage, tighten caps completely.

10. Seal caps with Parafilm for longer shelf life.

To inoculate:

1. Sterilely scrape up cell mass from a single colony of desired genotype.

2. Stab inoculating loop several times into agar.

3. Loosely replace cap, and incubate stab culture overnight at 37°C.

4. Following incubation, tighten cap, and store in the dark at room temperature. Wrap the cap with Parafilm for long-term storage.

II. DNA RESTRICTION

1 M Tris (pH 8.0)

Makes 100 ml.
Store at room temperature (indefinitely).

1. Dissolve 12.1 g of Tris base (m.w. = 121.10) in 70 ml of deionized or distilled water.

2. Adjust pH by slowly adding approximately 5.0 ml of concentrated hydrochloric acid (HCl); monitor with a pH meter. (If a pH meter is not available, adding 5.0 ml of concentrated HCl will yield a solution of approximately pH 8.0.)

3. Add deionized or distilled water to make a total volume of 100 ml of solution.

> **CAUTION**
>
> Avoid inhaling Tris powder; wear a mask over nose and mouth.

Notes:

  i. A yellow-colored solution indicates poor-quality Tris. Discard, and obtain from a different source.
 ii. Many types of electrodes do not accurately measure the pH of Tris solutions; check with manufacturer to obtain a suitable one.
iii. The pH of Tris solutions is temperature-dependent; measure pH at room temperature.

5 M Sodium Chloride (NaCl)

Makes 100 ml.
Store at room temperature (indefinitely).

1. Dissolve 29.2 g of NaCl (m.w. = 58.44) in 70 ml of deionized or distilled water.

2. Add deionized or distilled water to make a total volume of 100 ml of solution.

1 M Magnesium Chloride ($MgCl_2$)

Makes 100 ml.
Store at room temperature (indefinitely).

1. Dissolve 20.3 g of MgCl2 (6-hydrate, m.w. = 203.30) in 80 ml of deionized or distilled water.

2. Add deionized or distilled water to make a total volume of 100 ml of solution.

1 M Dithiothreitol (DTT)

Makes 10 ml.
Store at −20°C (indefinitely).

1. Dissolve 1.5 g of DTT (m.w. = 154.25) in 8 ml of deionized or distilled water.

2. Add deionized or distilled water to make a total volume of 10 ml of solution.

3. Dispense into 1-ml aliquots in 1.5-ml tubes.

Note: Do not autoclave DTT or solutions containing it.

10× Compromise Restriction Buffer

Makes 1 ml.
Store at −20°C (indefinitely).

Mix ingredients in a 1.5-ml tube.

100 µl of 1 M Tris (pH 8.0)	(100 mM)
100 µl of 1 M MgCl$_2$	(100 mM)
200 µl of 5 M NaCl	(1 M)
7 µl of 14.3 M β-mercaptoethanol (BME)	(100 mM)
593 µl of deionized water	

or

100 µl of 1 M Tris (pH 8.0)	(100 mM)
100 µl of 1 M MgCl$_2$	(100 mM)
200 µl of 5 M NaCl	(1 M)
10 µl of 1 M DTT	(10 mM)
590 µl of deionized water	

Notes:

i. β-Mercaptoethanol, also called 2-mercaptoethanol, is a 14.3 M liquid at room temperature.
ii. Compromise buffer provides salt conditions that allow relatively high activity of a number of different restriction enzymes, including all those used in these laboratories. Consider using the specific buffer provided by the manufacturer when using expensive enzymes in critical experiments.
iii. DTT used in the second recipe may precipitate from the solution with repeated freezing and thawing. Vortex vigorously to redissolve.

2× Restriction Buffer

Makes 1 ml.
Store at −20°C (indefinitely).

Mix 200 µl of 10× restriction buffer with 800 µl of deionized or distilled water.

30 mM S-Adenosyl Methionine (SAM)

Makes 1 ml.
Store at −20°C (1 year).

1. Obtain 1 M sulfuric acid (H_2SO_4) or prepare by *carefully* mixing 1 part concentrated acid (18 M) with 17 parts of deionized or distilled water.

2. Prepare a 5 mM solution by adding 5 µl of 1 M H_2SO_4 to 995 µl of deionized or distilled water.

3. Add 900 µl of 5 mM H_2SO_4 solution to 100 µl of ethanol (100%).

4. Dissolve 15.8 mg of SAM (iodide salt, grade I, m.w. = 526.30) in 1 ml of 5 mM H_2SO_4/10% ethanol solution.

5. Dispense 100-µl aliquots in 1.5-ml tubes. Use each solution once, and discard.

Notes:

i. SAM is very unstable, and activity diminishes rapidly with freezing and thawing. Store SAM solution on ice at all times; SAM solutions have half-lives on the order of 30 minutes at room temperature.
ii. 30 mM SAM is often supplied with M.*Eco*RI methylase.

2× Restriction Buffer/SAM

Makes 1 ml.
Hold on ice or at −20°C, and use within several hours of mixing.

1. Mix in a 1.5-ml tube:

    10 µl of 30 mM SAM
    200 µl of 10× restriction buffer
    790 µl of deionized water

2. Use within several hours, and discard.

Note: Store restriction buffer/SAM on ice at all times during manipulations; SAM solutions have half-lives on the order of 30 minutes at room temperature.

0.05% Glacial Acetic Acid

Makes 50 ml.
Store at room temperature (indefinitely).

1. Add 5 µl of the glacial acetic acid to 50 ml of deionized or distilled water.

2. The solution should be at approximately pH 4.0.

5 mg/ml RNase A (Pancreatic RNase)

Makes 20 ml.
Store at −20°C (indefinitely).

1. Dissolve 100 mg of RNase A in 20 ml of 0.05% glacial acetic acid, and transfer to a 50-ml conical tube.

2. Place the tube in a boiling-water bath for 15 minutes.

3. Cool the solution, and neutralize by adding 120 µl of 1 M Tris (pH 8.0).

4. Dispense 1-ml aliquots in 1.5-ml tubes.

Notes:
  i. Use only RNase A from bovine pancreas.
  ii. Dissolving RNase in the acetic acid prevents subsequent precipitation of the RNase. The solution can be prepared by simply dissolving RNase in deionized or distilled water; however, the RNase will occasionally precipitate from the solution and activity will be lost.

5× Restriction Buffer/RNase

Makes 1 ml.
Store at −20°C (several months).

Mix in a 1.5-ml tube:

  500 µl of 10× restriction buffer
  100 µl of 5 mg/ml RNase
  400 µl of water

III. GEL ELECTROPHORESIS

10× Tris/Borate/EDTA (TBE) Electrophoresis Buffer

Makes 1 liter.
Store at room temperature (indefinitely).

1. Add the following dry ingredients to 700 ml of deionized or distilled water in a 2-liter flask.

   1 g of NaOH (m.w. = 40.00)
   108 g of Tris base (m.w. = 121.10)
   55 g of boric acid (m.w. = 61.83)
   7.4 g of EDTA (disodium salt, m.w. = 372.24)

2. Stir to dissolve, preferably using a magnetic stir bar.

3. Add deionized water to bring total solution to 1 liter.

1× Tris/Borate/EDTA (TBE) Electrophoresis Buffer

Makes 5 liters.
Store at room temperature (indefinitely).

1. Into a spigeted carboy, add 9 liters of deionized or distilled water to 1 liter of 10× TBE electrophoresis buffer.

2. Stir to mix.

0.8% Agarose

Makes 200 ml.
Use fresh or store solidified agarose at room temperature (several weeks).

1. Add 1.6 g of agarose (low EEO electrophoresis grade) to 200 ml of 1× TBE electrophoresis buffer in a 600-ml beaker or Erlenmeyer flask.

2. Stir to suspend agarose.

3. Cover beaker with aluminum foil, and heat in boiling-water bath (double boiler) or on a hot plate until all agarose is dissolved (approximately 10 minutes).

   or

   Heat *uncovered* in a microwave oven at high setting until all agarose is dissolved (3–5 minutes per beaker).

4. Swirl solution and check bottom of beaker to make sure that all agarose has dissolved. (Just prior to complete dissolution, particles of agarose appear as translucent grains.) Reheat for several minutes if necessary.

5. Cover with aluminum foil, and hold in a hot-water bath (at about 60°C) until ready for use. Remove any "skin" of solidified agarose from surface prior to pouring.

Notes:

  i. 1.6-g samples of agarose powder can be preweighed and stored in capped test tubes until ready for use.
  ii. Solidified agarose can be stored at room temperature and then remelted over a boiling-water bath (15–20 minutes) or in a microwave oven (5–7 minutes per beaker) prior to use. Always loosen cap when remelting agarose in a bottle.

Loading Dye

Makes 100 ml.
Store at room temperature (indefinitely).

1. Dissolve the following ingredients in 60 ml of deionized or distilled water.

    0.25 g of bromophenol blue (m.w. = 669.96)
    0.25 g of xylene cyanol (m.w. = 538.60)
    50.00 g of sucrose (m.w. = 342.30)
    1.00 ml of 1 M Tris (pH 8.0)

2. Add deionized or distilled water to make a total volume of 100 ml of solution.

Note: Glycerol may be used instead of sucrose. Dissolve the xylene cyanol, bromophenol blue, and Tris in 49 ml of deionized or distilled water and stir in 50 ml of glycerol to make a total volume of 100 ml of solution.

1 μl/ml Ethidium Bromide Staining Solution

Makes 500 ml.
Store in dark at room temperature (indefinitely).

> **CAUTION**
>
> Ethidium bromide is a mutagen by the Ames microsome assay and a suspected carcinogen. Wear rubber gloves when preparing and using ethidium bromide solutions. Review Responsible Handling of Ethidium Bromide on page 256.

1. Add 100 µl of 5 mg/ml ethidium bromide to 500 ml of deionized or distilled water.

2. Store in unbreakable bottles (preferably opaque). Label bottle **CAUTION: Ethidium Bromide. Mutagen and cancer-suspect agent. Wear rubber gloves when handling.**

Note: Ethidium bromide is light-sensitive; store in dark container or wrap container in aluminum foil.

1% Methylene Blue

Makes 50 ml.
Store at room temperature (indefinitely).

Dissolve 0.5 g of methylene blue in 50 ml of deionized or distilled water.

0.025% Methylene Blue Staining Solution

Makes 400 ml.
Store at room temperature (indefinitely).

Add 10 ml of 1% methylene blue to 390 ml of deionized or distilled water.

IV. BACTERIAL TRANSFORMATION

1 M Calcium Chloride (CaCl$_2$)

Makes 100 ml.
Store at room temperature (indefinitely).

1. Dissolve 11.1 g of the anhydrous CaCl$_2$ (m.w. = 110.99) or 14.7 g of the dihydrate (m.w. = 146.99) in 80 ml of deionized or distilled water.

2. Add deionized or distilled water to make a total volume of 100 ml of solution.

50 mM Calcium Chloride (CaCl$_2$)

Makes 500 ml.
Store at 4°C or room temperature (indefinitely).

1. Mix 50 ml of 1 M CaCl$_2$ with 950 ml of deionized water.

2. Prerinse a 0.45- or 0.22-µm sterile filter by drawing through 50–100 ml of deionized or distilled water.

3. Pass CaCl$_2$ solution through prerinsed filter.

4. Dispense aliquots into presterilized 50-ml conical tubes or autoclaved 150–250-ml bottles.

Notes:
 i. Alternately, dispense 100-ml aliquots into 150–250-ml bottles; autoclave 15 minutes at 121°C.
 ii. With storage at 4°C, the solution is precooled and ready for making competent cells.

V. PLASMID MINIPREPARATION

0.5 M Ethylenediaminetetraacetic Acid (EDTA, pH 8.0)

Makes 100 ml.
Store at room temperature (indefinitely).

1. Add 18.6 g of EDTA (disodium salt, m.w. = 372.24) to 80 ml of deionized or distilled water.

2. Adjust pH by slowly adding approximately 2.2 g of sodium hydroxide pellets (m.w. = 40.00). (If a pH meter is not available, adding 2.2 g of NaOH pellets will make a solution of approximately pH 8.0.)

3. Mix vigorously with a magnetic stirrer or by hand. EDTA will only dissolve when the pH has reached 8.0 or higher.

Note: Use only the disodium salt of EDTA.

Tris/EDTA (TE) Buffer

Makes 100 ml.
Store at room temperature (indefinitely).

Mix:

1 ml of 1 M Tris (pH 8.0)	(10 mM)
200 µl of 0.5 M EDTA	(1 mM)
99 ml of deionized water	

Glucose/Tris/EDTA (GTE)

Makes 100 ml.
Store at 4°C or room temperature (indefinitely).

Mix:

0.9 g of glucose (m.w. = 180.16)	(50 mM)
2.5 ml of 1 M Tris (pH 8.0)	(25 mM)
2 ml of 0.5 M EDTA	(10 mM)
94.5 ml of deionized water	

Note: With storage at 4°C, the solution is precooled and ready for minipreparations.

5 M Potassium Acetate (KOAc)

Makes 200 ml.
Store at room temperature (indefinitely).

1. Add 98.1 g of potassium acetate (m.w. = 98.14) to 160 ml of deionized water.

2. Add deionized or distilled water to make a total volume of 200 ml of solution.

Potassium Acetate/Acetic Acid

Makes 100 ml.
Store at 4°C or room temperature (indefinitely).

Add 60 ml of 5 M potassium acetate and 11.5 ml of glacial acetic acid to 28.5 ml of deionized or distilled water.

Notes:

i. The sharp odor of the acetic acid distinguishes the finished KOAc/acetic acid solution from the KOAc stock. The two are easily confused.
ii. With storage at 4°C, the solution is precooled and ready for minipreparations.

10% Sodium Dodecyl Sulfate (SDS)

Makes 100 ml.
Store at room temperature (indefinitely).

1. Dissolve 10 g of electrophoresis-grade SDS (m.w. = 288.37) in 80 ml of deionized water.

2. Add deionized or distilled water to make a total volume of 100 ml of solution.

Notes:

i. Avoid inhaling SDS powder; wear a mask that covers both nose and mouth.
ii. SDS is the same as sodium lauryl sulfate.

1% SDS/0.2 N NaOH

Makes 10 ml.
Store at room temperature (several days).

Mix 1 ml of 10% SDS and 0.5 ml of 4 N NaOH into 8.5 ml of distilled water.

Notes:

i. Always use fresh SDS/NaOH solution.
ii. A precipitate may form at colder temperature. Warm solution, and shake gently to dissolve precipitate.

VI. DNA LIGATION

1 M Tris (pH 7.6)

Makes 100 ml.
Store at room temperature (indefinitely).

1. Dissolve 12.1 g of Tris base (m.w. = 121.10) in 70 ml of deionized or distilled water.

2. Adjust pH by slowly adding approximately 6.3 ml of concentrated hydrochloric acid (HCl); monitor with a pH meter. (If a pH meter is not available, adding 6.3 ml of concentrated HCl will yield a solution of approximately pH 7.6.)

3. Add deionized or distilled water to make a total volume of 100 ml of solution.

> **CAUTION**
>
> Avoid inhaling Tris powder; wear a mask over nose and mouth.

Notes:

i. A yellow-colored solution indicates poor-quality Tris. Discard, and obtain from a different source.
ii. Many types of electrodes do not accurately measure the pH of Tris solutions; check with manufacturer to obtain a suitable one.
iii. The pH of Tris solutions is temperature-dependent; measure pH at room temperature.

10 × Ligation Buffer

Makes 900 μl.
Store at −20°C (indefinitely).

Mix the following ingredients in a 1.5-ml tube:

    600 μl of 1 M Tris (pH 7.6)   (600 mM)
    100 μl of 1 M MgCl$_2$ (100 mM)
    70 μl of 1 M DTT   (70 mM)
    130 μl of deionized water

Note: DTT may precipitate from the solution with repeated freezing and thawing. Vortex vigorously to redissolve.

0.1 M Adenosine Triphosphate (ATP)

Makes 5 ml.
Store at −20°C (1 year).

1. Dissolve 0.3 g of ATP (disodium salt, m.w. = 605.19) in 5 ml of deionized or distilled water.

2. Dispense 500-μl aliquots into 1.5-ml tubes.

Note: ATP looses activity with repeated freezing and thawing. Discard aliquots that have been thawed several times.

2 × Ligation Buffer + ATP

Makes 500 μl.
Store at −20°C (1 month).

1. Mix 100 μl of 10 × ligation buffer and 10 μl of 0.1 M ATP in a 1.5-ml tube.

2. Add 390 μl of deionized or distilled water.

Note: ATP looses activity in dilute solution; use a fresh solution.

APPENDIX 3
Restriction Map Data for pAMP, pKAN, and Bacteriophage λ

I. Restriction Maps of pAMP and pKAN

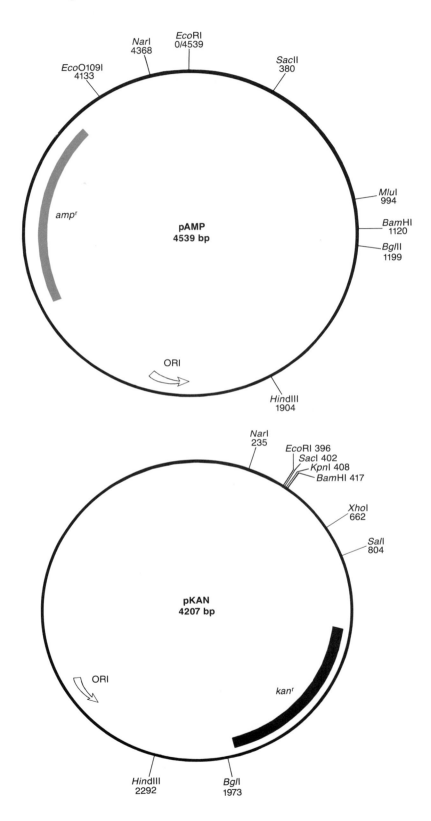

II. Restriction Enzymes That Cut Once in pAMP or pKAN and Corresponding Sites in Bacteriophage λ

Enzyme	Recognition sequence	Location pAMP	pKAN	bacteriophage λ
AatII	GACGT˅C	4078	4138	5105; 9394; 11,243; 14,974; 29,036; 40,806; 41,113; 42,247; 45,563; 45,592
AflIII	A˅CACGT A˅CGTGT	—	2651	none
AlwNI	CAGNNN˅CTG	2676	3062	none
BalI	TGG˅CCA	34	1727	18 sites
BamHI	G˅GATCC	1120	418	5505; 22,346; 27,972; 34,499; 41,732
BanII	GPuGCPy˅C	344	402	581; 10,086; 19,763; 21,570; 24,772; 25,877; 39,453
BglII	A˅GATCT	1199	1973	415; 22,425; 35,711; 38,103; 38,754; 38,814
BspMII	T˅CCGGA	1117	—	24 sites
Cfr10I	Pu˅CCGGPy	3238	3624	none
EcoO109I	PuG˅GNCCPy	4133	—	2815; 28,797; 48,473;
EcoRI	G˅AATTC	0/4539	396	21,226; 26,104; 31,747; 39,168; 44,972
HaeII	PuGCGC˅Py	—	1902	none
HindIII	A˅AGCTT	1904	2293	23,130; 25,157; 27,479; 36,895; 37,459; 37,584; 44,141
KpnI	GGTAC˅C	—	408	17,053; 18,556
MluI	A˅CGCGT	994	—	458; 5548; 15,372; 17,791; 19,996; 20,952; 22,220
NarI	GG˅CGCC	4368	235	45,679
NdeI	CA˅TATG	4316	183	27,630; 29,883; 33,679; 36,112; 36,668; 38,357; 40,131
RsrII	CG˅GACCG CG˅GTCCG	1016	—	3800; 6041; 13,983; 19,288; 22,242
SacI	GAGCT˅C	—	402	24,772; 25,877
SacII	CCGC˅GG	380	—	20,320; 20,530; 21,606; 40,386
SalI	G˅TCGAC	—	804	32,745; 33,244
SmaI	CCC˅GGG	—	412	19,397; 31,617; 39,888
SphI	GCATG˅C	—	1405	2212; 12,002; 23,942; 24,371; 27,374; 39,418
XhoI	C˅TCGAG	—	662	33,498

III. Construction of pAMP

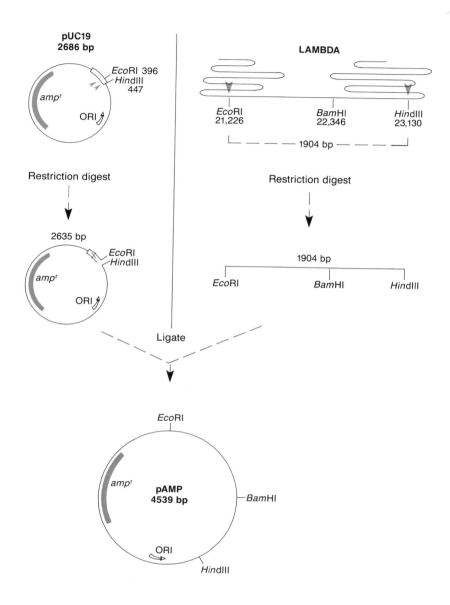

IV. Construction of pKAN

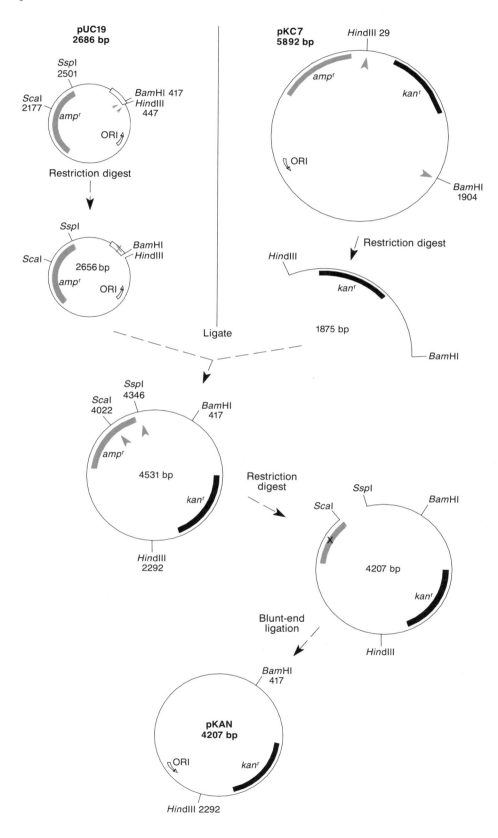

BIBLIOGRAPHY

Alberts, Bruce, Dennis Bray, Julian Lewis, Martin Raff, Keith Roberts, and James D. Watson. 1989. *Molecular biology of the cell.* Garland Publishing Inc., New York.

Angier, Natalie. 1988. *Natural obsessions: The search for the oncogene.* Houghton Mifflin Co., Boston.

Antebi, Elizabeth and David Fishlock. 1986. *Biotechnology: Strategies for life.* The Cambridge: MIT Press, Cambridge, Massachusetts.

Arnold, Caroline. 1986. *Genetics: From Mendel to gene splicing.* Franklin Watts, New York.

Bachman, Barbara. 1987. Derivations and genotypes of some mutant derivatives of *Escherichia coli* K-12. In Escherichia coli *and* Salmonella typhimurium: *Cellular and molecular biology* (ed. F.C. Neidhardt), vol 2, p. 1190. American Society for Microbiology, Washington, D.C.

Barrett, James, ed. 1986. *Contemporary classics in the life sciences.* ISI Press, Philadelphia.

Berridge, Michael J. 1985. The molecular basis of communication within the cell. *Sci. Am.* 253/4: 142–151.

Birnboim, H.C. and Janine Doly. 1979. A rapid alkaline extraction method for screening recombinant plasmid DNA. *Nucleic Acids Res.* 7: 1513.

Bishop, Michael J. 1982. Oncogenes. *Sci. Am.* 246/3: 81–89.

Bretscher, Mark S. 1985. The molecules of the cell membrane. *Sci. Am.* 253/4: 100–109.

Cairns, John. 1985. The treatment of diseases and the war against cancer. *Sci. Am.* 253/5: 51–59.

Cairns, John, Gunther S. Stent, and James D. Watson, eds. 1966. *Phage and the origins of molecular biology.* Cold Spring Harbor Laboratory, Cold Spring Harbor, New York.

Cech, Thomas R. 1986. RNA as an enzyme. *Sci. Am.* 255/5: 64–75.

Cherfas, Jeremy. 1982. *Man-made life.* Pantheon Books, New York.

Chilton, Mary-Dell. 1983. A vector for introducing new genes into plants. *Sci. Am.* 248/6: 51–59.

Cohen, Leonard A. 1987. Diet and cancer. *Sci. Am.* 257/5: 42–49.

Cohen, Stanley N., Annie C.Y. Chang, and Herbert W. Boyer. 1973. Construction of biologically function bacterial plasmids in vitro. *Proc. Natl. Acad. Sci.* 70: 3240.

Cohen, Stanley N., Annie C.Y. Chang, and Leslie Hsu. 1972. Nonchromosomal antibiotic resistance in bacteria: Genetic transformation of *Escherichia coli* by R-factor DNA. *Proc. Natl. Acad. Sci.* 69: 2110.

Corrick, James A. 1987. *Recent revolutions in biology*. Franklin Watts, New York.
Crick, Francis. 1988. *What mad pursuit*. Basic Books Inc., New York.
Croce, Carlo M. and George Klein. 1985. Chromosome translocations and human cancer. *Sci. Am. 252/3:* 54–73.
Crosson, Pierre and Norman J. Rosenberg. 1989. Strategies for agriculture. *Sci. Am. 261/3:* 128–135.
Dagert, M. and Stanislav D. Ehrich. 1979. Prolonged incubation in calcium chloride improves the competence of *Escherichia coli* cells. *Gene 6:* 23.
Darnell, James E., Jr. 1985. RNA. *Sci. Am. 253/4:* 68–87.
Darnell, James, Harvey Lodish, and David Baltimore. 1986. *Molecular cell biology*. Scientific American Books, New York.
Dawkins, Richard. 1978. *The selfish gene*. Oxford University Press, New York.
Dickerson, Richard E. 1983. The DNA helix and how it is read. *Sci. Am. 249/6:* 94–111.
Doolittle, Russell F. 1985. Proteins. *Sci. Am. 253/4:* 88–99.
Drlica, Karl. 1984. *Understanding DNA and gene cloning*. John Wiley and Sons, New York.
Dulbecco, Renato. 1987. *The design of life*. Yale University, New Haven.
Edelman, Gerald M. 1988. *Topobiology: An introduction to molecular embryology*. Basic Books Inc., New York.
Edelman, Gerald M. 1989. Topobiology. *Sci. Am. 260/5:* 76–88.
Edey, Maitland and Donald Johanson. 1989. *Blueprints: Solving the mystery of evolution*. Little, Brown, and Co., Boston.
Erlich, Henry A., ed. 1989. *PCR technology*. Stockton Press, New York.
Essex, Max and Phyllis J. Kanki. 1988. The origins of the AIDS virus. *Sci. Am. 259/4:* 64–71.
Fedoroff, Nina V. 1984. Transposable genetic elements in maize. *Sci. Am. 250/6:* 85–98.
Feldman, Michael and Lea Eisenbach. 1988. What makes a tumor cell metastatic? *Sci. Am. 259/5:* 60–87.
Felsenfeld, Gary. 1985. DNA. *Sci. Am. 253/4:* 58–67.
Fischer, Ernst Peter and Carol Lipson. 1988. *Thinking about science: Max Delbrück and the origins of molecular biology*. W.W. Norton and Co., New York.
Fox-Keller, Evelyn. 1983. *A feeling for the organism: The life and work of Barbara McClintock*. W.H. Freeman and Co., New York.
Fuhrmann, Walter and Friedrich Vogel. 1983. *Genetic counseling*. Springer-Verlag, New York.
Gallo, Robert C. 1986. The first human retrovirus. *Sci. Am. 255/6:* 88–101.
Gallo, Robert C. 1987. The AIDS virus. *Sci. Am. 256/1:* 46–73.
Gallo, Robert C. and Luc Montagnier. 1988. AIDS in 1988. *Sci. Am. 259/4:* 40–51.
Gehring, Walter J. 1985. The molecular basis of development. *Sci. Am. 253/4:* 152–163.
Gold, Michael. 1986. *A conspiracy of cells*. State University of New York Press, Albany.
Goldberg, Marshall. 1988. *Cell wars: The immune system's newest weapons against cancer*. Farrar-Straus-Giroux, New York.
Gonick, Larry and Mark Wheelis. 1983. *The cartoon guide to genetics*. Barnes and Noble Books, New York.
Greene, Patricia J., Mohindar S. Poonian, Alexander L. Nussbaum, L. Tobias, David E. Garfin, Herbert W. Boyer, and Howard M. Goodman. 1975.

Restriction and modification of a self-complementary octanucleotide containing the *Eco*RI substrate. *J. Mol. Biol.* 99: 237.

Grey, Howard M., Alessandro Sette, and Soren Buus. 1989. How T cells see antigen. *Sci. Am.* 261/5: 66–73.

Gribbin, John. 1987. *In search of the double helix*. Bantam Books, New York.

Grobstein, Clifford. 1977. The recombinant-DNA debate. *Sci. Am.* 237/1: 22–33.

Hall, Stephen S. 1987. *Invisible frontiers: The race to synthesize a human gene*. Tempus Books, Redmond, Washington.

Hanahan, Douglas. 1983. Studies on transformation of *Escherichia coli* with plasmids. *J. Mol. Biol.* 166: 557.

Hanahan, Douglas. 1987. Techniques for transformation of *E. coli*. In *DNA cloning: A practical approach* (Ed. D.M. Glover), vol. 1. IRL Press, Oxford.

Haseltine, William A. and Flossie Wong-Staal. 1988. The molecular biology of the AIDS virus. *Sci. Am.* 259/4: 52–63.

Helling, Robert B., Howard M. Goodman, and Herbert W. Boyer. 1974. Analysis of *Eco*RI fragments of DNA from lambdoid bacteriophages and other viruses by agarose-gel electrophoresis. *J. Virol.* 14: 1235.

Heyward, William L. and James W. Curran. 1988. The epidemiology of AIDS in the U.S. *Sci. Am.* 259/4: 72–81.

Hirsch, Martin S. and Joan C. Kaplan. 1987. Antiviral therapy. *Sci. Am.* 256/4: 76–85.

Hogle, James M., Marie Chow, and David J. Filman. 1987. The structure of poliovirus. *Sci. Am.* 256/3: 42–49.

Holliday, Robin. 1989. A different kind of inheritance. *Sci. Am.* 260/6: 60–73.

Ish-Horowicz, David and Julian F. Burke. 1981. Rapid and efficient cosmid cloning. *Nucleic Acids Res.* 9: 2989.

Kartner, Norbert and Victor Ling. 1989. Multidrug resistance in cancer. *Sci. Am.* 260/3: 44–51.

Jacob, Francois. 1988. *The statue within*. Basic Books Inc., New York.

Judson, Horace Freeland. 1979. *The eighth day of creation*. Simon and Schuster, New York.

Kessel, Richard G. and Randy H. Kardon. 1979. *Tissues and organs: A text atlas of scanning electron microscopy*. W.H. Freeman and Co., New York.

Kevles, Daniel J. 1985. *In the name of eugenics: Genetics and the uses of human heredity*. University of California Press, Berkeley.

Korey, Kenneth, ed. 1984. *The essential Darwin*. Little, Brown and Company, Boston.

Kornberg, Arthur. 1989. *For the love of enzymes*. Harvard University Press, Cambridge, Massachusetts.

Laurence, Jeffrey. 1985. The immune system in AIDS. *Sci. Am.* 253/6: 84–93.

Lawn, Richard M. and Gordon A. Vehar. 1986. The molecular genetics of hemophilia. *Sci. Am.* 254/3: 48–65.

Lewin, Benjamin. 1987. *Genes III*. John Wiley and Sons, New York.

Luria, Salvador E. and Joan L. Suit. 1987. Colicins and Col plasmids. In Escherichia coli *and* Salmonella typhimurium: *Cellular and molecular biology* (ed. F.C. Neidhardt), vol. 2, pp. 1620–1621. American Society for Microbiology, Washington, D.C.

Mandel, Manley and Akiko Higa. 1970. Calcium-dependent bacteriophage DNA infection. *J. Mol. Biol.* 53: 159.

Mann, Jonathan M., James Chin, Peter Piot, and Thomas Quinn. 1988. The international epidemiology of AIDS. *Sci. Am.* 259/4: 82–89.

McCarty, Maclyn. 1985. *The transforming principle: Discovering that genes are made of DNA*. W.W. Norton and Co., New York.

McIntosh, Richard and Kent L. McDonald. 1989. The mitotic spindle. *Sci. Am. 261/4:* 48–56.

Micklos, David, with Susan Zehl, Daniel Schechter, and Ellen Skaggs. 1988. *The first hundred years: A history of man and science at Cold Spring Harbor.* Cold Spring Harbor Laboratory, Cold Spring Harbor, New York.

The molecules of life: Readings from Scientific American. (Special issue.) 1985. 253/4.

Moore, John A., ed. 1986. *Science as a way of knowing. III—Genetics.* American Society of Zoologists, Washington, D.C.

Moore, Walter. 1990. *Schrodinger: Life and thought.* Cambridge University Press, New York.

Morrison, Philip, Phylis Morrison, and the office of Charles and Ray Eames. 1982. *Powers of ten: About the relative size of things in the universe.* Scientific American Library, New York.

Moses, Phyllis B. and Nam-Hai Chua. 1988. Light switches for plant genes. *Sci. Am. 258/4:* 64–69.

Muench, Karl H. 1988. *Genetic medicine.* Elsevier, New York.

Mullis, Kary, B. 1990. The unusual origin of the polymerase chain reaction. *Sci. Am. 262/4:* 56–65.

Murray, Andrew W. and Jack W. Szostak. 1987. Artificial chromosomes. *Sci. Am. 257/5:* 62–87.

Nathans, Jeremy. 1989. The genes for color vision. *Sci. Am. 260/2:* 42–49.

National Institute of General Medical Sciences. 1984. *The new human genetics: How gene splicing helps researchers fight inherited disease.* National Institutes of Health, Washington, D.C.

National Institutes of Health. 1982. Guidelines for research involving recombinant DNA molecules. *Federal Register 47/167:* 38051.

National Research Council. 1988. *Mapping and sequencing the human genome.* National Academy Press, Washington, D.C.

Neidhardt, Frederick C., ed. 1987. Escherichia coli *and* Salmonella typhimurium*: Cellular and molecular biology,* volumes I and II. American Society for Microbiology, Washington, D.C.

Nilsson, Lennart. 1985. *The body victorious.* Delacorte Press, New York.

Novick, Richard P. 1980. Plasmids. *Sci. Am. 243/6:* 103–127.

Oldstone, Michael B.A. 1989. Viral alteration of cell function. *Sci. Am. 261/2:* 42–48.

Oliver, Stephen and John Ward. 1985. *A dictionary of genetic engineering.* Cambridge University Press, New York.

Olson, Steve. 1986. *Biotechnology: An industry comes of age.* National Academy Press, Washington, D.C.

Patterson, David. 1987. The causes of Down Syndrome. *Sci. Am. 257/2:* 52–61.

Pestka, Sidney. 1983. The purification and manufacture of human interferons. *Sci. Am. 249/2:* 36–43.

Piller, Charles and Keith R. Yamamoto. 1988. *Gene wars: Military control over the new genetic technologies.* Beech Tree Books, New York.

Ptashne, Mark. 1989. How gene activators work. *Sci. Am. 260/1:* 41–47.

Quillardet, Phillipe and Maurice Hofnung. 1988. Ethidium bromide and safety— Readers suggest alternative solutions. Letter to editor. *Trends Genet.* 4: 89.

Radman, Miroslav and Robert Wagner. 1988. The high fidelity of DNA duplication. *Sci. Am. 259/2:* 40–47.

Rawn, David J. 1989. *Biochemistry.* Neil Patterson Publishers, Burlington, North Carolina.

Ross, Jeffrey. 1989. The turnover of messenger RNA. *Sci. Am. 260/4:* 48–55.

Sachs, Leo. 1986. Growth differentiation and the reversal of malignancy. *Sci. Am.* 254/1: 40–47.

Sambrook, Joseph, Edward F. Fritsch, and Thomas Maniatis. 1989. *Molecular cloning: A laboratory manual*, volumes I, II, and III. Cold Spring Harbor Laboratory Press, New York.

Sayre, Anne. 1975. *Rosalind Franklin and DNA*. W.W. Norton and Co., New York.

Schrödinger, Erwin. 1944. *What is life? The physical aspect of the living cell & mind and matter*. Cambridge University Press, New York.

Shapiro, James A. 1988. Bacteria as multicellular organisms. *Sci. Am.* 258/6: 82–89.

Sharp, Philip A., Bill Sugden, and Joe Sambrook. 1973. Detection of two restriction endonuclease activities in *Haemophilus parainfluenzaea* using analytical agarose-ethidium bromide electrophoresis. *Biochemistry 12:* 3055.

Snyder, Solomon H. 1985. The molecular basis of communication between cells. *Sci. Am.* 253/4: 132–141.

Stahl, Franklin W. 1987. Genetic recombination. *Sci. Am.* 256/2: 90–101.

Stebbins, G. Ledyard and Francisco J. Ayala. 1985. The evolution of Darwinism. *Sci. Am.* 253/1: 72–85.

Sylvester, Edward J. 1987. *Target: Cancer*. Charles Scribner's Sons, New York.

Sylvester, Edward J. and Lynn C. Klotz. 1987. *The gene age*. Charles Scribner's Sons, New York.

Tonegawa, Susumu. 1985. The molecules of the immune system. *Sci. Am.* 253/4: 122–131.

Tramontano, Alfonzo and Richard Lerner. 1988. Catalytic antibodies. *Sci. Am.* 258/3: 58.

Varmus, Harold. 1987. Reverse transcription. *Sci. Am.* 257/3: 56–65.

Wambaugh, Joseph. 1989. *The blooding*. William Morrow and Co., New York.

Watson, James D. 1980. *The double helix: A personal account of the discovery of the structure of DNA*. W.W. Norton and Co., New York.

Watson, James D. and John Tooze. 1981. *The DNA story: A documentary history of gene cloning*. W.H. Freeman and Co., San Francisco.

Watson, James D., John Tooze, and David T. Kurtz. 1983. *Recombinant DNA: A short course*. Scientific American Books, New York.

Watson, James D., Nancy H. Hopkins, Jeffrey W. Roberts, Joan Argetsinger Steitz, and Alan M. Weiner. 1987. *Molecular biology of the gene*, volumes I and II. Benjamin/Cummings Publishing Co. Inc., Menlo Park, California.

Weber, Klaus and Mary Osborn. 1985. The molecules of the cell matrix. *Sci. Am.* 253/4: 110–121.

Weinberg, Robert A. 1983. A molecular basis of cancer. *Sci. Am.* 249/5: 126–142.

Weinberg, Robert A. 1985. The molecules of life. *Sci. Am.* 253/4: 48–57.

Weinberg, Robert A. 1988. Finding the anti-oncogene. *Sci. Am.* 259/3: 44–53.

Weintraub, Harold M. 1990. Antisense RNA and DNA. *Sci. Am.* 262/1: 40–46.

What Science Knows About AIDS. (Special issue.) 1988. *Sci. Am.* 259/4.

White, Ray and Jean-Marc Lalouel. 1988. Chromosome mapping with DNA markers. *Sci. Am.* 258/2: 40–49.

Wilson, Allan C. 1985. The molecular basis of evolution. *Sci. Am.* 253/4: 164–175.

Witt, Steven C. 1982. *Genetic engineering of plants*. California Agricultural Lands Project, San Francisco.

Witt, Steven C. 1985. *Biotechnology and genetic diversity*. California Agricultural Lands Project, San Francisco.

Wolpert, Lewis and Alison Richards. 1988. *A passion for science.* Oxford University Press, New York.

Young, John Ding and Zanvil A. Cohn. 1988. How killer cells kill. *Sci. Am. 258/1:* 38–45.

Zimmerman, Burke K. 1984. *Biofuture: Confronting the genetic era.* Plenum Publishers, New York.

NAME INDEX

Altman, Sid, 194
Anderson, W. French, 160
Arber, Werner, 41
Avery, Oswald T., 6, 19–21, *21*, 55, 115
Axel, Richard, 122
Barbacid, Mariano, 125
Beachy, Roger, 173
Beadle, George W., 6, *18*, 18–19, 32, 55
Benzer, Seymour, *23*, 33
Berg, Paul, 42–44, *44*, 61, 84
Bernal, J. Desmond, 5, 26
Bertani, Giuseppe, 41
Bishop, J. Michael, *117*, 117, 118
Bohr, Niels, 3,
Botstein, David, 143, 149–150
Boyer, Herb, 42–44, *44*, 61, 84
Bragg, William Henry, 26
Bragg, William Lawrence, 26
Brenner, Sydney, *34*, 34
Bridges, Calvin, 14–17, *16*
Brinster, Ralph, 177
Cantor, Charles, 142–143
Capecchi, Mario, 163–164
Cech, Thomas, 194
Chang, Annie, 43–44, *44*, 61, 84
Chargaff, Erwin, 26, *29*, 29
Chase, Martha, 6, 23–25, *24*
Cohen, Stanley, 42–44, *44*, 61, 84
Corey, R.B., 28
Correns, Carl, 14
Creighton, Harriet, 17
Crick, Francis, 6, *8*, 9, 10, 26, 28–30, 32, 34, 36
d'Herelle, Felix, 22
Darwin, Charles, 6, *10*, 10–12, 26
Davis, Ronald, 149–150
de Vries, Hugo, *14*, 14,
Delbruck, Max, 2, 3, 22–23, *23*
Edelman, Gerald, 97
Franklin, Rosalind, 28, *29*
Gallo, Robert, 120
Garrod, Sir Archibald, *18*, 18–19, 32, 149
Gilbert, Walter, 80
Griffith, Fred, 19–20, 55, 115
Gusella, James, 153–155
Hall, Benjamin, 34, 68
Hanafusa, Hidesaburo, 117

Harlow, Edward, 132
Harvey, William, *4*, 4
Hershey, Alfred, 6, 22, 23–25, *24*
Herskowitz, Ira, 97
Hicks, James, 97
Higa, A., 43
Holley, Robert, 34
Hood, Leroy, 142–143, 147
Jeffreys, Alec, 165–166, *166*
Jenner, Edward, 182
Jones, Donald, *170*, 170
Khorana, H. Gobind, 35
Kornberg, Arthur, *31*, 31
Leder, Philip, 61–62
Lederberg, Esther, 69
Lederberg, Joshua, 69
Lerner, Richard, 191–194
Levene, Phoebus, 21
Linn, Stuart, 41
Linnaeus, Carolus, 9
Luria, Salvador, 22, *23*, 41
Lyell, Sir Charles, 11–12
MacLeod, Colin, 20
Mandel, Morton, 43
Maniatis, Tom, 62
Matthei, Heinrich, 35
Maxam, Allan, 80
McCarty, Maclyn, 20
McClintock, Barbara, 17, *94*, 94
McKusick, Victor, 148
Mendel, Gregor, 6, *12*, 12, 26
Meselson, Matthew, *30*, 30–31, 41, 45
Miescher, Friedrich, 26
Milstein, Cesar, 78
Monroe, Miles (alias Woody Allen), *39*, 39
Morgan, Thomas Hunt, 14–17, *16*, 141, 150
Muller, Hermann, 14–17, 18
Nathans, Daniel, 41–42, *44*, 48
Nirenberg, Marshall, *35*, 35
Nixon, Richard, *110*, 111
Olsen, Maynard, 142–143
Palmiter, Richard, 177
Pauling, Linus, 5, 26, 28, *29*, 191
Rous, Peyton, *113*, 113
Sambrook, Joseph, 48–49
Sanger, Fred, 80, 82

Italic numbers indicate photos.

Schleiden, Matthias, 4,
Schlosser, Luna (alias Diane Keaton), *39*, 39
Schrodinger, Erwin, 3,
Schwann, Theodor, 4,
Scolnick, Mark, 149–150
Shope, Richard E., 114
Shull, George, 169–170, *170*
Smith, Hamilton, 41–42, *44*
Southern, Ed, 83
Spiegelman, Sol, 34, 68
Stahl, Frank, *30*, 30–31
Stern, Curt, 17
Stevens, Nettie, 14,
Strathern, Jeffrey, 97
Sturdevant, Alfred, 14–17,
Sutton, Walter, 6, 12, 14
Tatum, Edward, 6, *18*, 18–19, 32, 55

Tonegawa, Susumu, 97
Usher, Archbishop James, 10,
van der Eb, Alex, 120
Varmus, Harold, *117*, 117, 118
von Laue, Max, 26
Wallace, Alfred, 6, 10
Watson, James D., 6, *8*, 9, 10, 26, 28–30, 139
Weigle, Jean, 41
Wexler, Nancy, *153*, 153–155
White, Ray, 141, 165
Wigler, Michael, *122*, 122
Wilcox, Kent, 41
Wilkins, Maurice, 28, *29*
Wilson, Edmund, *14*, 14
Wyman, Arlene, 165
Yanofsky, Charles, 32
Yuan, Robert, 41, 45

ments of 100,000 or more nucleotides. 145–146

molecular cloning. The biological amplification of a specific DNA sequence through mitotic division of a host cell into which it has been transformed or transfected. 62

subcloning. The process of transferring a cloned DNA fragment from one vector to another. 71–73, 72, 141, 146

Clotting factor. Any of several proteins involved in the formation of a blood clot, the absence of any of which leads to the bleeding disorder hemophilia. 178–179

cM. *See* CentiMorgan.

Codon. A group of three nucleotides that specifies addition of one of the 20 amino acids during translation of an mRNA into a polypeptide. 34–37, *35*

initiation codon. The mRNA sequence AUG, coding for methionine, which initiates translation of mRNA. 75

termination (stop) codon. Any of three mRNA sequences (UGA, UAG, UAA) that do not code for an amino acid and thus signal the end of protein synthesis. 81, 90–91

Coenzyme (cofactor). An organic molecule, such as a vitamin, that binds to an enzyme and is required for its catalytic activity. 88, 253, 322

Colony. A group of identical cells (clones) derived from a single progenitor cell. 53, 63, 215, 231, 293, 311, 380, 383, 416

Competency. An ephemeral state, induced by treatment with cold cations, during which bacterial cells are capable of uptaking foreign DNA. 55–57, 215, 239, 293, 361

Complementarity-determining region (CDR). A portion of an antibody molecule that forms a structure complementary to a region (epitope) of an antigen molecule. 98, 195

Complementary DNA. *See* DNA.

Complementary nucleotides. *See* Nucleotide.

Concatemer. A DNA segment composed of repeated sequences linked end to end. 122, 338

Condensation reaction. A reaction in which a water molecule is liberated when two reacting molecules are joined. 51

Consensus sequence. A DNA sequence that is highly conserved in different species. 100–101

Constitutive promoter. *See* Promoter.

Contiguous (contig) map. *See* Mapping.

Cosmid. A plasmid vector containing a COS site that enables it to be packaged into infective bacteriophage λ particles. Used for cloning DNA sequences of 35,000–45,000 bp. *64,* 71, 146–147

Cross-hybridization. The hydrogen bonding of a single-stranded DNA sequence that is partially but not entirely complementary to a single-stranded substrate. Often, this involves hybridizing a DNA probe for a specific DNA sequence to the homologous sequences of different species. 106

Crossing-over. The exchange of DNA sequences between chromatids of homologous chromosomes during meiosis. *17*

Crown gall disease. A tumorous growth of certain dicot plants caused by *Agrobacterium tumefaciens*. 171

CTF. *See* CCAAT-binding transcription factor.

Cyclic AMP (cyclic adenosine monophosphate). A second messenger that regulates many intracellular reactions by transducing signals from extracellular growth factors to cellular metabolic pathways. 133

Cytogenetics. Study that relates the appearance and behavior of chromosomes to genetic phenomenon. 16

Dalton. A unit of measurement equal to the mass of a hydrogen atom, 1.67×10^{-24} gram 1/Avogadro's number. 62–64, *63*

Death phase. *See* Growth phase.

Denature. To induce structural alterations that disrupt the biological activity of a molecule. Often refers to breaking hydrogen bonds between base pairs in double-stranded nucleic acid molecules to produce in single-stranded polynucleotides or altering the secondary and tertiary structure of a protein, destroying its activity. 34, *59,* 68–69, 74, 80, *83*–84, 143, 189, 322

Density gradient centrifugation. *See* Centrifugation.

Deoxyribonucleic acid. *See* DNA.

Diabetes. A disease associated with the absence or reduced levels of insulin, a hormone essential for the transport of glucose to cells. 178

Dicotyledon (dicot). A flowering plant whose embryo possesses two cotyledons, or seed leaves. 171–172

Dideoxynucleotide. *See* Nucleotide.

Digest. To cut DNA molecules with one or more restriction endonucleases. 41, 50, 65, 247, 277, 327, 341, 416

Diploid. The condition when the genome of an organism consists of two copies of each chromosome. 95–96

Directional cloning. *See* Cloning.

Dissociator (*Ds*). A transposable element in maize, whose mobility is dependent on another element, Activator (*Ac*). 94–95

Diversity (D) segment. A small DNA segment that, along with a Joining segment, links a constant and variable gene to yield a functional gene encoding an immunoglobulin heavy chain. *98*

DNA (deoxyribonucleic acid). An organic acid and polymer composed of four nitrogenous bases—adenine, thymine, cytosine, and guanine—linked via intervening units of phosphate and the pentose sugar deoxyribose. DNA is the genetic material of most organisms and usually exists as a double-stranded molecule in which two antiparallel strands are held together by hydrogen bonds between adenine-thymine and cytosine-guanine. 1, 9, 20–32, *28,* 199, 215, 247, 277, 293, 311, 327, 355, 361, 379, 383, 393, 413

DNA (deoxyribonucleic acid) (*continued*)
β-DNA. The normal form of DNA found in biological systems, which exists as a right-handed helix. 101
cDNA (copy DNA). DNA synthesized from an RNA template using reverse transcriptase. 70–77, *71, 74,* 79–81, 98, 104, 122, 143, 148
cDNA library. *See* Library.
complementary DNA or RNA. The matching strand of a DNA or RNA molecule to which its bases pair. 30–34, 42, 46–47, 50, 67–74, 84, 107, 143
DNA fingerprint. The unique pattern of DNA fragments identified by Southern hybridization (using a probe that binds to a polymorphic region of DNA) or by polymerase chain reaction (using primers flanking the polymorphic region). 164–167, *165*
DNA ligase. *See* Ligase.
DNA polymerase. *See* Polymerase.
DNA polymorphism. One of two or more alternate forms (alleles) of a chromosomal locus that differ in nucleotide sequence or have variable numbers of repeated nucleotide units. 151–*154, 152,* 157–158, 164
DNase (deoxyribonuclease). *See* Nuclease.
DNA sequencing. Procedures for determining the nucleotide sequence of a DNA fragment. 80–83, *82,* 139–148, *142,* 184
double-stranded complementary DNA (dscDNA). A duplex DNA molecule copied from a cDNA template. 73–74
duplex DNA. Double-stranded DNA. 46, 68, 81–84
Z-DNA. A region of DNA that is "flipped" into a left-handed helix, characterized by alternating purines and pyrimidines, and which may be the target of a DNA-binding protein. 101
DNA diagnosis. The use of DNA polymorphisms to detect the presence of a disease gene. 157–160, *159*
chronic myelogenous leukemia. *159*
Huntington's disease. *154*
sickle cell anemia. *152*
Dominant(-acting) oncogene. *See* Oncogene.
Dominant gene. *See* Gene.
Double helix. Describes the coiling of the antiparallel strands of the DNA molecule, resembling a spiral staircase in which the paired bases form the steps and the sugar-phosphate backbones form the rails. 9, 26–29
Double minutes. Chromosome fragments that are the result of chromosome duplication and in which oncogenes may be amplified. A common cytological feature of tumor cells. *126*
Double-stranded complementary DNA (dscDNA). *See* DNA.
Downstream. The region extending in a 3′ direction from a gene. 75, *101,* 102
Drosophila melanogaster. The fruit fly whose common use in genetic studies was introduced by Thomas Hunt Morgan in the early 1900s. *14–17,* 105, 107
Ds. *See* Dissociator.
Duchenne muscular dystrophy (DMD). An X-linked disease, caused by the dystrophin gene, characterized by gradual muscle wasting beginning in infancy. 155–158, *157*

Electrophoresis. The technique of separating charged molecules in a matrix to which is applied an electrical field. 48–50, 65, 145, 247, 345
agarose gel electrophoresis. A matrix composed of a highly purified form of agar is used to separate larger DNA and RNA molecules ranging from 100 to 20,000 nucleotides. *48–50,* 83–84, 247
field-inversion electrophoresis (orthogonal-field-alternation gel electrophoresis, OFAGE). A longer forward current is alternated with a shorter reverse current to separate very large DNA molecules of up to 10 million nucleotides. 145
polyacrylamide gel electrophoresis. Electrophoresis through a matrix composed of a synthetic polymer, used to separate proteins, small DNA, or RNA molecules of up to 1000 nucleotides. Used in DNA sequencing. 48–49, 81, 267
pulse-field electrophoresis (PFE). Current is alternated between pairs of electrodes set at angles to one another to separate very large DNA molecules of up to 10 million nucleotides. 144–145
Electroporation. A method for transforming DNA, especially useful for plant cells, in which high-voltage pulses of electricity are used to open pores in the cell membrane through which foreign DNA can pass. 164, 172
Embryology. The study of the morphological and biochemical development of the fertilized egg into an adult organism. 86–87, 106–*107*
Endonuclease. *See* Nuclease.
Endoplasmic reticulum (ER). A system of membranes in the cytoplasm of a cell that is involved in the synthesis of glycoproteins and lipids. Rough ER is associated with ribosomes; smooth ER is ribosome-free. 88
Enhancer. A regulatory DNA sequence that greatly increases transcription of a gene. An enhancer retains its influence regardless of orientation and can be located several thousand base pairs upstream or downstream from the gene it influences. 92, *101*–102, 147
Enzyme. A protein that functions as a biological catalyst to increase the rate of a specific biochemical reaction. 18, 21, 32, *88,* 186–187, 191–194, *192,* 215, 247, 277, 311, 327, 341, 355, 416
polyvalent enzyme. A polypeptide that has been genetically engineered to contain active sites, each of which catalyzes different reactions or successive steps of a reaction cascade. 191
Epitope. The structural feature of an antigen molecule—in proteins, consisting of several amino acids—to which a specific antibody binds. 78–79, 97
Escherichia coli (E. coli). A commensal bacterium inhabiting the human colon that is widely used in

biology, both as a simple model of cell biochemical function and as a host for molecular cloning experiments. 51–53, 64, 147, 179–181, 199, 215, 239, 253, 293, 311, 337, 355, 361, 373, 383, 393, 416

Ethidium bromide. A fluorescent dye used to stain DNA and RNA. The ethidium bromide molecule intercalates (stacks) between nucleotides, and the ethidium bromide–nucleic acid complex fluoresces when exposed to ultraviolet light. 49–50, 256, 337

Eukaryote. An organism whose cells possess a nucleus and other membrane-bound vesicles, including all members of the protist, fungi, plant, and animal kingdoms. 36, 61, 83, 103, 289

Evolution. The long-term process through which a population of organisms accumulates genetic changes that enable its members to successfully adapt to environmental conditions and to better exploit food resources. 10–12, 87

Exon. The portion of a gene found in mature mRNA. Exons of a gene are linked together by mRNA splicing. 92–93, 103–105

Exponential phase. See Growth phase.

Expression library. See Library.

Fibrin. A fibrous protein that forms a scaffold for the formation of a blood clot. The enzyme thrombin catalyzes the reaction to convert the plasma protein fibrinogen into fibrin. 181

Fibroblast. A precursor cell of connective tissue that is relatively easy to maintain in cell culture. 104, 115

Field-inversion electrophoresis. See Electrophoresis.

Flanking region. The DNA sequences extending on either side of a specific locus or gene. See also 3' or 5'. 94–96, 102, 155–158, 163, 184, 195

Focus. A clump of transformed cultured cells displaying morphological and biochemical properties that are the in vitro equivalent of a solid tumor. 115–116

focus assay. A screening procedure to detect certain transforming oncogenes. Viral or cellular DNA sequences are transfected into cultured cells, which are then observed for focus formation. 122

Fourier transformations. The mathematical equations that are used to convert raw X-ray crystallographic data into an electron density map. 189

Fusion gene. See Gene.

Gamete. A haploid sex cell, egg or sperm, that contains a single copy of each chromosome. 12–14, 160

GC box. A guanine- and cytosine-rich sequence present in the promoter of many eukaryotic housekeeping genes, which is the binding site for the SP1 protein. 101–102

Gene. A locus on a chromosome that encodes a specific protein or several related proteins. 12, 14, 18–20, 215, 293, 311, 340, 358, 361, 383, 393, 416

dominant gene. A gene whose phenotype is expressed when it is present in a single copy. 12–13

fusion gene. A hybrid gene created by joining portions of two different genes (to produce a new protein) or by joining a gene to a different promoter (to alter or regulate gene transcription). 109, 130, 175–176, 182

gene amplification. The presence of multiple copies of a gene. Amplification is one mechanism through which proto-oncogenes are activated in malignant cells. See also Amplify. 126–127, 132, 174

gene expression. The process of producing a protein from its DNA- and mRNA-coding sequences. 87, 92–94, 103, 107–109, 147, 175

gene targeting (gene trapping). A method to select for transfected cells in which a DNA sequence has integrated at a homologous site on the chromosomes. 163

gene therapy. See also Germ cell gene therapy; Somatic cell gene therapy. 160–164, 161, 163

gene translocation. The movement of a gene or gene fragment from one chromosomal location to another, which often alters or abolishes expression. 125–126, 158

recessive gene. Characterized as having a phenotype expressed only when both copies of the gene are mutated or missing. 12–13

regulatory gene. A gene whose protein controls the activity of other genes or metabolic pathways. 105–108, 107, 120, 291

split gene. A configuration common to eukaryotic genes where exons (sequences in mRNA) are interrupted by introns (sequences removed from precursor mRNA during mRNA splicing). 103

Genetic code. The three-letter code that translates nucleic acid sequence into protein sequence. 32–37, 79, 105

Genetic disease. A disease that has its origin in changes to the genetic material, DNA. Usually refers to diseases that are inherited in a Mendelian fashion, although noninherited forms of cancer also result from DNA mutation. 112, 140, 148–150, 160

Genetic engineering. The manipulation of an organism's genetic endowment by introducing or eliminating specific genes. A broad definition of genetic engineering also includes selective breeding and other means of artificial selection. 1, 120, 172, 175, 181–184, 189

Genetic map. See Mapping.

Genome. The genetic complement contained in the chromosomes of a given organism, usually the haploid chromosome state. 52, 62, 139–141, 140, 148, 171, 271

Genomic library. See Library.

Genotype. The structure of DNA that determines the expression of a trait (phenotype). 12, 311, 383, 393

Germ cell (germ line). Gametes and the progenitor tissues from which they are derived. 160–162, *161*
  germ cell (germ line) gene therapy. The repair or replacement of a defective gene within the gamete-forming tissues, which produces a heritable change in an organism's genetic constitution. 160–162, *161*
Globin. The protein constituent of hemoglobin molecules that bind oxygen and carbon dioxide in red blood cells. 62–64, *63*, 70–71, 102–103, 141, *152*–153
Glycosylation. The addition of sugar side groups to a polypeptide. 88
Golgi apparatus. The cellular organelle responsible for modifying proteins prior to transport from the cell. 88, 180
G protein (guanine-nucleotide-binding protein). A membrane-anchored protein that hydrolyzes GTP and transduces a hormonal signal from a membrane receptor to an intracellular second messenger. 133–134
Green revolution. Advances in genetics, petrochemicals, and machinery that culminated in a dramatic increase in crop productivity during the third quarter of the 20th century. *169*–170
Growth factor. A serum protein that stimulates cell division when it binds to its cell-surface receptor. 133–*135*, *134*
  growth factor receptor. A membrane-spanning protein that selectively binds its growth factor and then transduces a signal for cell division to other molecules in the cytoplasm and nucleus. *134*–135
  platelet-derived growth factor (PDGF). A protein secreted by platelets to stimulate the replication of connective tissue cells and smooth muscle cells at wound sites. Also stimulates growth of fibroblasts in cell culture. 133–135
Growth phase (curve). The characteristic periods in the growth of a bacterial culture, as indicated by the shape of a graph of viable cell number versus time. *52*, 215, 235, 392
  death phase. The final growth phase, during which nutrients have been depleted and cell number decreases. *52*, 235
  lag phase. The initial growth phase, during which cell number remains relatively constant prior to rapid growth. *52*, 215, 235, 246
  logarithmic phase (log or exponential growth phase). The steepest slope of the growth curve—the phase of vigorous growth during which cell number doubles every 20–30 minutes. *52*–53, 55–56, 219, 235, 239, 309, 361
  stationary phase. The plateau of the growth curve after log growth, during which cell number remains constant. New cells are produced at the same rate as older cells die. *52*, 215, 235, 246, 308, 311

Hairpin loop. A loop formed when single-stranded DNA or RNA base pairs to itself, as when synthesizing dscDNA. 73–*74*, 413
Haploid. The chromosome number equal to one complete set of the genetic endowment of a eukaryotic organism. *52*, 62, 95, 139, 171
Helix-turn-helix. A secondary polypeptide structure, found in several DNA-binding proteins, in which two adjacent α helices are oriented at approximately right angles to one another. 100, 107, 187
Hemophilia. An X-linked recessive genetic disease, caused by a mutation in the gene for clotting factor VIII (hemophilia A) or clotting factor IX (hemophilia B), which leads to abnormal blood clotting. 16, 178–179, 182
Herbicide. Any substance that is toxic to plants; usually used to kill specific unwanted plants. 170, 173–175, *174*
Heteroduplex. A double-stranded DNA molecule or DNA-RNA hybrid, where each strand is of a different origin. 162
Heterogeneous nuclear RNA (hnRNA). *See* RNA.
Histone. Any of five related proteins, composed primarily of basic amino acids, which are the scaffold around which DNA is wound to form the chromatin structure of eukaryotic chromosomes. 139, 337
Homeobox (homeo domain). A consensus 180-bp sequence found within the coding region of homeotic genes that is thought to encode a DNA-binding region in the expressed protein. 105–108, *106*
Homeotic mutation. Any mutation that disrupts a developmental pathway involved in determining spatial relationships of tissues and body parts. 105
Homogeneously staining region (HSR). A dark-staining region of a chromosome that arises from adjacent units of a repeated chromosomal fragment and in which oncogenes may be amplified. A common cytological feature of tumor cells. *126*
Homolog. Any member of a set of genes or DNA sequences from different organisms whose nucleotide sequences show a high degree of one-to-one correspondence. 117, 124
Homologous chromosomes. Chromosomes that have the same linear arrangement of genes—a pair of matching chromosomes in a diploid organism. 14, 17
Homologous recombination. The exchange of DNA fragments between two DNA molecules or chromatids of paired chromosomes (during crossing over) at the site of identical nucleotide sequences. 162–164, *163*, 338
HSR. *See* Homogeneously staining region.
Human Genome Project. A project coordinated by the National Institutes of Health (NIH) and the Department of Energy (DOE) to determine the entire nucleotide sequence of the human chromosomes. 139–141, 147–148
Human growth hormone (HGH, somatotrophin). A protein produced in the pituitary gland that stimulates the liver to produce somatomedins,

which stimulate growth of bone and muscle. *177–179*, 181

Humoral immunity. The production of antibodies (immunoglobulin) by B lymphocytes in the blood. 183–184

Huntington's disease (Huntington's chorea). An autosomal dominant genetic disease, whose symptoms of involuntary muscle movements (chorea) and mental dysfunction usually do not appear until the age of 35–45, 153–155, *154*

Hybrid. The offspring of two parents differing in at least one genetic characteristic (trait). Also, heteroduplex DNA or DNA-RNA molecule. 17, 64, 169–171, 311, 358

Hybridization. The hydrogen bonding of complementary DNA and/or RNA sequences to form a duplex molecule. 68–69, 83–84, 103, 127, 151, 322

Northern hybridization (Northern blotting). A procedure in which RNA fragments are transferred from an agarose gel to a nitrocellulose filter, where the RNA is then hybridized to a radioactive probe. 83–85

Southern hybridization (Southern blotting). A procedure in which DNA restriction fragments are transferred from an agarose gel to a nitrocellulose filter, where the denatured DNA is then hybridized to a radioactive probe. *82–85*

Hybridoma. A hybrid cell, composed of a B lymphocyte fused to a tumor cell, which grows indefinitely in tissue culture and is selected for the secretion of a specific antibody of interest. 78–79, 192–194

Hydrogen bond. A relatively weak bond formed between a hydrogen atom (which is covalently bound to a nitrogen or oxygen atom) and a nitrogen or oxygen with an unshared electron pair. *28–30*, 34, 47, 68, 189, 272, 341

Hydrolysis. A reaction in which a molecule of water is added at the site of cleavage of a molecule to two products. 47, 134, 157, 191–*192*, 234

Imaginal disc. Any of 19 clusters of cells in *Drosophila* larvae that develop into major body parts of the adult fly (imago). *105*–106

Immortalizing oncogene. *See* Oncogene.

Inbreed. To mate closely related individuals. 149, 169–170

Inositol lipid. *See* Phospholipid.

In situ. Refers to performing assays or manipulations with intact tissues. *107*, 126–*127*, 174

Insulin. A peptide hormone secreted by the islets of Langerhans of the pancreas that regulates the level of sugar in the blood. 109, 130, 135, 178–181, *180*

preproinsulin. A polypeptide precursor of proinsulin, from which a 24-amino-acid signal peptide sequence is responsible for extracellular secretion. 179–*180*

proinsulin. A polypeptide precursor of insulin, from which a 33-amino-acid sequence is removed enzymatically to form active insulin. 179–181, *180*

Interferon. A family of small proteins that stimulate viral resistance in cells. 182

Intergenic regions. DNA sequences located between genes that comprise a large percentage of the human genome with no known function. 148

Interleukin-2 (IL-2). An autostimulating protein produced by activated helper T cells that causes the proliferation of helper T cells. 160, 182

Intron. A nucleotide sequence that intervenes between exons, which is excised from pre-mRNA during RNA processing. 83, 92–*93*, *103*–105, 124, 148, 179–181

In vitro. Literally "in glass," refers to the recreation of biological processes in an artificial environment. 3, 35, 65, 75, 100, 114–116, 120, 154, 189

In vivo. Refers to biological processes that take place within a living organism or cell. 116, 120, 337

Ion. A charged particle. 45, 55–57, 252, 299

Islets of Langerhans. A small patch of endocrine cells in the pancreas composed of α and β cells. The β cells secrete insulin, and the α cells secrete glucagon. 109, 130, 179

Isoform. Related proteins encoded by different genes or generated by alternative RNA splicing of transcripts encoded by the same gene. 92, 104

Isotope. One of two or more forms of an element that have the same number of protons (atomic number) but differing numbers of neutrons (mass numbers). Radioactive isotopes are commonly used to make DNA probes and metabolic tracers. 24–*25*, 30–*31*

Joining (J) segment. A small DNA segment that links constant and variable genes to yield a functional gene encoding an immunogobulin heavy or light chain. 98–99

Kanamycin. An antibiotic of the aminoglycoside family that poisons translation by binding to the ribosomes. 43, 54–*55*, 225, 299, 341, 358, 361, 373, 383

Kinase. An enzyme that catalyzes the transfer of phosphate groups to substrate molecules. *See also* Protein kinase. 79, 88, 122, 133, *135*, 163, 184–185

Lag phase. *See* Growth phase.

Lawn. A uniform and uninterrupted layer of bacterial growth, in which individual colonies cannot be observed. 66, 307, 379

LB broth. *See* Luria-Bertani broth.

Legume. A member of the pea family that possesses root nodules containing nitrogen-fixing bacteria. 171

Leucine zipper. A secondary protein structure in which projecting leucine residues on two polypeptide chains interdigitate to form a stable dimer. 100, 136, 187

Leukemia. A liquid tumor characterized by the overproduction of nonfunctional or immature white blood cells. 112, 118, 120, 127

Library. A collection of cells, usually bacteria or yeast, that have been transformed with recombinant vectors carrying DNA inserts from a single species. 62–84

    cDNA library. A library composed of complementary copies of cellular mRNAs. 73–74, 79, 104, 143

    expression library. A library of cDNAs whose encoded proteins are expressed by specialized vectors. 74–77, *76*, 194

    genomic library. A library composed of fragments of genomic DNA. 62, 64–*69*, 83, 104, 143, 146–148, 154–155, 291

Ligase (DNA ligase). An enzyme that catalyzes a condensation reaction that links two DNA molecules via the formation of a phosphodiester bond between the 3' hydroxyl and 5' phosphate of adjacent nucleotides. 43, *43*, 51, *51*, 62, 73, 341, 355

Ligate. The process of joining two or more DNA fragments. 50–51, 65, *66*, 73–75, *74*, 109, 122, *123*, 163, 194, 199, 252, 298, 341, 355, 361, 379, 383, 393, 413

Lineage. A chart that traces the flow of genetic information from generation to generation. 149

Linkage. The frequency of coinheritance of a pair of genes and/or genetic markers, which provides a measure of their physical proximity to one another on a chromosome. 16–*17*, 141, 146, *150*, 153–155, 158

    disequilibrium. The situation when a pair of genes and/or genetic markers are so closely linked that their chance of coinheritance is 100%. 158

    linkage map. *See* Mapping.

Linked genes/markers. Genes and/or markers that are so closely associated on the chromosome that they are coinherited in 80% or more of cases. *150*, 154, 157–158

Linker. A short, double-stranded oligonucleotide containing a restriction endonuclease recognition site, which is ligated to the ends of a DNA fragment. 73–74

Liquid tumor. Any of the cancers of the circulating cells of the blood. *112*

Locus (plural, loci). A specific location or site on a chromosome. 94–97, 151, *155*–156, 164–165

Log phase. *See* Growth phase.

Luria-Bertani (LB) broth. A rich nutrient medium—containing carbohydrates, amino acids, nucleotide phosphates, salts, and vitamins—commonly used to propagate *E. coli*. 52, 215, 239, 293, 361, 391

Lymphoblastoid cell line. A lymphocyte cell line that grows indefinitely in culture, derived from lymphocytes that are immortalized by fusion with a cancer cell or transformation by an oncogene. 154

Lymphoma. Any of the cancers of the lymphatic tissue. 112, 120, 125, 129

Lysis. The destruction of the cell membrane. *59*, 65, 69–70, 76–77, 115, 144, 166

Lysogen. A bacterial cell whose chromosome contains integrated viral DNA. 65, 253

    lysogenic. A type or phase of the virus life cycle during which the virus integrates into the host chromosome of the infected cell, often remaining essentially dormant for some period of time. 113, 115

Lytic. A phase of the virus life cycle during which the virus replicates within the host cell, releasing a new generation of viruses when the infected cell lyses. 65, 113, 115

Malignant. Having the properties of cancerous growth. 90, 111–112, 114, *116*–117, 120, 125, 129–130, 132

Mapping. Determining the physical location of a gene or genetic marker on a chromosome.

    contiguous (contig) map. The alignment of sequence data from large, adjacent regions of the genome to produce a continuous nucleotide sequence across a chromosomal region. 141, 146–147

    genetic (linkage) map. A linear map of the relative positions of genes along a chromosome. Distances are established by linkage analysis, which determines the frequency at which two gene loci become separated during chromosomal recombination. 141, 146

    physical map. A map showing physical locations on a DNA molecule, such as restriction sites, RFLPs, and sequence-tagged sites. 141

    restriction map. A physical map showing the locations of restriction enzyme recognition sites. 42, 71–73, *72*, 141, 147

Mating-type switch. The cycling of yeast cells between **a** and α mating types, which involves the transposition of gene "cassettes" into an active site of expression. 95–97, *96*

Megabase cloning. *See* Cloning.

Meiosis. The reduction division process by which haploid gametes and spores are formed, consisting of a single duplication of the genetic material followed by two mitotic divisions. 12–14, 17, 97, *150*–151

Messenger RNA (mRNA). *See* RNA.

Metabolism. The biochemical processes that sustain a living cell or organism. 18, 149, 178, 235

Metallothionein promoter. *See* Promoter.

Metastasis. The ability of cancerous cells to invade surrounding tissues, enter the circulatory system, and establish new malignancies in body tissues distant from the site of the original tumor. 160

Methionine. The amino acid encoded by the sequence AUG. 35, 81, 181, 277

Methylase (modifying enzyme). An enzyme that adds a methyl group to a specific nucleotide within a restriction enzyme recognition sequence, thus blocking binding and cutting by the restriction enzyme at that site. 47, 74, 277

Methylate. The addition of one or more methyl groups (CH$_3$) to a molecule. 41, 44, *46–47*, 277
Microbe. A microorganism. 171, 179, 226
Microinjection. A means to introduce a solution of DNA, protein, or other soluble material into a cell using a fine microcapillary pipet. 108–*109*, 160, 173, 176–177
Mitosis. The replication of a cell to form two daughter cells with identical sets of chromosomes. 40, 146
Modifying enzyme. See Methylase. 41, 47
Molecular biology. The study of the biochemical and molecular interactions within living cells. 1–7, *3*, 9, 19, 23, 51, 53, 171
Molecular cloning. See Cloning.
Molecular genetics. The study of the flow and regulation of genetic information between DNA, RNA, and protein molecules. 2, 36, 61, 87–88, 156
Monoclonal antibodies. See Antibody.
Monocotyledon (monocot). A flowering plant whose embryo possesses one cotyledon, or seed leaf, including all the staple cereal crops, grasses, lilies, and palms. 171, 172
Movable genetic element. See Transposon.
Multi-locus probe. See Probe.
Mutation. An alteration in DNA structure or sequence of a gene. 16–18, 32, 90–91, 94, 99, *105*, 115, 124–*125*, 150–154, *152*, 156, *159*, 171, 186, 189–191, *190*
  point mutation. A change in a single base pair of a DNA sequence in a gene. 99, 124, 151–153, *152*, *159*, 189
Myeloma. A tumor of the cells of the bone marrow. 78–79, 192

Natural selection. The differential survival and reproduction of organisms with genetic characteristics that enable them to better utilize environmental resources. 11
Nicked circle. See Plasmid.
Nick translation. A procedure for making a DNA probe in which a DNA fragment is treated with DNase to produce single-stranded nicks, followed by incorporation of radioactive nucleotides from the nicked sites by DNA polymerase I. 84
Nitrocellulose. A membrane used to immobilize DNA, RNA, or protein, which can then be probed with a labeled sequence or antibody. 68, 74–77, *83–84*
Nitrogen fixation. The conversion of atmospheric nitrogen (N$_2$) to biologically usable nitrates (NO$_3^{-2}$). 187
Nitrogenous bases. The purines (adenine and guanine) and pyrimidines (thymine, cytosine, and uracil) that comprise DNA and RNA molecules. 27
Northern hybridization (blotting). See Hybridization.
Nuclease. A class of enzymes that degrades DNA and/or RNA molecules by cleaving the phosphodiester bonds that link adjacent nucleotides. 44, 146, 253, 327
  deoxyribonuclease (DNase). Substrate is DNA. 20, 84, 327
  endonuclease. Cleaves at internal sites in the substrate molecule. See also Restriction endonuclease. 41, 44–47, *45*, *46*, 62, 71, 73, 151, 247, 277, 327, 345
  exonuclease. Progressively cleaves from the end of the substrate molecule. 44, 56, 89
  ribonuclease (RNase). Substrate is RNA. 20, 59, 327
  S1 nuclease. Substrate is single-stranded DNA or RNA. 73–74
Nuclein. The term used by Friedrich Miescher to describe the nuclear material he discovered in 1869, which today is known as DNA. 26
Nucleoside. A building block of DNA and RNA, consisting of a nitrogenous base linked to a five-carbon sugar. 143, 184–186, *185*, 234, 289
  nucleoside analog. A synthetic molecule that resembles a naturally occurring nucleoside, but that lacks a bond site needed to link it to an adjacent nucleotide. 164, 184–186, *185*
Nucleotide. A building block of DNA and RNA, consisting of a nitrogenous base, a five-carbon sugar, and a phosphate group. 21, 26–29, *28*, 52, 67–69, 80, 100, 139–140, 271, 289, 311, 328
  complementary nucleotides. Members of the pairs adenine-thymine, adenine-uracil, and guanine-cytosine that have the ability to hydrogen bond to one another. 50
  dideoxynucleotide (*did*N). A deoxynucleotide that lacks a 3′ hydroxyl group, and is thus unable to form a 3′–5′ phosphodiester bond necessary for chain elongation. Dideoxynucleotides are used in DNA sequencing and the treatment of viral diseases. *80*, 184–*185*
  oligonucleotide. A DNA polymer composed of only a few nucleotides. 79, 84, 143–144, 159, 189
  polynucleotide. A DNA polymer composed of multiple nucleotides. 31
Nucleus. The membrane-bound region of a eukaryotic cell that contains the chromosomes. 33, 92, 103, 133, 135, 139, 172

Oncogene. A gene that contributes to cancer formation when mutated or inappropriately expressed. 88, 90, 100, 111–137, *125*, *130*
  cellular oncogene (proto-oncogene). A normal gene that when mutated or improperly expressed contributes to the development of cancer. *118–119*, 122, 125, 130
  dominant(-acting) oncogene. A gene that stimulates cell proliferation and contributes to oncogenesis when present in a single copy. 129–132
  *erb*B. Membrane receptor that binds epidermal growth hormone. 133–*135*, *134*
  immortalizing oncogene. A gene that upon transfection enables a primary cell to grow indefinitely in culture. 129, 154
  myc. A nuclear oncogene involved in immortalizing cells. *118–119*, *125–130*, *126*, *127*, *128*, 135–136

Oncogene (*continued*)
  ras. Converts GTP to GDP; a step in signal transduction. 124–125, 128–129, 132, 134–135, 159
  recessive(-acting) oncogene or anti-oncogene. A single copy of this gene is sufficient to suppress cell proliferation; the loss of both copies of the gene contributes to cancer formation. 130–132
  transforming oncogene. A gene that upon transfection converts a previously immortalized cell to the malignant phenotype. 129, 154
  viral oncogene. A viral gene that contributes to malignancies in vertebrate hosts. 117–118
Oncogenesis. The progression of cytological, genetic, and cellular changes that culminate in a malignant tumor. 118–120, 124, 128–130, 132, 162
  multi-step oncogenesis. The involvement of multiple oncogenes in tumor expression. 129–130
Oncoprotein. The protein produced by an oncogene. 88, 125–126, 132–136, 134
Open pollination. Pollination by wind, insects, or other natural mechanisms. 169
Open reading frame. *See* Reading frame.
Operator. A prokaryotic regulatory element that interacts with a repressor to control the transcription of adjacent structural genes. 107
Origin of replication. The nucleotide sequence at which DNA synthesis is initiated. 43, 53, 146, 340, 358, 361, 393
Overlapping reading frame. *See* Reading frame.
Ovum. A female gamete. 39–40

Paleontology. The study of the fossil record of past geological periods and of the phylogenetic relationships between ancient and contemporary plant and animal species. 11–12
Palindromic sequence. A DNA locus whose 5′-to-3′ sequence is identical on each DNA strand. Recognition sites of many restriction enzymes are palindromic. 46
Pedigree. A diagram mapping the genetic history of a particular family. 153–154, 158
Pesticide. A substance that kills harmful organisms (for example, an insecticide or fungicide). 170, 173–174, 187
Phage (particle). *See* Bacteriophage.
Phagocytosis. The cellular process of engulfing a particle or organism through ameboid-like movement of the cell membrane. 121–122
Phenotype. The expression of gene alleles (genotype) as a detectable physical or biochemical trait. 12, 57, 129, 149, 164, 293, 311, 383, 393
Phenylketonuria. A genetic disease caused by lack of an enzyme needed to convert excess phenylalanine to tyrosine. Accumulated phenylpyruvate interferes with normal brain development, leading to severe mental retardation. 149
Pheromone. A hormone-like substance that is secreted into the environment. 97
Phosphatase. An enzyme that hydrolyzes esters of phosphoric acid, removing a phosphate group. 88
Phosphate ($^{-}PO_4$). 26–29, 27–28, 46, 48, 51, 56–57, 70, 79, 88, 135, 175, 185, 297
  pyrophosphate. A diphosphate molecule that may be liberated when adenosine triphosphate (ATP) is hydrolyzed to adenosine monophosphate (AMP). 355
Phosphodiester bond. A bond in which a phosphate group joins adjacent carbons through ester linkages. A condensation reaction between adjacent nucleotides results in a phosphodiester bond between 3′ and 5′ carbons in DNA and RNA. 44, 47, 51, 80, 84, 341
Phospholipid. A class of lipid molecules in which a phosphate group is linked to glycerol and two fatty acyl groups. A chief component of biological membranes. 56–57, 59
  inositol phospholipid. A membrane-anchored phospholipid that transduces hormonal signals by stimulating the release of any of several chemical messengers. 133
Phosphorylation. The addition of a phosphate group to a compound. 88, 135, 185
Physical map. *See* Mapping. 141
Plaque. A clear spot on a lawn of bacteria or cultured cells where cells have been lysed by viral infection. 66, 69
Plasmid (p). A circular DNA molecule, capable of autonomous replication, which typically carries one or more genes encoding antibiotic resistance proteins. 42–43, 50–60, 53, 62–64, 74–75, 84, 122, 146, 163, 171–172, 177, 181, 184, 189–190, 199, 215, 231, 253, 290, 293, 311, 327, 341, 358, 361, 379, 383, 393, 413
  nicked circle (relaxed circle). During extraction of plasmid DNA from the bacterial cell, one strand of the DNA becomes nicked. This relaxes the torsional strain needed to maintain supercoiling, producing the familiar form of plasmid. 338
  pAMP. Ampicillin-resistant plasmid developed for this laboratory course. 72, 215, 290, 293, 311, 336, 341, 342, 358, 361, 379, 393, 405
  pBR322. A derivation of ColE1, one of the first plasmid vectors widely used. 298, 315, 340
  pKAN. Kanamycin-resistant plasmid developed for this laboratory course. *Labs*
  pSC101. A plasmid under stringent growth control, developed by Stanley Cohen. 43
  pUC. A widely used expression plasmid containing a β-galactosidase gene. 315, 340
  relaxed plasmid. A plasmid that replicates independently of the main bacterial chromosome and is present in 10–500 copies per cell. 53
  stringent plasmid. A plasmid that only replicates along with the main bacterial chromosome and is present as a single copy, or at most several copies, per cell. 53
  supercoiled plasmid. The predominant in vivo form of plasmid, in which the plasmid is coiled around histone-like proteins. Supporting proteins are stripped away during extraction from the bacterial cell, causing the plasmid molecule to supercoil around itself in vitro. 298, 337–338, 361, 393, 415–416

Platelet-derived growth factor (PDGF). *See* Growth factor.
Point mutation. *See* Mutation.
Poly(A) tail. A polymer of 100–200 adenine residues found at the 3′ end of most eukaryotic mRNAs. 70–71, 88–89, 92
Polyclonal antibodies. *See* Antibody. 78
Polylinker. A short DNA sequence containing several restriction enzyme recognition sites that is contained in cloning vectors. 75–76
Polymer. A molecule composed of repeated subunits. 21, 46, 66
Polymerase. An enzyme that catalyzes the addition of multiple subunits to a substrate molecule. 195
  DNA polymerase. Synthesizes a double-stranded DNA molecule using a primer and DNA as a template. 31, *43*, 53, 73, 80, 84, 184–186, 189
  poly(A) polymerase. Catalyzes the addition of adenine residues to the 3′ end of pre-mRNAs to form the poly(A) tail. 91–92
  polymerase chain reaction (PCR). A procedure that enzymatically amplifies a target DNA sequence of up to several thousand base pairs through repeated replication by DNA polymerase. 141–*144*, 159, 164, 166, 186, *193*–194
  RNA polymerase. Transcribes RNA from a DNA template. *33*, 75, 92, 100–*101*
  Taq polymerase. A heat-stable DNA polymerase isolated from the bacterium *Thermus aquaticus,* used in PCR. 143–144
Polynucleotide. *See* Nucleotide.
Polypeptide (protein). A polymer composed of multiple amino acid units linked by peptide bonds. 28, 35, *35*, 37, 88–90, *89*, 97, 99, 106, *134*, 136, 157, 179–181, *180*, 187, 191, 194
Polyploid. A multiple of the haploid chromosome number that results from chromosome replication without nuclear division. 171
Polysaccharide. A polymer composed of multiple units of monosaccharide (simple sugar). 19–*20*, 97
Polyvalent enzyme. *See* Enzyme.
Polyvalent vaccine. *See* Vaccine.
Porcine. Relating to or derived from pigs. 179
Preenzyme. *See* Zymogen.
Preproinsulin. *See* Insulin.
Primary cell. A cell or cell line taken directly from a living organism, which is not immortalized. 114, 129–*130*
Primer. A short DNA or RNA fragment annealed to single-stranded DNA, from which DNA polymerase extends a new DNA strand to produce a duplex molecule. 70, 73–*74*, 80, 84, 143–*144*, 159, 189
Prion. *See* Proteinaceous infectious particle.
Probe. A single-stranded DNA that has been radioactively labeled and is used to identify complementary sequences in genes or DNA fragments of interest. 68–71, 73, 79, 83–*84*, 97, 104, 106–*107*, 126–*127*, 143, 151–155, *152*, 164–167, *165*
  multilocus probe. A probe that hybridizes to a number of different sites in the genome of an organism. 164–166, *165*
  single-locus probe. A probe that hybridizes to a single site in the genome of an organism. 164–*167*
Proinsulin. *See* Insulin.
Prokaryote. An organism whose cell(s) lacks a nucleus and other membrane-bound vesicles. Prokaryotes include all members of the Kingdom Monera. 45, 94, 107, 171, 289
Promoter. A region of DNA extending 150–300 bp upstream from the transcription start site that contains binding sites for RNA polymerase and a number of proteins that regulate the rate of transcription of the adjacent gene. 75–76, 92, 94–95, 100–*101*, 107, 109, 118, 120, 130, 135, 147, 172–176
  constitutive promoter. An unregulated promoter that allows for continual transcription of its associated gene. 173, 176
  metallothionein. A protective protein that binds heavy metals, such as cadmium and lead. 109
  rat insulin promoter (RIP). 109
Pronucleus. Either of the two haploid gamete nuclei just prior to their fusion in the fertilized ovum. 108, 177
Protease. An enzyme that cleaves peptide bonds that link amino acids in protein molecules. 181, 194
Protein. A polymer of amino acids linked via peptide bonds and which may be composed of two or more polypeptide chains. 21, 23–25, 32–37, *35*, 56, 59, 75, 78, 81, 87–89, 92, 97, *104*, 117, 161, 178–182, *180*, 186–189, *188*, 226, 234, 297, 311, 337, 373, 391
Proteinaceous infectious particle (prion). A proposed pathogen composed only of protein with no detectable nucleic acid and which is responsible for Creutzfeldt-Jakob disease and kuru in humans and scrapie in sheep. 179
Protein kinase. An enzyme that adds phosphate groups to a protein molecule at serine, threonine, or tyrosine residues. 88, 133, 135
Proteolytic. The ability to break down protein molecules. 88
Proto-oncogene. *See* Oncogene.
Provirus. *See* Virus.
Pyrophosphate. *See* Phosphate.

Reading frame. A series of triplet codons beginning from a specific nucleotide. Depending on where one begins, each DNA strand contains three different reading frames. 37, 81, 147, 175
  open reading frame. A long DNA sequence that is uninterrupted by a stop codon and encodes part or all of a protein. 81, 156
  overlapping reading frames. Start codons in different reading frames generate different polypeptides from the same DNA sequence. 37
Recessive(-acting) oncogene. *See* Oncogene.
Recognition sequence (site). A nucleotide sequence—composed typically of 4, 6, or 8 nucleotides—that is recognized by a restriction endonuclease. Type II enzymes cut (and their corresponding modification enzymes methylate) within or very near the recognition se-

Recognition sequence (site) (*continued*)
quence. 41–42, 44–47, *46*, 145, 247, *271*, 277, 289, 338, 341, 358, 413
Recombinant DNA. The process of cutting and recombining DNA fragments as a means to isolate genes or to alter their structure and function. 42–44, 160, 341, 359, 361, 373, 393, 414
Recombination frequency. The frequency at which crossing over occurs between two chromosomal loci—the probability that two loci will become unlinked during meiosis. 151, 157–158
Regulatory gene. *See* Gene.
Relaxed circle. *See* Plasmid.
Relaxed plasmid. *See* Plasmid.
Renature. The reannealing (hydrogen bonding) of single-stranded DNA and/or RNA to form a duplex molecule. 59, 68, 322
Repressor. A DNA-binding protein in prokaryotes that blocks gene transcription by binding to the operator. 107, 126
Reptation. The end-on movement of a DNA molecule through a gel matrix during electrophoresis. 145
Restriction endonuclease (enzyme). A class of endonucleases that cleaves DNA after recognizing a specific sequence. 41–47, *45*, 247, 327
*Bam*HI (GGATCC). *45*, 47, 71–72, 75–76, 247, *271*, 290, 341, *342*, 358, 361, 383, 393, 414
*Eco*RI (GAATTC). 42–47, *46*, 65–66, 71–76, *72*, 247, *271*, 277
*Hin*dIII (AAGCTT). *45*, 47, *72*, 75–76, 247, *271*, 277, 341, *342*, 355, 361, 383, 393, 413
Type I. Cuts nonspecifically a distance greater than 1000 bp from its recognition sequence and contains both restriction and methylation activities. 44–47
Type II. Cuts at or near a short, and often symmetrical, recognition sequence. A separate enzyme methylates the same recognition sequence. 44–47
Type III. Cuts 24–26 bp downstream from a short, asymmetrical recognition sequence. Requires ATP and contains both restriction and methylation activities. 44–47
Restriction-fragment-length polymorphism (RFLP). Differences in nucleotide sequence between alleles at a chromosomal locus result in restriction fragments of varying lengths detected by Southern analysis. 84, 141, 151–155, 157
Restriction map. *See* Mapping.
Retrovirus. A member of a class of RNA viruses that utilizes the enzyme reverse transcriptase to reverse copy its genome into a DNA intermediate, which integrates into the host-cell chromosome. Many naturally occurring cancers of vertebrate animals are caused by retroviruses. 36, 113–*114*, 117–120, *118*–*119*,154, *161*–*162*, 186
Reverse genetics. Using linkage analysis and polymorphic markers to isolate a disease gene in the absence of a known metabolic defect, then using the DNA sequence of the cloned gene to predict the amino acid sequence of its encoded protein. 149–151, 156

Reverse transcriptase (RNA-dependent DNA polymerase). An enzyme isolated from retrovirus-infected cells that synthesizes a complementary (c)DNA strand from an RNA template. 36, 70, 73, 186
Ribosome-binding site. The region of an mRNA molecule that binds the ribosome to initiate translation. 75–76, 90
Ribozyme. *See* Catalyst.
RNA (ribonucleic acid). An organic acid composed of repeating nucleotide units of adenine, guanine, cytosine, and uracil, whose ribose components are linked by phosphodiester bonds. 1, 26–27, 32–37, *33*, *34*, *35*, *36*, 322, 327, 413
alternative mRNA splicing. The inclusion or exclusion of different exons to form diiferent mRNA transcripts. 102–105, *103*, *104*
antisense RNA. A complementary RNA sequence that binds to a naturally occurring (sense) mRNA molecule, thus blocking its translation. 175–176
heterogeneous nuclear RNA (hnRNA). The name originally given to large RNA molecules found in the nucleus, which are now known to be unedited mRNA transcripts, or pre-mRNAs. 102
messenger RNA (mRNA). The class of RNA molecules that copies the genetic information from DNA, in the nucleus, and carries it to ribosomes, in the cytoplasm, where it is translated into protein. *35*, 88–90, 92–93, 102–105, 327
mRNA cap structure. An inverted and methylated guanosine residue that is added posttranscriptionally to the 5' end of eukaryotic mRNAs, where it presumably helps prevent degradation by exonucleases. 90–92, *91*
mRNA half-life. The time during which one-half of the transcripts of a specific mRNA degrades. 88–90
mRNA splicing (processing). The excision of introns and subsequent joining of exons of pre-mRNA to form a mature mRNA transcript. 92–93, 102–105
pre-mRNA. The initial mRNA transcript prior to any mRNA processing. 92–93, 103, 179
ribosomal RNA (rRNA). The RNA component of the ribosome. 33, 327
RNA polymerase. *See* Polymerase.
small nuclear RNA (snRNA). Short RNA transcripts of 100–300 bp that associate with proteins to form small nuclear ribonucleoprotein particles (snRNPs), which participate in RNA processing. 100
transfer RNA (tRNA). The class of small RNA molecules that transfer amino acids to the ribosome during protein synthesis. *35*, 81, 90, 194, 327

Saccharomyces cerevisiae. Brewer's yeast. 94–97, *96*
Salmonella. A genus of rod-shaped, gram-negative bacteria that are a common cause of food poisoning. 94–97, *96*

Sarcoma. A tumor arising from mesoderm, or connective, tissue. *112–113*, 124, 133

Satellite RNA (viroids). A small, self-splicing RNA molecule that accompanies several plant viruses, including tobacco ringspot virus. 195

Selectable marker. A gene whose expression allows one to identify cells that have been transformed or transfected with a vector containing the marker gene. 58

    amp^r. (β-Lactamase) ampicillin resistance gene. *54*, 76, 383, 393, 414

    kan^r. Kanamycin resistance gene. *55*, 383, 393, 414

    tk. *See* Thymidine kinase.

Semiconservative replication. During DNA duplication, each strand of a parent DNA molecule is a template for the synthesis of its new complementary strand. Thus, one half of a preexisting DNA molecule is conserved during each round of replication. 30–*32*, *31*

Sequence hypothesis. Francis Crick's seminal concept that genetic information exists as a linear DNA code; DNA and protein sequence are colinear. 32–33

Sequence-tagged site (STS). A unique (single-copy) DNA sequence used as a mapping landmark on a chromosome. *See also* Mapping. 141, 143

Sickle cell anemia. A genetic disease in which a point mutation in the β globin gene leads to production of abnormal hemoglobin and misshaped red blood cells with diminished oxygen transport. 149, 151–153, *152*

Signal peptide (sequence). A 16–30 amino acid sequence, located at the amino terminus of secreted polypeptides, that conducts the protein into the endoplasmic reticulum for posttranslational modifications. *180*, 182

Signal transduction. The biochemical events that conduct the signal of a hormone or growth factor from the cell exterior, through the cell membrane, and into the cytoplasm. This involves a number of molecules, including receptors, G proteins, and second messengers. 133

Site-directed mutagenesis. The process of introducing specific base-pair mutations into a gene. 189–191, *190*

Somatic cell. Any nongerm cell that composes the body of an organism and which possesses a set of multiploid chromosomes (diploid in most organisms). 39, 160–162

    somatic cell gene therapy. The repair or replacement of a defective gene within somatic tissue. 160–162, *161*

Somatotrophin. *See* Human growth hormone.

Southern hybridization (blotting). *See* Hybridization.

SP1 protein. A DNA-binding protein that binds to the GC box of eukaryotic promoters. *101*, 107

Species. A classification of related organisms that can freely interbreed. 9–12, 45, 61, 102, 114, 117

Split gene. *See* Gene.

Spore. An asexual reproductive cell, often that has the ability to resist unfavorable environment conditions. 18–*19*, 95

Stationary phase. *See* Growth phase.

Stem cell. An undifferentiated cell, usually of embryonic origin, that gives rise to one or more types of specialized cells. 161–162

Sticky end. A protruding, single-stranded nucleotide sequence produced when a restriction endonuclease cleaves off center in its recognition sequence. 42–43, 50–51, 271, 341, *342*, 359

Stop codon. *See* Codon.

Stringency. Reaction conditions—notably temperature, salt, and pH—that dictate the annealing of single-stranded DNA/DNA, DNA/RNA, and RNA/RNA hybrids. At high stringency, duplexes form only between strands with perfect one-to-one complementarity; lower stringency allows annealing between strands with some degree of mismatch between bases. 68, 79

Stringent plasmid. *See* Plasmid.

Structure-functionalism. The scientific tradition that stresses the relationship between a physical structure and its function, for example, the related disciplines of anatomy and physiology. 4–5

Subcloning. *See* Cloning.

Subunit vaccine. *See* Vaccine.

Supercoiled plasmid. *See* Plasmid.

Supernatant. The soluble liquid fraction of a sample after centrifugation or precipitation of insoluble solids. 24, 59, 70

Synapsis. The pairing of homologous chromosome pairs during prophase of the first meiotic division, when crossing over occurs. *150–151*

TATA box. An adenine- and thymine-rich promoter sequence located 25–30 bp upstream of a gene, which is the binding site of RNA polymerase. 92, 100–102, *101*

T-DNA (tumor-DNA). The transforming region of DNA in the Ti plasmid of *Agrobacterium tumefaciens*. 171–*173*, 176

Telomere. The end of a chromosome. 146, 155

Template. An RNA or single-stranded DNA molecule upon which a complementary nucleotide strand is synthesized. 30–*31*, 33–35, 70, 143

Termination codon. *See* Codon.

Tetracycline. An antibiotic that interferes with protein synthesis in prokaryotes. 43, 54

Thymidine kinase (*tk*). An enzyme that allows a cell to utilize an alternate metabolic pathway for incorporating thymidine into DNA. Used as a selectable marker to identify transfected eukaryotic cells. 122, *163–164*, 184

Ti (tumor-inducing) plasmid. A giant plasmid of *Agrobacterium tumefaciens* that is responsible for tumor formation in infected plants. Ti plasmids are used as vectors to introduce foreign DNA into plant cells. 171–173, *172*

T lymphocyte. A white blood cell responsible for the cell-mediated immune response. 120

Trait. *See* Phenotype.

Transcription. The process of creating a complementary RNA copy of DNA. 33–34, 36, 92, 100–102

Transduction. The transfer of DNA sequences from one bacterium to another via lysogenic infection by a bacteriophage (transducing phage). 117

Transfection. The uptake and expression of a foreign DNA sequence by cultured eukaryotic cells. 64, 120–122

Transformation. In prokaryotes, the natural or induced uptake and expression of a foreign DNA sequence—typically a recombinant plasmid in experimental systems. In higher eukaryotes, the conversion of cultured cells to a malignant phenotype—typically through infection by a tumor virus or transfection with an oncogene. 19–20, 43, 51–53, 55–58, 57, 90, 115–117, 116, 171–172, 199, 215, 231, 239, 293, 311, 340, 358, 361, 373, 390, 393, 415

transformant. In prokaryotes, a cell that has been genetically altered through the uptake of foreign DNA. In higher eukaryotes, a cultured cell that has acquired a malignant phenotype, 19–20, 43, 51–53, 55–57, 116, 293, 361, 379, 390, 416

transformation efficiency. The number of bacterial cells that uptake and express plasmid DNA divided by the mass of plasmid used (in transformants/microgram). 55–56, 307, 365, 379

Transforming oncogene. *See* Oncogene.

Transgenic. A vertebrate organism in which a foreign DNA gene (a transgene) is stably incorporated into its genome early in embryonic development. The transgene is present in both somatic and germ cells, is expressed in one or more tissues, and is inherited by offspring in a Mendelian fashion. 108–110, 173, 176–178, 177

Transition-state intermediate. In a chemical reaction, an unstable and high-energy configuration assumed by reactants on the way to making products. Enzymes are thought to bind and stabilize the transition state, thus lowering the energy of activation needed to drive the reaction to completion. 191

Translation. The process of converting the genetic information of an mRNA on ribosomes into a polypeptide. Transfer RNA molecules carry the appropriate amino acids to the ribosome, where they are joined by peptide bonds. 34–36, 35, 90–91

Translocation. The movement or reciprocal exchange of large chromosomal segments, typically between two different chromosomes. 125–126

Transposition. The movement of a DNA segment within the genome of an organism. 92–97, 95, 96

Transposon (transposable, or movable genetic element). A relatively small DNA segment that has the ability to move from one chromosomal position to another. 36, 92–97, 95, 96

Trypanosomes. A genus of flagellated microorganisms that is responsible for African sleeping sickness. 94–97, 96

Trypsin. A proteolytic enzyme that hydrolyzes peptide bonds on the carboxyl side of the amino acids arginine and lysine. 20, 88

Trypsinogen. The inactive precursor (zymogen) of trypsin synthesized by the acinar cells of the pancreas. 88

Tumor virus. *See* Virus.

Ultrabithorax (*Ubx*). A homeotic gene that acts on the development of thoracic segments T2 and T3 in *Drosophila*. 108

Upstream. The region extending in a 5' direction from a gene. 100–102, 101, 118, 120

Vaccine. A preparation of dead or weakened pathogen, or of derived antigenic determinants, that is used to induce formation of antibodies against the pathogen. 173, 182–184

polyvalent vaccine. A recombinant organism into which has been cloned antigenic determinants from a number of different disease-causing organisms. 182–184, 183

subunit vaccine. A vaccine composed of a purified antigenic determinant that is separated from the virulent organism. 183–184

Vaccinia. The cowpox virus used to vaccinate against smallpox and, experimentally, as a carrier of genes for antigenic determinants cloned from other disease organisms. 182–184, 183

Variable surface glycoprotein (VSG). One of a battery of antigenic determinants expressed by a microorganism to elude immune detection. 95–96

Vector. An autonomously replicating DNA molecule into which foreign DNA fragments are inserted and then propagated in a host cell. 40, 50–51, 53–54, 64–66, 145–147, 146, 172, 189–191, 190, 298, 315

Viral oncogene. *See* Oncogene.

Viroid. A plant pathogen that consists of a naked RNA molecule of approximately 250–350 nucleotides, whose extensive base pairing results in a nearly correct double helix. 173, 195

Virulence. The ability of an organism to cause disease. 19–21, 20

Virus. An infectious particle composed of a protein capsule and a nucleic acid core, which is dependent on a host organism for replication.

provirus. A double-stranded DNA copy of an RNA virus genome that is integrated into the host chromosome during lysogenic infection. 116–120, 118, 119, 161–162

tumor virus. A virus capable of transforming a cell to a malignant phenotype. 111–121, 114, 118, 119

X-linked disease. A genetic disease caused by a mutation on the X chromosome. In X-linked recessive conditions, a normal female "carrier" passes on the mutated X chromosome to an affected son. 15–16

X-ray crystallography. The diffraction pattern of X-rays passing through a pure crystal of a substance used to deduce its atomic structure. 2, 186–189, *188*

Yeast artificial chromosome (YAC). A vector used to clone DNA fragments of up to 400,000 bp, which contains the minimum chromosomal sequences needed to replicate in yeast—telomeres, centromere, and origins of replication. 145–147, *146*

Z-DNA. *See* DNA.

Zinc fingers. A structural motif of DNA-binding proteins in which finger-like loops of amino acids are stabilized by interactions with zinc atoms. 100

Zymogen (preenzyme). A nonfunctional precursor of an enzyme, which is converted to an active form through biochemical modification. *88*